GROUND WATER HANDBOOK

To Jay Lehr
with best regards,
and appreciation for
your encouragement!
Keith Anderson

By

Keith E. Anderson
Senior Consultant
Montgomery Watson

National Ground Water Association
Dublin, Ohio U.S.A. 43017

Ground Water Handbook
First Edition

National Ground Water Association
6375 Riverside Dr.
Dublin, OH 43017
(614) 761-1711

Printed in the United States of America

Library of Congress Cataloging-in-Publication Data

Anderson, Keith E. (Keith Elliott), 1920–
Ground water handbook / by Keith E. Anderson.
p. cm.
Includes index.
ISBN 1-56034-052-5
1. Wells – Handbooks, manuals, etc. 2. Water, Underground-
–Handbooks, manuals, etc. I. Title.
TD407.A55 1991
551.49—dc20 90-43108
CIP

Cover design: Janelle McClary
Typesetting: Merry Pryor
Louise Hambel

Preface

The enormous growth of the ground water industry over the past half-century has prompted the preparation of this new, updated handbook. The goal in its preparation has been to provide as much useful, basic information as possible to assist the many persons throughout the world who are involved with the industry—including those who explore for, develop, analyze, investigate, and are the ultimate users of this priceless natural resource.

The choice of material to be included (or not) in the book has been a somewhat subjective exercise, based largely upon the experience and judgement of the author. Because of space limitations, much of the material presented has had to be condensed. Every effort has been made, however, to provide sufficient detail so that the book will be useful for work in both field and office environments.

The author greatly appreciates and acknowledges the encouragement of his many friends and colleagues who have made valuable suggestions for material to be included in the handbook and who have reviewed much of the manuscript.

Keith E. Anderson, P.E., P.G.
Boise, Idaho, U.S.A.

Introduction

Every user of this handbook must recognize that this is not a "textbook" for persons working with ground water. In all cases, it is assumed that the reader is familiar with the particular problem at hand and how to solve it—but wants a single source for the formulas, standards, tables, graphs, and other reference materials that are constantly needed.

Every user of the handbook is encouraged to browse through each chapter to get an overview of the material that has been assembled—keeping in mind that some of the data could appear under more than one chapter title (such as flow measurement and irrigation, or pumps and electrical data). The index has been made as detailed as possible to make the book easy to use.

It is anticipated that the handbook will be useful to those working with ground water throughout the world. Unfortunately the world is not, as yet, on a single system of units of measurement. Most of the material presented is in the units commonly in use in the United States; a complete table of conversions to metric (SI) units is included in Chapter 1, however. Certain tables or graphs are presented with dual systems of measurement units.

Every effort has been made to include the latest information on the type of material that is constantly subject to change—such as water-quality standards. Readers are urged, however, to keep informed of current regulatory standards.

An attempt has been made to make the data in the handbook as accurate as possible. The author and publisher, while appreciating being notified of any errors or omissions that may be noted by readers, will in no event be liable for direct, indirect, or consequential damages resulting from use of the handbook.

Dedication

This handbook is dedicated to O.E. Meinzer and R.M. Leggette who introduced me to my career in ground water; to the newer generation(s) of professionals exemplified by such friends as P.K. Chaturjee of India, Roberto Antonio of the Philippines, and Terry Scanlan and Bern Hinckley of the United States; to my long-time friend and partner Jack Kelly; and especially to my wife, Dorothy, without whose encouragement, support, and patience the book could not have been completed.

Table of Contents

Chapter	Title	Page
	Preface	iii
	Introduction	iv
1	Mathematics and Conversions	1
2	Drilling	61
3	Pipe and Casing	99
4	Hydraulics and Water Data	123
5	Pumps and Water Systems	157
6	Flow Measurement	189
7	Electrical Data	223
8	Well Design	235
9	Irrigation	249
10	Water Quality	275
11	Hydrogeology and Ground Water Flow	291
12	Miscellaneous	335
	Index	395

Chapter 1

Mathematics and Conversions

LENGTH UNIT		EQUIVALENT							
		Centimeters	Inches	Feet	Yards	Meters	Rods	Kilometers	Miles
1 Centimeter	=	ONE	0.3937	0.0328	0.01093	0.01	0.00199	0.00001	—
1 Inch	=	2.54	ONE	0.08333	0.0278	0.0254	0.00505	0.0000254	0.0000158
1 Foot	=	30.48	12	ONE	0.3333	0.3048	0.0606	0.000305	0.000189
1 Yard	=	91.44	36	3	ONE	0.9144	0.18181	0.000915	0.000568
1 Meter	=	100	39.37	3.2808	1.0936	ONE	0.1988	0.001	0.000621
1 Rod	=	502.9	198	16.5	5.5	5.0292	ONE	0.00503	0.00312
1 Kilometer	=	100,000	39,370	3,280.8	1,093.6	1,000	198.83	ONE	0.62137
1 Mile	=	160,934	63,360	5,280	1,760	1,609.3	320	1.6093	ONE

AREA UNIT		EQUIVALENT							
		Square Centimeters	Square Inches	Square Feet	Square Yards	Square Meters	Acres	Hectares	Square Miles
1 Sq. Centimeter	=	ONE	0.155	0.001076	0.0001196	0.0001	—	—	
1 Sq. Inch	=	6.452	ONE	0.00694	0.0007716	0.0006452	—	—	—
1 Sq. Foot	=	929	144	ONE	0.1111	0.0929	0.00002296	—	—
1 Sq. Yard	=	8,361	1,296	9	ONE	0.8361	0.0002066	0.0000836	—
1 Sq. Meter	=	10,000	1,550	10.76	1.196	ONE	0.0002471	0.0001	—
1 Acre	=	40,465,284	6,272,640	43,560	4,840	4,047	ONE	0.4047	0.001563
1 Hectare	=	—	—	107,639	11,960	10,000	2.471	ONE	0.003861
1 Sq. Mile	=	—	—	27,878,400	3,097,600	2,589,998	640	259	ONE

VOLUME UNIT		EQUIVALENT							
		Cubic	Cubic Centimeters	Liters Inches	U.S. Gallons	Imperial	Cubic Feet Gallons	Cubic Yards	Cubic Meters
1 Cubic Centimeter	=	ONE	0.06102	0.001	0.0002642	0.00022	0.00003561	0.00000132	0.000001
1 Cubic Inch	=	16.387	ONE	0.01639	0.004329	0.003605	0.0005787	0.00002143	0.00001639
1 Liter	=	1,000	61.0234	ONE	0.26417	0.22008	0.03531	0.001308	0.001
1 U.S. Gallon	=	3,785.4	231	3.7854	ONE	0.83267	0.13368	0.004951	0.003785
1 Imperial Gallon	=	4,542.5	277.42	4.543	1.20095	ONE	0.1605	0.00595	0.00454
1 Cubic Foot	=	28,317	1,728	28.317	7.48052	6.2321	ONE	0.03704	0.02832
1 Cubic Yard	=	764,560	46,656	764.56	201.974	168.18	27	ONE	0.76456
1 Cubic Meter	=	1,000,000	61,023	1,000	264.17	220	35.315	1.308	ONE

WEIGHT UNIT		EQUIVALENT							
		Grams	Ounces (Avoirdupois)	Ounces (Troy)	Pounds (Avoirdupois)	Kilograms	Tons (Short)	Tons (Long)	Tons (Metric)
1 Gram	=	ONE	0.03527	0.03215	0.002205	0.001	0.000001102	0.000000984	0.000001
1 Ounce (Avoir.)	=	28.3495	ONE	0.9115	0.0625	0.0283	0.00003125	0.0000279	0.00002835
1 Ounce (Troy)	=	31.1035	1.0971	ONE	0.08333	0.0311	0.00003428	0.00003061	0.0000311
1 Pound (Avoir.)	=	453.5924	16	14.5833	ONE	0.4535	0.0005	0.0004464	0.0004535
1 Kilogram	=	1,000	35.274	32.1507	2.205	ONE	0.001102	0.0009842	0.001
1 Ton (Short)	=	907,185	32,000	29,168	2,000	907.185	ONE	0.89286	0.90718
1 Ton (Long)	=	1,016,047	35,840	32,668	2,240	1,016.047	1.12	ONE	1.016
1 Ton (Metric)	=	1,000,000	35,274	32,151	2,205	1,000	1.1023	0.9843	ONE

FLOW UNIT	EQUIVALENT						
	U.S. Gallons Per Minute	Imperial Gal. Per Minute	Cubic Feet Per Second	Acre-Feet Per Day	Liters Per Second	Cubic Meters Per Second	Million U.S. Gallons/Day
1 U.S. Gallon/Minute =	ONE	0.8327	0.002228	0.00442	0.06309	0.000063	0.00144
1 Imperial Gal/Min =	1.201	ONE	0.00267	0.00531	0.0757	0.000076	0.00173
1 Cubic Foot/Second =	448.83	373.72	ONE	1.9835	28.317	0.02832	0.646
1 Acre-Foot/Day =	226.24	188.38	0.5042	ONE	14.28	0.01428	0.3259
1 Liter/Second =	15.85	13.20	0.0353	0.07	ONE	0.001	0.0228
1 Cubic Meter/Second =	15,848	13,196	35.31	70.04	1,000	ONE	22.827
1 U.S. Million Gals/Day =	694.44	578.24	1.547	3.068	43.81	0.0438	ONE

POWER UNIT	EQUIVALENT						
	Watts	Kilowatts	Horsepower	Foot Pounds per Minute	Joules per Second	U.S. Gal/Min per 100′ Lift	BTU per Hour
1 Watt =	ONE	0.001	0.001341	44.254	1	0.05308	3.415
1 Kilowatt =	1,000	ONE	1.341	44,254	1,000	53.078	3,415
1 Horsepower =	745.7	0.7457	ONE	33,000	745.7	39.588	2,546.4
1 Foot Pound/Min =	0.022597	0.0000226	0.0000303	ONE	0.022597	0.00199	0.0772
1 Joule per Second =	1	0.001	0.001341	44.254	ONE	0.05308	3.415
1 US GPM/100′ Lift =	18.839	0.01884	0.02526	833.58	18.839	ONE	64.295
1 BTU per Hour =	0.2928	0.000293	0.0003927	12.953	0.2928	0.01555	ONE

Unit Conversions

Multiply	By	To Obtain
Absolute viscosity (poise)	1	Gram/second centimeter
Absolute viscosity (centipoise)	0.01	Poise
Acceleration due to gravity (g)	32.17405	Feet/second squared
	980.665	Centimeters/second squared
Acres	0.40469	Hectares
	10	Square chains
	43,560	Square feet
	4,046.86	Square meters
	0.001562	Square miles
	160	Square rods
	5,645.38	Square varas
	4,480	Square yards
Acre-feet	43,560	Cubic feet
	1233.482	Cubic meters
	325,851.45	Gallons (US)
	1,233,482	Liters
Acre-inch	3,630	Cubic feet
	102.79	Cubic meters
	2,7154.29	Gallons
Acre-feet/hour	726	Cubic feet/minute
	12.1	Cubic feet/second/
	5,430.86	Gallons (US)/minute
Acre-feet/day	0.50417	Cubic feet/second
	0.01428	Cubic meters/second
	226.24	Gallons (US)/minute
Angstroms	0.0001	Microns
	0.003937	Millionths of inch
Ares	0.02471	Acres
	0.01	Hectares
	1,076.39	Square feet
Atmospheres	76.0	Centimeters Hg @ 32F
	33.94	Feet of water @ 62F
	29.921	Inches of Hg @ 32F
	10,333	Kilograms/square meter
	101.330	Kilopascals
	1,013.25	Millibars
	760.0	Millimeters Hg @ 32F
	235.1408	Ounces/square inch
	14.6963	Pounds/square inch
	1.058	Tons/square foot

Unit Conversions (Cont'd)

Multiply	By	To Obtain
Bags (sacks) of cement	94	Pounds of cement
Barrels of oil	42	Gallons (US) of oil
Barrels (42 gal)/day	0.0292	Gallons (US)/minute
Barrels of cement	376	Pounds of cement
Bars	0.98692	Atmospheres
	100	Kilopascals
	1,000	Millibars
	14.504	Pounds/square inch
Board feet	144	Cubic inches
Boiler horsepower	33,479	BTU/hour
	9.803	Kilowatts
	34.5	Pounds of water evaporated per hour @ 212F
BTU (British Thermal Unit)	252.016	Calories (gm)
(Heat required to raise	0.252	Calories (Kg)
temperature of one pound	777.54	Foot pounds
of water one degree F)	0.0003927	Horsepower hours
	1,054.8	Joules
	107.5	Kilogram meters
	0.0002928	Kilowatt hours
BTU/cubic foot	8.89	Calories (Kg)/cu meter @ 32F
BTU/minute	12.96	Foot pounds/second
	0.02356	Horsepower
	0.01757	Kilowatts
	17.57	Watts
BTU/pound	0.556	Calories (Kg)/Kilogram
Bushels	2,150.4	Cubic inches
	35.239	Liters
	4	Pecks
	32	Quarts (dry)
Cables	120	Fathoms
	240	Yards
	219.456	Meters
Calories (gram)	0.003968	BTU
	0.001	Calories (Kg)
	3.088	Foot pounds
	4.185	Joules
	0.4265	Kilogram meters
	0.0011628	Watt hours
Calories (Kg)	3.968	BTU
	1,000	Calories (gm)
	3,088	Foot pounds

Unit Conversions (Cont'd)

Multiply	By	To Obtain
Calories (Kg)	0.001558	Horsepower hours
	4,185	Joules
	426.5	Kilogram meters
	1.1628	Watt hours
Calories (Kg)/cubic meter	0.1124	BTU/cubic foot at 32F
Calories (Kg)/Kg	1.8	BTU/pound
Calories (Kg)/minute	51.43	Foot pounds/second
	0.09351	Horsepower
	0.06972	Kilowatts
Carats (diamond)	200	Milligrams
Centigrams	0.01	Grams
Centiliters	0.01	Liters
Centimeters	0.3937	Inches
	0.032808	Feet
	0.01	Meters
	10	Millimeters
Centimeters of mercury @ 32F	0.01316	Atmospheres
	0.4461	Feet of water at 62F
	136	Kgs/square meter
	27.85	Pounds/square foot
	0.1934	Pounds/square inch
Centimeters/second	2,834.65	Feet/day
	0.032808	Feet/second
	1,034,647	Feet/year
	1,417.31	Inches/hour
Centipoise	0.00067197	Pounds/second foot
	2.4191	Pounds/hour foot
	0.01	Poise
Chains	4	Rods
	66	Feet
	100	Links
Cheval-vapeur (CV)	1	Metric horsepower
	75	Kilogram meters/second
	0.98632	Horsepower
Circular inches	1,000,000	Circular mils
	0.7854	Square inches
	785,400	Square mils
Circular mils	0.000001	Circular inches
	0.7854	Square mils
	0.0000007854	Square inches

Unit Conversions (Cont'd)

Multiply	By	To Obtain
Cubic centimeters	0.00003531	Cubic Feet
	0.06102	Cubic inches
	0.0002642	Gallons (US)
	0.001	Liters
	0.002113	Pints (liquid US)
	0.001057	Quarts (liquid US)
	0.03381	Ounces (fluid)
Cubic feet	0.00002296	Acre feet
	0.803564	Bushels
	28,317	Cubic centimeters
	1,728	Cubic inches
	0.02832	Cubic meters
	0.03704	Cubic yards
	7.48052	Gallons (US)
	6.229	Gallons (Imperial)
	28.32	Liters
	59.84	Pints (liquid US)
	29.92	Quarts (liquid US)
Cubic feet of water	62.4266	Pounds @ 39.2F
	62.3554	Pounds @ 62F
Cubic feet/minute	472	Cubic centimeters/sec
	0.1247	Gallons (US)/second
	0.472	Liters/second
	62.36	Pounds water/min @ 62F
	7.48052	Gallons (US)/minute
	6.2321	Gallons (Imperial)/min
	10,772	Gallons (US)/day
	0.033058	Acre feet/day
Cubic feet/second	1.98347	Acre feet/day
	723.97	Acre feet/year (365 days)
	60	Cubic feet/minute
	86,400	Cubic feet/day
	0.02832	Cubic meters/second
	1.6992	Cubic meters/minute
	101.95	Cubic meters/hour
	2,446.8	Cubic meters/day
	646,317	Gallons (US)/day
	448.83117	Gallons (US)/minute
	374.03	Gallons (Imperial)/minute
	28.31685	Liters/second

Unit Conversions (Cont'd)

Multiply	By	To Obtain
Cubic feet/square foot/day	0.3048	Cubic meters/sq meter/day (=meters/day)
	0.0006944	Cubic feet/sq foot/minute (=feet/minute)
	0.5	Cubic inches/sq inch/hour (=inches/hour)
	0.3646	Darcies
	7.48052	Gallons/sq. foot/day
	304.8	Liters/sq. meter/day
Cubic inches	16.387	Cubic centimers
	0.0005787	Cubic feet
	0.004329	Gallons (US)
	0.01639	Liters
	0.03463	Pints (liquid US)
	0.01732	Quarts (liquid US)
Cubic meters	0.0008106	Acre feet
	1,000,000	Cubic centimeters
	35.31	Cubic feet
	61,023	Cubic inches
	1.308	Cubic yards
	264.2	Gallons (US)
	220	Gallons (Imperial)
	1,000	Liters
	2,113	Pints (liquid US)
	1,057	Quarts (liquid US)
Cubic meters/second (cumec)	35.31	Cubic feet/second
	2,419,200	Cubic meters/28 days
	2,592,000	Cubic meters/30 days
	2,678,400	Cubic meters/31 days
	31,536,000	Cubic meters/365 days
Cubic meters/hour	0.00981	Cubic feet/second
	4.39	Gallons (US)/minute
	264.2	Gallons (US)/hour
	220	Gallons (Imperial)/hour
	0.2778	Liters/second
Cubic yards	764,600	Cubic centimeters
	27	Cubic feet
	46,656	Cubic inches
	0.7646	Cubic meters
	201.97	Gallons (US)
	168.18	Gallons (Imperial)
	764.6	Liters

Unit Conversions (Cont'd)

Multiply	By	To Obtain
Cubic yards	1,616	Pints (liquid US)
	807.9	Quarts (liquid US)
Cubit	18	Inches
Darcy (Water @ 20C)	0.000966	Centimeters/second
	2.743	Feet/day
	20.53	Gallons (US)/square foot/day
	0.836	Meters/day
Days	24	Hours
	1,440	Minutes
	86,400	Seconds
Decigrams	0.1	Grams
Deciliters	0.1	Liters
Decimeters	0.1	Meters
Degrees (angle)	60	Minutes
	0.01745	Radians
	3,600	Seconds
Degrees F (temp.) (--less 32)	0.5556	Degrees C
Degrees C (temp.)	1.8 (plus 32)	Degrees F
Degrees/second	0.01745	Radians/sec
	0.1667	Revolutions/min
	0.002778	Revolutions/sec
Dekagrams	10	Grams
Dekaliters	10	Liters
Dekameters	10	Meters
Diameter (circle)	3.1415926536 (=pi)	Circumference
Diameter (circle)	0.88623	Side of equal square
Diameter cubed (sphere)	0.5236	Volume (sphere)
Diam (major) x diam (minor)	0.7854	Area of ellipse
Diameter squared (sphere)	3.1416	Surface (sphere)
Diameter (inches) x RPM	0.262	Belt speed in ft/min
Digits	0.75	Inches
Drams (avoirdupois)	27.34375	Grains
	1.771845	Grams
	0.0625	Ounces (avoir.)
Fathoms	6	Feet
Feet	30.48	Centimeters
	12	Inches
	0.3048	Meters
	304.8	Millimeters
	0.06061	Rods
	0.33333	Yards

Unit Conversions (Cont'd)

Multiply	By	To Obtain
Feet of water @ 62F	0.029465	Atmospheres
	0.88162	Inches of Hg @ 32F
	304.44	Kilograms/square meter
	62.3554	Pounds/square foot
	0.43302	Pounds/square inch
Feet/day	0.0003528	Centimeters/second
	0.0000116	Feet/second
	0.000694	Feet/minutes
	365	Feet/year
	0.5	Inches/hour
	0.3048	Meters/day
Feet/minute	0.5080	Centimeters/second
	1,440	Feet/day
	0.01667	Feet/second
	0.01829	Kilometers/hour
	0.3048	Meters/minute
	0.01136	Miles/hour
Feet/second	30.48	Centimeters/second
	1.0973	Kilometers/hour
	0.5921	Knots
	18.29	Meters/minute
	0.3048	Meters/second
	0.68182	Miles/hour
	0.01136	Miles/minute
Feet/second squared	30.48	Centimeters/second squared
Foot pounds	0.0012861	BTU
	0.32412	Calories (gm)
	0.0003241	Calories (Kg)
	1.3558	Joules
	0.13826	Kilogram meters
	0.0003766	Watt hours
Foot pounds/minute	0.0012861	BTU/minute
	0.01667	Foot pounds/second
	0.0000303	Horsepower
Foot pounds/second	0.07717	BTU/minute
	0.001818	Horsepower
	0.01945	Calories (Kg)/minute
	0.001356	Kilowatts

Unit Conversions (Cont'd)

Multiply	By	To Obtain
Furlong	40	Rods
	220	Yards
	660	Feet
	0.125	Miles
	0.2012	Kilometers
Gallons (Imperial)	277.42	Cubic inches
	4.543	Liters
	1.20095	Gallons (US)
Gallons (US)	3,785	Cubic centimeters
	0.13368	Cubic feet
	231	Cubic inches
	0.003785	Cubic meters
	0.004951	Cubic yards
	3.78534	Liters
	8	Pints (liquid US)
	4	Quarts (liquid US)
	0.83267	Gallons (Imperial)
Gallons (US)/day/square foot	0.13368	Feet/day
	48.79	Feet/year
	0.0668	Inches/hour
	0.04075	Meters/day
	14.9	Meters/year
	0.0000472	Centimeter/second
	0.0487	Darcies
	48.71	Millidarcies
Gallons (US) of water @ 62F	8.3357	Pounds
Gallons (US) of water/minute	6.002	Tons of water/day
Gallons (US)/hour	24	Gallons/day
	0.01667	Gallons/minute
Gallons (US)/minute	0.0044192	Acre feet/day
	34.286	Barrels (42gal.)/day
	0.002228	Cubic feet/second
	0.13368	Cubic feet/minute
	8.0208	Cubic feet/hour
	0.000063	Cubic meters/second
	60	Gallons (US)/hour
	1,440	Gallons (US)/day
	0.06309	Liters/second
	3.78533	Liters/minute

Unit Conversions (Cont'd)

Multiply	By	To Obtain
Gallons (US)/minute/acre	0.156	Liters/second/hectare
Gallons (US)/minute/foot	0.207	Liters/second/meter
Grains	1	Grains (avoirdupois)
	1	Grains (apothecary)
	1	Grains (troy)
	0.0648	Grams
	0.0020833	Ounces (troy)
	0.0022857	Ounces (avoirdupois)
Grains/gallon (US)	17.118	Parts/million
	17.118	Milligrams/liter
	142.86	Pounds/million gals. (US)
Grams (force)	980.665	Dynes
Grams (mass)	15.43	Grains
	0.001	Kilograms
	1,000	Milligrams
	0.03527	Ounces (avoirdupois)
	0.03215	Ounces (troy)
	0.0022046	Pounds
Grams/centimeter	0.0056	Pounds/inch
Grams/cubic centimeter	1,000	Kilograms/cubic meter
	62.43	Pounds/cubic foot
	0.03613	Pounds/cubic inch
Grams/liter	58.417	Grains/gallon (US)
	8.345	Pounds/100 gals. (US)
	0.062427	Pounds/cubic foot
	1,000	Parts/million
	1,000	Milligrams/liter
Gravity (g)	980.665	Centimeters/sec. squared
	32.17405	Feet/second squared
Hand	10.16	Centimeters
	4	Inches
Hectares	2,471	Acres
	107,639	Square feet
	10,000	Square meters
	0.01	Square kilometers
Hectograms	100	Grams
Hectoliters	100	Liters
Hectometers	100	Meters

Unit Conversions (Cont'd)

Multiply	By	To Obtain
Hectowatts	100	Watts
Hogshead	63	Gallons (US)
	238.4759	Liters
Horsepower	42.44	BTU/minute
	10.17	Calories (Kg)/minute
	33,000	Foot pounds/minute
	550	Foot pounds/second
	0.7457	Kilowatts
	1.014	Metric horsepower (CV)
	745.7	Watts
Horsepower hours	2,546.5	BTU
	641,700	Calories (gm)
	641.7	Calories (Kg)
	1,980,000	Foot pounds
	2,684,500	Joules
	273,740	Kilogram meters
	0.7457	Kilowatt hours
	745.7	Watt hours
Inches	2.540	Centimeters
	0.08333	Feet
	0.0254	Meters
	25.4	Millimeters
	1,000	Mils
	72	Points (printing type)
	0.02778	Yards
Inches of Hg @ 32F	0.03342	Atmospheres
	1.1343	Feet of water @ 62F
	13.6114	Inches of water @ 62F
	345.3	Kilograms/sq. meter
	3.386	Kilopascals
	7.85872	Ounces/square inch
	70.727	Pounds/square foot
	0.49117	Pounds/square inch
Inches of water @ 62F	0.002455	Atmospheres
	0.07347	Inches of Hg @ 32F
	25.37	Kilograms/sq. meter
	0.5771	Ounces/square inch
	5.1963	Pounds/square foot
	0.03609	Pounds/square inch
Inches/hour	0.000706	Centimeters/second
	24	Inches/day

Unit Conversions (Cont'd)

Multiply	By	To Obtain
Joules	0.00094805	BTU
	0.239	Calories (gm)
	0.000239	Calories (Kg)
	0.73756	Foot pounds
	0.10197	Kilogram meters
	0.0002778	Watt hours
	1	Watt second
Kilograms	980,665	Dynes
	1,000	Grams
	35.274	Ounces (avoirdupois)
	32.1507	Ounces (troy)
	2.205	Pounds
	0.06852	Slugs
	0.001102	Tons (short)
	0.001	Tons (metric)
Kilogram meters	0.009302	BTU
	2.344	Calories (gm)
	0.002344	Calories (Kg)
	7.233	Foot pounds
	9.806	Joules
	0.002724	Watt hours
Kilograms/cubic meter	0.06243	Pounds/cubic foot
Kilograms/square centimeter	1	Metric atmosphere
	14.223	Pounds/square inch
Kilogram/square meter	0.000096784	Atmospheres
	0.007356	Centimeters of Hg @ 32F
	0.003285	Feet of water at 62F
	0.1	Gram/sq. centimeter
	0.002896	Inches of Hg @ 32F
	0.20482	Pounds/square foot
	0.001422	Pounds/square inch
Kiloliters	1,000	Liters
Kilometers	100,000	Centimeters
	3,280.84	Feet
	1,000	Meters
	0.62137	Miles
	1,093.6	Yards
Kilometers/hour	27.778	Centimeters/second
	54.68	Feet/minute
	0.9113	Feet/second
	0.53959	Knots

Unit Conversions (Cont'd)

Multiply	By	To Obtain
Kilometers/hour	16.67	Meters/minute
	0.27778	Meters/second
	0.62137	Miles/hour
Kilopascals	0.00987	Atmospheres
	0.2953	Inches of Hg @ 32F
	1,000	Pascals
	20.886	Pounds/square foot
	0.1450	Pounds/square inch
Kilowatts	56.92	BTU/minute
	14.34	Calories (Kg)/minute
	44,250	Foot pounds/minute
	737.6	Foot pounds/second
	1.341	Horsepower
	1,000	Watts
Kilowatt Hours	3,413	BTU
	860,500	Calories (gm)
	860.5	Calories (Kg)
	2,655,200	Foot pounds
	1.341	Horsepower hours
	3,600,000	Joules
	367,100	Kilogram meters
	1,000	Watt hours
Knots	1.85325	Kilometers/hour
	0.51479	Meters/second
	1.15155	Miles/hour
	1	Nautical miles/hour
Leagues	3	Miles
Links	7.92	Inches
Liters	1,000	Cubic centimeters
	0.03531	Cubic feet
	61.02	Cubic inches
	0.001	Cubic meters
	0.001308	Cubic yards
	0.264172	Gallons (US)
	0.22	Gallons (Imperial)
	2.113	Pints (liquid US)
	2.2018	Pounds of water @ 62F
	1.057	Quarts (liquid US)
Liters/minute	0.0005886	Cubic feet/second
	0.004403	Gallons (US)/second
	0.264172	Gallons (US)/minute

Unit Conversions (Cont'd)

Multiply	By	To Obtain
Liters/second	15.8508	Gallons (US)/minute
	0.001	Cubic meters/second
	8.64	Millimeters/hectare/day
	259.2	Mm/hectare/30 days
Liters/second/hectare	6.415	Gallons (US)/min/acre
Liters/second/meter	4.82	Gallons (US)/min/foot
Meters	100	Centimeters
	3.28084	Feet
	39.37008	Inches
	0.001	Kilometers
	0.0006214	Miles
	1,000	Millimeters
	1.0936	Yards
Meters/minute	1,667	Centimeters/second
	3.2808	Feet/minute
	0.05468	Feet/second
	0.06	Kilometers/hour
	0.03728	Miles/hour
Meters/second	100	Centimeters/second
	196.85	Feet/minute
	3.28084	Feet/second
	3.6	Kilometers/hour
	0.06	Kilometers/minute
	2.23694	Miles/hour
	0.03728	Miles/minute
Meters/second squared	3.28084	Feet/second squared
Microns	0.00003937	Inches
	0.000001	Meters
	0.001	Millimeters
	0.03937	Mils
Miles	80	Chains
	5,280	Feet
	63,360	Inches
	1.6093	Kilometers
	1,609.34	Meters
	0.8684	Nautical miles
	320	Rods
	1,760	Yards
Miles/hour	44.704	Centimeters/second
	88	Feet/minute

Unit Conversions (Cont'd)

Multiply	By	To Obtain
Miles/hour	1.467	Feet/second
	1.6093	Kilometers/hours
	0.86839	Knots
	26.82	Meters/minute
	0.44704	Meters/second
Miles/minute	2,682	Centimeters/second
	88	Feet/second
	1.6093	Kilometers/minute
	60	Miles/hour
Millibars	0.0009869	Atmospheres
	0.001	Bars
	10.197	Kilograms/square meter
	0.1	Kilopascals
	0.0145	Pounds/square inch
Milliers	1,000	Kilograms
Milligram	0.01543	Grains
	0.001	Grams
Milligrams/liter	1	Parts per million
Milliliters	0.001	Liters
Million gallons (US)/day	1.54723	Cubic feet/second
	3,785	Cubic meters/day
	694.45	Gallons (US)/minute
	44.780	Liters/second
Millimeters	0.1	Centimeters
	0.0032808	Feet
	0.03937	Inches
	0.001	Meters
	1,000	Microns
	39.37	Mils
Mils	0.001	Inches
	25.4	Microns
	0.0254	Millimeters
Miner's inches	0.02	Cubic feet/sec (in ID, KS, NE, NV, NM, ND, SD, UT, WA)
	0.025	Cubic feet/sec (in AZ, CA, MT, OR)
	0.026	Cubic feet/sec (in CO)
	0.028	Cubic feet/sec (in Brit. Columbia)
	8.977	Gallons (US)/ min (in ID, KS, NE, NV, NM, ND, SD, UT, WA)

Unit Conversions (Cont'd)

Multiply	By	To Obtain
Miner's inches	11.221	Gallons (US)/min (in AZ, CA, MT, OR)
	11.688	Gallons (US)/min (in CO)
	12.572	Gallons (US)/min (in B. C.)
Minutes (angle)	0.0167	Degrees
	0.0002909	Radians
	60	Seconds
Months (28 days)	672	Hours
	40,320	Minutes
Months (30 days)	720	Hours
	43,200	Minutes
Months (31 days)	744	Hours
	44,640	Minutes
Nautical miles	6,080.2	Feet
	1.8533	Kilometers
	1.1516	Miles
	2,026.8	Yards
Newtons	7.231	Poundals
	0.2248	Pounds (force)
Ounces (avoirdupois)	16	Drams (avoirdupois)
	437.5	Grains
	28.349527	Grams
	0.9115	Ounces (troy)
	0.0625	Pounds (avoirdupois)
Ounces (fluid)	29.57	Cubic centimeters
	1.805	Cubic inches
	0.25	Gills
	0.02957	Liters
	29.57	Milliliters
	0.0625	Pints (liquid US)
	0.03125	Quarts (liquid US)
Ounces (troy)	480	Grains
	31.103481	Grams
	1.09714	Ounces (avoirdupois)
	20	Pennyweights (troy)
	0.08333	Pounds
Ounces/square inch	0.004253	Atmospheres
	4.39	Centimeters of water @ 62F
	1.732	Inches of water @ 62F
	0.12725	Inches of Hg @ 32F

Unit Conversions (Cont'd)

Multiply	By	To Obtain
Palms	3	Inches
Parts per million	0.0584	Grains/gallon (US)
	0.07016	Grains/gallon (Imperial)
	1	Milligrams/liter
	8.345	Pounds/million gals (US)
Pascals	0.001	Kilopascals
Pennyweights (troy)	24	Grains
	1.55517	Grams
	0.05	Ounces (troy)
Pints (liquid US)	473.1	Cubic centimeters
	28.875	Cubic inches
	4	Gills
	16	Ounces (fluid)
	0.5	Quarts (liquid US)
Poise	100	Centipoise
	242	Pounds/hour foot
	0.0672	Pounds/second foot
Poundals	0.1383	Newtons
Pounds (avoirdupois)	256	Drams (avoirdupois)
	7,000	Grains
	453.5924	Grams
	0.4535	Kilograms
	16	Ounces (avoirdupois)
	14.5833	Ounces (troy)
	1.21528	Pounds (troy)
	0.03108	Slugs
	0.0005	Tons (short)
Pounds (troy)	5,760	Grains
	373.24177	Grams
	13.1657	Ounces (avoirdupois)
	12	Ounces (troy)
	240	Pennyweights (troy)
	0.822857	Pounds (avoirdupois)
Pounds of water @ 62F	0.01604	Cubic feet
	27.72	Cubic inches
	0.120	Gallons (US)
Pounds (force)	4.448	Newtons
Pounds/cubic foot	0.01601846	Grams/cubic centimeter
	16.01846	Kilograms/cubic meter
	0.0005787	Pounds/cubic inch
	0.13368	Pounds/gallon (US)
	0.03108	Slugs/cubic foot

Unit Conversions (Cont'd)

Multiply	By	To Obtain
Pounds/foot	1.488	Kilograms/meter
Pounds/inch	178.6	Grams/centimeter
Pounds/hour foot	0.4132	Centipoise
	0.004132	Poise grams/second cm
Pounds/second foot	1,488.1	Centipoise
	14.881	Poise grams/second cm
Pound/square foot	0.0004725	Atmospheres
	0.016037	Feet of water @ 62F
	0.014139	Inches of Hg @ 32F
	4.882	Kilograms/square meter
	0.04788	Kilopascals
	0.35913	Millimeters of Hg @ 32F
	0.006944	Pounds/square inch
Pounds/square inch	0.068044	Atmospheres
	2.30934	Feet of water @ 62F
	2.0360	Inches of Hg @ 32F
	27.712	Inches of water @ 62F
	703.067	Kilograms/square meter
	6.8948	Kilopascals
	51.7151	Millimeters of Hg @ 32F
	144	Pounds/square foot
Quadrants (angular)	90	Degrees
	5,400	Minutes
	1.571	Radians
	324,000	Seconds
Quarts (dry)	67.20	Cubic inches
Quarts (liquid US)	946.3	Cubic centimeters
	57.75	Cubic inches
	0.9463	Liters
	32	Ounces (fluid)
	2	Pints (liquid US)
Quires	25	Sheets
Radians	57.30	Degrees
	3,438	Minutes
	0.637	Quadrants
	206,265	Seconds
Radians/second	57.30	Degrees/second
	0.1592	Revolutions/second
	9.549	Revolutions/minute
Reams	500	Sheets

Unit Conversions (Cont'd)

Multiply	By	To Obtain
Revolutions	360	Degrees
	4	Quadrants
	6.283	Radians
Revolutions/minute	6	Degrees/second
	0.1047	Radians/second
	0.01667	Revolutions/second
Revolution/second	360	Degrees/second
	6.283	Radians/second
	60	Revolutions/minute
Rods	16.5	Feet
	5.5	Yards
Seconds (angle)	0.000004848	Radians
	0.01667	Minutes
Sections (land survey)	640	Acres
	259	Hectares
	1	Square miles
Slugs	14.5939	Kilograms
	32.18	Pounds
Span	9	Inches
Square centimeters	0.0010764	Square feet
	0.1550	Square inches
	0.0001	Square meters
	100	Square millimeters
Square feet	0.000022957	Acres
	929.0	Square centimeters
	144	Square inches
	0.092903	Square meters
	0.1111	Square yards
Square inches	1.27324	Circular inches
	1,273,239	Circular mils
	6.4516	Square centimeters
	0.0069444	Square feet
	645.16	Square millimeters
	1,000,000	Square mils
Square kilometers	247.1054	Acres
	100.0	Hectares
	10,763,349	Square feet
	1,000,000	Square meters
	0.3861	Square miles
	1,195,930	Square yards

Unit Conversions (Cont'd)

Multiply	By	To Obtain
Square meters	0.0002471	Acres
	10,000	Square centimeters
	10.76391	Square feet
	1,550.0	Square inches
	1.19599	Square yards
Square miles	640	Acres
	258.999	Hectares
	1	Section (land survey)
	27,878,400	Square feet
	2.58999	Square kilometers
	102,400	Square rods
	3,097,600	Square yards
Square millimeters	1,973	Circular mils
	0.01	Square centimeters
	0.00155	Square inches
	1,550	Square mils
Square mils	1.27324	Circular mills
	0.0006452	Square millimeters
Square yards	0.00020661	Acres
	9	Square feet
	0.83613	Square meters
Stere	1	Cubic meters
Stone	6.35029	Kilograms
	14	Pounds
Tons (long)	1,016.047	Kilograms
	2,240	Pounds
	1.120	Tons (short)
Tons (short)	907.1847	Kilograms
	32,000	Ounces
	2,000	Pounds
	0.90718	Tons (metric)
	0.89286	Tons (long)
Tons (metric)	1,000	Kilograms
	2,204.62	Pounds
	1.1023	Tons (short)
Tons of refrigeration	12,000	BTU/hour
	288,000	BTU/day
Tons of water/day @ 62F	1.3263	Cubic feet/hour
	0.16510	Gallons (US)/minute
	83.33	Pounds of water/hour

Unit Conversions (Cont'd)

Multiply	By	To Obtain
Watts	0.05692	BTU/minute
	0.01434	Calories (kg)/minute
	44.26	Foot pounds/minute
	0.7376	Foot pounds/second
	0.001341	Horsepower
	1	Joule/second
	0.001	Kilowatts
Watt hours	3.413	BTU
	860.5	Calories (gm)
	0.8605	Calories (Kg)
	2,655	Foot pounds
	0.001341	Horsepower hours
	3,600	Joules
	367.1	Kilogram meters
	0.001	Kilowatt hours
Yards	91.44	Centimeters
	3	Feet
	36	Inches
	0.9144	Meters
	0.1818	Rods
Year (365 days)	8,760	Hours
	525,600	Minutes
	31,536,000	Seconds

Metric System Multipliers

Symbol	Designation	Factor	
T	tera-	10^{12}	1 000 000 000 000
G	giga-	10^{9}	1 000 000 000
M	mega-	10^{6}	1 000 000
k	kilo-	10^{3}	1 000
h	hecto-	10^{2}	100
da	deca-	10	10
d	deci-	10^{-1}	0.1
c	centi-	10^{-2}	0.01
m	milli-	10^{-3}	0.001
μ	micro-	10^{-6}	0.000 001
n	nano-	10^{-9}	0.000 000 001
p	pico-	10^{-12}	0.000 000 000 001

Temperature Conversions

C	C or F	F	C	C or F	F	C	C or F	F	C	C or F	F	C	C or F	F	C	C or F	F	C	C or F	F
−101	−150	−238	−22.8	−9	15.8	−8.9	16	60.8	5.0	41	105.8	18.9	66	150.8	32.8	91	195.8	46.7	116	240.8
−95.6	−140	−220	−22.2	−8	17.6	−8.3	17	62.6	5.6	42	107.6	19.4	67	152.6	33.3	92	197.6	47.2	117	242.6
−90.0	−130	−202	−21.7	−7	19.4	−7.8	18	64.4	6.1	43	109.4	20.0	68	154.4	33.9	93	199.4	47.8	118	244.4
−84.4	−120	−184	−21.1	−6	21.2	−7.2	19	66.2	6.7	44	111.2	20.6	69	156.2	34.4	94	201.2	48.3	119	246.2
−78.9	−110	−166	−20.6	−5	23.0	−6.7	20	68.0	7.2	45	113.0	21.1	70	158.0	35.0	95	203.0	48.9	120	248.0
−73.3	−100	−148	−20.0	−4	24.8	−6.1	21	69.8	7.8	46	114.8	21.7	71	159.8	35.6	96	204.8	54.4	130	266.0
−67.8	−90	−130	−19.4	−3	26.6	−5.6	22	71.6	8.3	47	116.6	22.2	72	161.6	36.1	97	206.6	60.0	140	284.0
−62.2	−80	−112	−18.9	−2	28.4	−5.0	23	73.4	8.9	48	118.4	22.8	73	163.4	36.7	98	208.4	65.6	150	302.0
−56.7	−70	−94.0	−18.3	−1	30.2	−4.4	24	75.2	9.4	49	120.2	23.3	74	165.2	37.2	99	210.2	71.1	160	320.0
−51.1	−60	−76.0	−17.8	0	32.0	−3.9	25	77.0	10.0	50	122.0	23.9	75	167.0	37.8	100	212.0	76.7	170	338.0
−45.6	−50	−58.0	−17.2	1	33.8	−3.3	26	78.8	10.6	51	123.8	24.4	76	168.8	38.3	101	213.8	82.2	180	356.0
−40.0	−40	−40.0	−16.7	2	35.6	−2.8	27	80.6	11.1	52	125.6	25.0	77	170.6	38.9	102	215.6	87.8	190	374.0
−34.4	−30	−22.0	−16.1	3	37.4	−2.2	28	82.4	11.7	53	127.4	25.6	78	172.4	39.4	103	217.4	93.3	200	392.0
−31.7	−25	−13.0	−15.6	4	39.2	−1.7	29	84.2	12.2	54	129.2	26.1	79	174.2	40.0	104	219.2	98.9	210	410.0
−28.9	−20	−04.0	−15.0	5	41.0	−1.1	30	86.0	12.8	55	131.0	26.7	80	176.0	40.6	105	221.0	100.0	212	413.6
−28.3	−19	−02.2	−14.4	6	42.8	−0.6	31	87.8	13.3	56	132.8	27.2	81	177.8	41.1	106	222.8	104.4	220	428.0
−27.8	−18	−00.4	−13.9	7	44.6	0.0	32	89.6	13.9	57	134.6	27.8	82	179.6	41.7	107	224.6	110.0	230	446.0
−27.2	−17	01.4	−13.3	8	46.4	0.6	33	91.4	14.4	58	136.4	28.3	83	181.4	42.2	108	226.4	115.6	240	464.0
−26.7	−16	03.2	−12.8	9	48.2	1.1	34	93.2	15.0	59	138.2	28.9	84	183.2	42.8	109	228.2	121.1	250	482.0
−26.1	−15	05.0	−12.2	10	50.0	1.7	35	95.0	15.6	60	140.0	29.4	85	185.0	43.3	110	230.0	126.7	260	500.0
−25.6	−14	06.8	−11.7	11	51.8	2.2	36	96.8	16.1	61	141.8	30.0	86	186.8	43.9	111	231.8	132.2	270	518.0
−25.0	−13	08.6	−11.1	12	53.6	2.8	37	98.6	16.7	62	143.6	30.6	87	188.6	44.4	112	233.6	137.8	280	536.0
−24.4	−12	10.4	−10.6	13	55.4	3.3	38	100.4	17.2	63	145.4	31.1	88	190.4	45.0	113	235.4	143.3	290	554.0
−23.9	−11	12.2	−10.0	14	57.2	3.9	39	102.2	17.8	64	147.2	31.7	89	192.2	45.6	114	237.2	148.9	300	572.0
−23.3	−10	14.0	−9.4	15	59.0	4.4	40	104.0	18.3	65	149.0	32.2	90	194.0	46.1	115	239.0			

$C = 5/9\,(F - 32)$ or $F = (9/5\,C) + 32$

Conversion — Inches of Mercury to Feet of Water

Inches of Mercury	0.0	0.1	0.2	0.3	0.4	0.5	0.6	0.7	0.8	0.9
0	0.00	0.11	0.23	0.34	0.45	0.57	0.68	0.79	0.91	1.02
1	1.13	1.25	1.36	1.47	1.59	1.70	1.81	1.93	2.04	2.15
2	2.27	2.38	2.49	2.61	2.72	2.83	2.95	3.06	3.17	3.29
3	3.40	3.51	3.63	3.74	3.85	3.97	4.08	4.19	4.31	4.42
4	4.53	4.65	4.76	4.87	4.99	5.10	5.21	5.33	5.44	5.55
5	5.67	5.78	5.89	6.00	6.12	6.23	6.34	6.46	6.57	6.68
6	6.80	6.91	7.02	7.14	7.25	7.36	7.48	7.59	7.70	7.82
7	7.93	8.04	8.16	8.27	8.38	8.50	8.61	8.72	8.84	8.95
8	9.06	9.18	9.29	9.40	9.52	9.63	9.74	9.86	9.97	10.08
9	10.20	10.31	10.42	10.54	10.65	10.76	10.88	10.99	11.10	11.22
10	11.33	11.44	11.56	11.67	11.78	11.90	12.01	12.12	12.24	12.35
11	12.46	12.58	12.69	12.80	12.92	13.03	13.14	13.26	13.37	13.48
12	13.60	13.71	13.82	13.94	14.05	14.16	14.28	14.39	14.50	14.62
13	14.73	14.84	14.96	15.07	15.18	15.30	15.41	15.52	15.64	15.75
14	15.86	15.98	16.09	16.20	16.32	16.43	16.54	16.66	16.77	16.88
15	17.00	17.11	17.22	17.33	17.45	17.56	17.67	17.79	17.90	18.01
16	18.13	18.24	18.35	18.47	18.58	18.69	18.81	18.92	19.03	19.15
17	19.26	19.37	19.49	19.60	19.71	19.83	19.94	20.05	20.17	20.28
18	20.39	20.51	20.62	20.73	20.85	20.96	21.07	21.19	21.30	21.41
19	21.53	21.64	21.75	21.87	21.98	22.09	22.21	22.32	22.43	22.55
20	22.66	22.77	22.89	23.00	23.11	23.23	23.34	23.45	23.57	23.68
21	23.79	23.91	24.02	24.13	24.25	24.36	24.47	24.59	24.70	24.81
22	24.93	25.04	25.15	25.27	25.38	25.49	25.61	25.72	25.83	25.95
23	26.06	26.17	26.29	26.40	26.51	26.63	26.74	26.85	26.97	27.08
24	27.19	27.31	27.42	27.53	27.65	27.76	27.87	27.99	28.10	28.21
25	28.33	28.44	28.55	28.66	28.78	28.89	29.00	29.12	29.23	29.34
26	29.46	29.57	29.68	29.80	29.91	30.02	30.14	30.25	30.36	30.48
27	30.59	30.70	30.82	30.93	31.04	31.16	31.27	31.38	31.50	31.61
28	31.72	31.84	31.95	32.06	32.18	32.29	32.40	32.52	32.63	32.74
29	32.86	32.97	33.08	33.20	33.31	33.42	33.54	33.65	33.76	33.88
30	33.99	34.10	34.22	34.33	34.44	34.56	34.67	34.78	34.90	35.01

Inches and Fractions As Decimals of a Foot

Inches	Fractions Of Inches							
	0	1/8	1/4	3/8	1/2	5/8	3/4	7/8
0	0.0000	0.0104	0.0208	0.0313	0.0417	0.0521	0.0625	0.0729
1	0.0833	0.0938	0.1042	0.1146	0.1250	0.1354	0.1458	0.1563
2	0.1667	0.1771	0.1875	0.1979	0.2083	0.2188	0.2292	0.2396
3	0.2500	0.2604	0.2708	0.2813	0.2917	0.3021	0.3125	0.3229
4	0.3333	0.3438	0.3542	0.3646	0.3750	0.3854	0.3958	0.4063
5	0.4167	0.4271	0.4375	0.4479	0.4583	0.4688	0.4792	0.4896
6	0.5000	0.5104	0.5208	0.5313	0.5417	0.5521	0.5625	0.5729
7	0.5833	0.5938	0.6042	0.6146	0.6250	0.6354	0.6458	0.6563
8	0.6667	0.6771	0.6875	0.6979	0.7083	0.7188	0.7292	0.7396
9	0.7500	0.7604	0.7708	0.7813	0.7917	0.8021	0.8125	0.8229
10	0.8333	0.8438	0.8542	0.8646	0.8750	0.8854	0.8958	0.9063
11	0.9167	0.9271	0.9375	0.9479	0.9583	0.9688	0.9792	0.9896
12	1.0000	—	—	—	—	—	—	—

Inches and Fractions in Centimeters

Inches	Fractions of Inches							
	0	1/8	1/4	3/8	1/2	5/8	3/4	7/8
0	0.0000	0.3175	0.6350	0.9525	1.2700	1.5875	1.9050	2.2225
1	2.5400	2.8575	3.1750	3.4925	3.8100	4.1275	4.4450	4.7625
2	5.0800	5.3975	5.7150	6.0325	6.3500	6.6675	6.9850	7.3025
3	7.6200	7.9375	8.2550	8.5725	8.8900	9.2075	9.5250	9.8425
4	10.1600	10.4775	10.7950	11.1125	11.4300	11.7475	12.0650	12.3825
5	12.7000	13.0175	13.3350	13.6525	13.9700	14.2875	14.6050	14.9225
6	15.2400	15.5575	15.8750	16.1925	16.5100	16.8275	17.1450	17.4625
7	17.7800	18.0975	18.4150	18.7325	19.0500	19.3675	19.6850	20.0025
8	20.3200	20.6375	20.9550	21.2725	21.5900	21.9075	22.2250	22.5425
9	22.8600	23.1775	23.4950	23.8125	24.1300	24.4475	24.7650	25.0825
10	25.4000	25.7175	26.0350	26.3525	26.6700	26.9875	27.3050	27.6225
11	27.9400	28.2575	28.5750	28.8925	29.2100	29.5275	29.8450	30.1625
12	30.4800	—	—	—	—	—	—	—

Decimal Equivalents of Common Fractions

8ths	16ths	32nds	64ths	Decimal	8ths	16ths	32nds	64ths	Decimal
			1	0.015625				33	0.515625
		1	2	0.03125			17	34	0.53125
			3	0.046875				35	0.546875
	1	2	4	0.0625		9	18	36	0.5625
			5	0.078125				37	0.578125
		3	6	0.09375			19	38	0.59375
			7	0.109375				39	0.609375
1	2	4	8	0.125	5	10	20	40	0.625
			9	0.140625				41	0.640625
		5	10	0.15625			21	42	0.65625
			11	0.171875				43	0.671875
	3	6	12	0.1875		11	22	44	0.6875
			13	0.203125				45	0.703125
		7	14	0.21875			23	46	0.71875
			15	0.234375				47	0.734375
2	4	8	16	0.25	6	12	24	48	0.75
			17	0.265625				49	0.765625
		9	18	0.28125			25	50	0.78125
			19	0.296875				51	0.796875
	5	10	20	0.3125		13	26	52	0.8125
			21	0.328125				53	0.828125
		11	22	0.34375			27	54	0.84375
			23	0.359375				55	0.859375
3	6	12	24	0.375	7	14	28	56	0.875
			25	0.390625				57	0.890625
		13	26	0.40625			29	58	0.90625
			27	0.421875				59	0.921875
	7	14	28	0.4375		15	30	60	0.9375
			29	0.453125				61	0.953125
		15	30	0.46875			31	62	0.96875
			31	0.484375				63	0.984375
4	8	16	32	0.5	8	16	32	64	1

Decimal Time Conversions

Seconds = Minutes or Minutes = Hours		Seconds = Minutes or Minutes = Hours		Hours = Days	
Seconds / Minutes	Minutes / Hours	Seconds / Minutes	Minutes / Hours	Hours	Days
1	0.017	31	0.517	1	0.042
2	0.033	32	0.533	2	0.083
3	0.050	33	0.550	3	0.125
4	0.067	34	0.567	4	0.167
5	0.083	35	0.583	5	0.208
6	0.100	36	0.600	6	0.250
7	0.117	37	0.617	7	0.292
8	0.133	38	0.633	8	0.333
9	0.150	39	0.650	9	0.375
10	0.167	40	0.667	10	0.417
11	0.183	41	0.683	11	0.458
12	0.200	42	0.700	12	0.500
13	0.217	43	0.717	13	0.542
14	0.233	44	0.733	14	0.583
15	0.250	45	0.750	15	0.625
16	0.267	46	0.767	16	0.667
17	0.283	47	0.783	17	0.708
18	0.300	48	0.800	18	0.750
19	0.317	49	0.817	19	0.792
20	0.333	50	0.833	20	0.833
21	0.350	51	0.850	21	0.875
22	0.367	52	0.867	22	0.917
23	0.383	53	0.883	23	0.958
24	0.400	54	0.900	24	1.000
25	0.417	55	0.917		
26	0.433	56	0.933		
27	0.450	57	0.950		
28	0.467	58	0.967		
29	0.483	59	0.983		
30	0.500	60	1.000		

Minutes and Seconds to Decimal Parts of Degrees

Min	Deg	Sec	Deg	Deg	Min	Sec	Deg	Min	Sec
0	0.00000	0	0.00000	0.00	0	00	0.60	36	
1	0.01667	1	0.00028	0.01	0	36	0.61	36	36
2	0.03333	2	0.00056	0.02	1	12	0.62	37	12
3	0.05000	3	0.00083	0.03	1	48	0.63	37	48
4	0.06667	4	0.00111	0.04	2	24	0.64	38	24
5	0.08333	5	0.00139	0.05	3		0.65	39	
6	0.10000	6	0.00167	0.06	3	36	0.66	39	36
7	0.11667	7	0.00194	0.07	4	12	0.67	40	12
8	0.13333	8	0.00222	0.08	4	48	0.68	40	48
9	0.15000	9	0.00250	0.09	5	24	0.69	41	24
10	0.16667	10	0.00278	0.10	6		0.70	42	
11	0.18333	11	0.00306	0.11	6	36	0.71	42	36
12	0.20000	12	0.00333	0.12	7	12	0.72	43	12
13	0.21667	13	0.00361	0.13	7	48	0.73	43	48
14	0.23333	14	0.00389	0.14	8	24	0.74	44	24
15	0.25000	15	0.00417	0.15	9		0.75	45	
16	0.26667	16	0.00444	0.16	9	36	0.76	45	36
17	0.28333	17	0.00472	0.17	10	12	0.77	46	12
18	0.30000	18	0.00500	0.18	10	48	0.78	46	48
19	0.31667	19	0.00528	0.19	11	24	0.79	47	24
20	0.33333	20	0.00556	0.20	12		0.80	48	
21	0.35000	21	0.00583	0.21	12	36	0.81	48	36
22	0.36667	22	0.00611	0.22	13	12	0.82	49	12
23	0.38333	23	0.00639	0.23	13	48	0.83	49	48
24	0.40000	24	0.00667	0.24	14	24	0.84	50	24
25	0.41667	25	0.00694	0.25	15		0.85	51	
26	0.43333	26	0.00722	0.26	15	36	0.86	51	36
27	0.45000	27	0.00750	0.27	16	12	0.87	52	12
28	0.46667	28	0.00778	0.28	16	48	0.88	52	48
29	0.48333	29	0.00806	0.29	17	24	0.89	53	24
30	0.50000	30	0.00833	0.30	18		0.90	54	
31	0.51667	31	0.00861	0.31	18	36	0.91	54	36
32	0.53333	32	0.00889	0.32	19	12	0.92	55	12
33	0.55000	33	0.00917	0.33	19	48	0.93	55	48
34	0.56667	34	0.00944	0.34	20	24	0.94	56	24
35	0.58333	35	0.00972	0.35	21		0.9	57	
36	0.60000	36	0.01000	0.36	21	36	0.96	57	36
37	0.61667	37	0.01028	0.37	22	12	0.97	58	12
38	0.63333	38	0.01056	0.38	22	48	0.98	58	48
39	0.65000	39	0.01083	0.39	23	24	0.99	59	24
40	0.66667	40	0.01111	0.40	24		1.00	60	
41	0.68333	41	0.01139	0.41	24	36			
42	0.70000	42	0.01167	0.42	25	12			
43	0.71667	43	0.01194	0.43	25	48			
44	0.73333	44	0.01222	0.44	26	24			
45	0.75000	45	0.01250	0.45	27				
46	0.76667	46	0.01278	0.46	27	36			
47	0.78333	47	0.01306	0.47	28	12			
48	0.80000	48	0.01333	0.48	28	48			
49	0.81667	49	0.01361	0.49	29	24			
50	0.83333	50	0.01389	0.50	30				
51	0.85000	51	0.01417	0.51	30	36			
52	0.86667	52	0.01444	0.52	31	12			
53	0.88333	53	0.01472	0.53	31	48			
54	0.90000	54	0.01500	0.54	32	24			
55	0.91667	55	0.01528	0.55	33				
56	0.93333	56	0.01556	0.56	33	36			
57	0.95000	57	0.01583	0.57	34	12			
58	0.96667	58	0.01611	0.58	34	48			
59	0.98333	59	0.01639	0.59	35	24			
60	1.00000	60	0.01667	0.60	36				

Degrees-Radians Conversion

Degrees	Radians	Degrees	Radians	Degrees	Radians	Degrees	Radians
10	0.174533	1	0.017453	0.1	0.001745	0.01	0.000175
20	0.349066	2	0.034907	0.2	0.003491	0.02	0.000349
30	0.523599	3	0.052360	0.3	0.005236	0.03	0.000524
40	0.698132	4	0.069813	0.4	0.006981	0.04	0.000698
50	0.872665	5	0.087266	0.5	0.008727	0.05	0.000873
60	1.047198	6	0.104720	0.6	0.010472	0.06	0.001047
70	1.221730	7	0.122173	0.7	0.012217	0.07	0.001222
80	1.396263	8	0.139626	0.8	0.013963	0.08	0.001396
90	1.570796	9	0.157080	0.9	0.015708	0.09	0.001571

Radians	Degrees	Radians	Degrees	Radians	Degrees	Radians	Degrees
1	57.2958	0.1	5.7296	0.01	0.5730	0.001	0.0573
2	114.5916	0.2	11.4592	0.02	1.1459	0.002	0.1146
3	171.8873	0.3	17.1887	0.03	1.7189	0.003	0.1719
4	229.1831	0.4	22.9183	0.04	2.2918	0.004	0.2292
5	286.4789	0.5	28.6479	0.05	2.8648	0.005	0.2865
6	343.7747	0.6	34.3775	0.06	3.4377	0.006	0.3438
7	401.0705	0.7	40.1070	0.07	4.0107	0.007	0.4011
8	458.3662	0.8	45.8366	0.08	4.5837	0.008	0.4584
9	515.6620	0.9	51.5662	0.09	5.1566	0.009	0.5157
10	572.9578	1.00	57.2958	0.10	5.7296	0.010	0.5730

Compounding and Discounting Formulas

The tables in this chapter were computed from the following standard formulas:

Compounding factor for 1 (what an initial amount becomes when growing at compound interest):

$$s = (1 + i)^n$$

where: s = sum at end of period
i = interest rate
n = number of periods

Compounding factor for 1 per annum (growth of equal year-end deposits all growing at compound interest):

$$s = \frac{(1 + i)^n - 1}{i}$$

Sinking fund factor (level deposit required each period to reach 1 by the end of n periods):

$$s = \frac{i}{(1 + i)^n - 1}$$

Discount factor (how much 1 at a future date is worth today):

$$v^n = \frac{1}{(1 + i)^n}$$

where: v^n = present worth of 1 at the end of period n with an interest rate of i per period

Present worth of annuity factor (how much 1 received or paid per period for n periods is worth today with interest rate of i per period):

$$a = \frac{1 - \dfrac{1}{(1 + i)^n}}{i}$$

where: a = present worth of an annuity of 1 per period at end of period n with interest rate of i per period

Capital recovery factor (payment per period that will repay a loan of 1 in n periods with compound interest on unpaid balance):

$$p = \frac{i}{1 - \dfrac{1}{(1 + i)^n}}$$

where: p = payment per period

NOTE: If formulas are used for periods of n *months,* then interest rate i must be *per month* (or annual interest rate ÷ 12).

Amount of Monthly Payment (Principal and Interest) to Repay $1,000 Loan — for Term and Annual Interest Rate Shown

Years	6%	7%	8%	9%	10%	11%	12%	13%	14%	15%	16%	17%	18%	19%	20%
1	$86.07	$86.53	$86.99	$87.45	$87.92	$88.38	$88.85	$89.32	$89.79	$90.26	$90.73	$91.20	$91.68	$92.16	$92.63
2	44.32	44.77	45.23	45.68	46.14	46.61	47.07	47.54	48.01	48.49	48.96	49.44	49.92	50.41	50.90
3	30.42	30.88	31.34	31.80	32.27	32.74	33.21	33.69	34.18	34.67	35.16	35.65	36.15	36.66	37.16
5	19.33	19.80	20.28	20.76	21.25	21.74	22.24	22.75	23.27	23.79	24.32	24.85	25.39	25.94	26.49
6	16.57	17.05	17.53	18.03	18.53	19.03	19.55	20.07	20.61	21.15	21.69	22.25	22.81	23.38	23.95
7	14.61	15.09	15.59	16.09	16.60	17.12	17.65	18.19	18.74	19.30	19.86	20.44	21.02	21.61	22.21
8	13.14	13.63	14.14	14.65	15.17	15.71	16.25	16.81	17.37	17.95	18.53	19.12	19.72	20.33	20.95
9	12.01	12.51	13.02	13.54	14.08	14.63	15.18	15.75	16.33	16.92	17.53	18.14	18.76	19.39	20.03
10	11.10	11.61	12.13	12.67	13.22	13.78	14.35	14.93	15.53	16.13	16.75	17.38	18.02	18.67	19.33
11	10.37	10.88	11.42	11.96	12.52	13.09	13.68	14.28	14.89	15.51	16.14	16.79	17.44	18.11	18.79
12	9.76	10.28	10.82	11.38	11.95	12.54	13.13	13.75	14.37	15.01	15.66	16.32	16.99	17.67	18.37
13	9.25	9.78	10.33	10.90	11.48	12.08	12.69	13.31	13.95	14.60	15.27	15.94	16.63	17.33	18.04
14	8.81	9.35	9.91	10.49	11.08	11.69	12.31	12.95	13.60	14.27	14.95	15.64	16.34	17.05	17.77
15	8.44	8.99	9.56	10.14	10.75	11.37	12.00	12.65	13.32	14.00	14.69	15.39	16.10	16.83	17.56
16	8.11	8.67	9.25	9.85	10.46	11.09	11.74	12.40	13.08	13.77	14.47	15.19	15.91	16.65	17.39
17	7.83	8.40	8.98	9.59	10.21	10.85	11.51	12.19	12.87	13.58	14.29	15.02	15.76	16.50	17.26
18	7.58	8.16	8.75	9.36	10.00	10.65	11.32	12.00	12.70	13.42	14.14	14.88	15.63	16.38	17.15
19	7.36	7.94	8.55	9.17	9.81	10.47	11.15	11.85	12.56	13.28	14.02	14.76	15.52	16.29	17.06
20	7.16	7.75	8.36	9.00	9.65	10.32	11.01	11.72	12.44	13.17	13.91	14.67	15.43	16.21	16.99
21	6.99	7.58	8.20	8.85	9.51	10.19	10.89	11.60	12.33	13.07	13.82	14.59	15.36	16.14	16.93
22	6.83	7.43	8.06	8.71	9.38	10.07	10.78	11.50	12.24	12.99	13.75	14.52	15.30	16.09	16.88
23	6.69	7.30	7.93	8.59	9.27	9.97	10.69	11.42	12.16	12.92	13.69	14.46	15.25	16.04	16.84
24	6.56	7.18	7.82	8.49	9.17	9.88	10.60	11.34	12.10	12.86	13.63	14.42	15.21	16.01	16.81
25	6.44	7.07	7.72	8.39	9.09	9.80	10.53	11.28	12.04	12.81	13.59	14.38	15.17	15.98	16.78
30	6.00	6.65	7.34	8.05	8.78	9.52	10.29	11.06	11.85	12.64	13.45	14.26	15.07	15.89	16.71

Discount Factor – How much 1 at a future date is worth today

Year	1%	3%	5%	6%	8%	10%	12%	14%	16%	18%	20%
1	0.990	0.971	0.952	0.943	0.926	0.909	0.893	0.877	0.862	0.847	0.833
2	0.980	0.943	0.907	0.890	0.857	0.826	0.797	0.769	0.743	0.718	0.694
3	0.971	0.915	0.864	0.840	0.794	0.751	0.712	0.675	0.641	0.609	0.579
4	0.961	0.888	0.823	0.792	0.735	0.683	0.636	0.592	0.552	0.516	0.482
5	0.951	0.863	0.784	0.747	0.681	0.621	0.567	0.519	0.476	0.437	0.402
6	0.942	0.837	0.746	0.705	0.630	0.564	0.507	0.456	0.410	0.370	0.335
7	0.933	0.813	0.711	0.665	0.583	0.513	0.452	0.400	0.354	0.314	0.279
8	0.923	0.789	0.677	0.627	0.540	0.467	0.404	0.351	0.305	0.266	0.233
9	0.914	0.766	0.645	0.592	0.500	0.424	0.361	0.308	0.263	0.225	0.194
10	0.905	0.744	0.614	0.558	0.463	0.386	0.322	0.270	0.227	0.191	0.162
11	0.896	0.722	0.585	0.527	0.429	0.350	0.287	0.237	0.195	0.162	0.135
12	0.887	0.701	0.557	0.497	0.397	0.319	0.257	0.208	0.168	0.137	0.112
13	0.879	0.681	0.530	0.469	0.368	0.290	0.229	0.182	0.145	0.116	0.093
14	0.870	0.661	0.505	0.442	0.340	0.263	0.205	0.160	0.125	0.099	0.078
15	0.861	0.642	0.481	0.417	0.315	0.239	0.183	0.140	0.108	0.084	0.065
16	0.853	0.623	0.458	0.394	0.292	0.218	0.163	0.123	0.093	0.071	0.054
17	0.844	0.605	0.436	0.371	0.270	0.198	0.146	0.108	0.080	0.060	0.045
18	0.836	0.587	0.416	0.350	0.250	0.180	0.130	0.095	0.069	0.051	0.038
19	0.828	0.570	0.396	0.331	0.232	0.164	0.116	0.083	0.060	0.043	0.031
20	0.820	0.554	0.377	0.312	0.215	0.149	0.104	0.073	0.051	0.037	0.026
21	0.811	0.538	0.359	0.294	0.199	0.135	0.093	0.064	0.044	0.031	0.022
22	0.803	0.522	0.342	0.278	0.184	0.123	0.083	0.056	0.038	0.026	0.018
23	0.795	0.507	0.326	0.262	0.170	0.112	0.074	0.049	0.033	0.022	0.015
24	0.788	0.492	0.310	0.247	0.158	0.102	0.066	0.043	0.028	0.019	0.013
25	0.780	0.478	0.295	0.233	0.146	0.092	0.059	0.038	0.024	0.016	0.010

Compounding Factor for 1

What an initial amount becomes when growing at compound interest

Year	1%	3%	5%	6%	8%	10%	12%	14%	16%	18%	20%
1	1.010	1.030	1.050	1.060	1.080	1.100	1.120	1.140	1.160	1.180	1.200
2	1.020	1.061	1.103	1.124	1.166	1.210	1.254	1.300	1.346	1.392	1.440
3	1.030	1.093	1.158	1.191	1.260	1.331	1.405	1.482	1.561	1.643	1.728
4	1.041	1.126	1.216	1.262	1.360	1.464	1.574	1.689	1.811	1.939	2.074
5	1.051	1.159	1.276	1.338	1.469	1.611	1.762	1.925	2.100	2.288	2.488
6	1.062	1.194	1.340	1.419	1.587	1.772	1.974	2.195	2.436	2.700	2.986
7	1.072	1.230	1.407	1.504	1.714	1.949	2.211	2.502	2.826	3.185	3.583
8	1.083	1.267	1.477	1.594	1.851	2.144	2.476	2.853	3.278	3.759	4.300
9	1.094	1.305	1.551	1.689	1.999	2.358	2.773	3.252	3.803	4.435	5.160
10	1.105	1.344	1.629	1.791	2.159	2.594	3.106	3.707	4.411	5.234	6.192
11	1.116	1.384	1.710	1.898	2.332	2.853	3.479	4.226	5.117	6.176	7.430
12	1.127	1.426	1.796	2.012	2.518	3.138	3.896	4.818	5.936	7.288	8.916
13	1.138	1.469	1.886	2.133	2.720	3.452	4.363	5.492	6.886	8.599	10.699
14	1.149	1.513	1.980	2.261	2.937	3.797	4.887	6.261	7.988	10.147	12.839
15	1.161	1.558	2.079	2.397	3.172	4.177	5.474	7.138	9.266	11.974	15.407
16	1.173	1.605	2.183	2.540	3.426	4.595	6.130	8.137	10.748	14.129	18.488
17	1.184	1.653	2.292	2.693	3.700	5.054	6.866	9.276	12.468	16.672	22.186
18	1.196	1.702	2.407	2.854	3.996	5.560	7.690	10.575	14.463	19.673	26.623
19	1.208	1.754	2.527	3.026	4.316	6.116	8.613	12.056	16.777	23.214	31.948
20	1.220	1.806	2.653	3.207	4.661	6.727	9.646	13.743	19.461	27.393	38.338
21	1.232	1.860	2.786	3.400	5.034	7.400	10.804	15.668	22.574	32.324	46.005
22	1.245	1.916	2.925	3.604	5.437	8.140	12.100	17.861	26.186	38.142	55.206
23	1.257	1.974	3.072	3.820	5.871	8.954	13.552	20.362	30.376	45.008	66.247
24	1.270	2.033	3.225	4.049	6.341	9.850	15.179	23.212	35.236	53.109	79.497
25	1.282	2.094	3.386	4.292	6.848	10.835	17.000	26.462	40.874	62.669	95.396

Compound Factor for 1 per Annum

Growth of equal year-end deposits all growing at compound interest

Year	1%	3%	5%	6%	8%	10%	12%	14%	16%	18%	20%
1	1.000	1.000	1.000	1.000	1.000	1.000	1.000	1.000	1.000	1.000	1.000
2	2.010	2.030	2.050	2.060	2.080	2.100	2.120	2.140	2.160	2.180	2.200
3	3.030	3.091	3.153	3.184	3.246	3.310	3.374	3.440	3.506	3.572	3.640
4	4.060	4.184	4.310	4.375	4.506	4.641	4.779	4.921	5.066	5.215	5.368
5	5.101	5.309	5.526	5.637	5.867	6.105	6.353	6.610	6.877	7.154	7.442
6	6.152	6.468	6.802	6.975	7.336	7.716	8.115	8.536	8.977	9.442	9.930
7	7.214	7.662	8.142	8.394	8.923	9.487	10.089	10.730	11.414	12.142	12.916
8	8.286	8.892	9.549	9.897	10.637	11.436	12.300	13.233	14.240	15.327	16.499
9	9.369	10.159	11.027	11.491	12.488	13.579	14.776	16.085	17.519	19.086	20.799
10	10.462	11.464	12.578	13.181	14.487	15.937	17.549	19.337	21.321	23.521	25.959
11	11.567	12.808	14.207	14.972	16.645	18.531	20.655	23.045	25.733	28.755	32.150
12	12.683	14.192	15.917	16.870	18.977	21.384	24.133	27.271	30.850	34.931	39.581
13	13.809	15.618	17.713	18.882	21.495	24.523	28.029	32.089	36.786	42.219	48.497
14	14.947	17.086	19.599	21.015	24.215	27.975	32.393	37.581	43.672	50.818	59.196
15	16.097	18.599	21.579	23.276	27.152	31.772	37.280	43.842	51.660	60.965	72.035
16	17.258	20.157	23.657	25.673	30.324	35.950	42.753	50.980	60.925	72.939	87.442
17	18.430	21.762	25.840	28.213	33.750	40.545	48.884	59.118	71.673	87.068	105.931
18	19.615	23.414	28.132	30.906	37.450	45.599	55.750	68.394	84.141	103.740	128.117
19	20.811	25.117	30.539	33.760	41.446	51.159	63.440	78.969	98.603	123.414	154.740
20	22.019	26.870	33.066	36.786	45.762	57.275	72.052	91.025	115.380	146.628	186.688
21	23.239	28.676	35.719	39.993	50.423	64.002	81.699	104.768	134.841	174.021	225.026
22	24.472	30.537	38.505	43.392	55.457	71.403	92.503	120.436	157.415	206.345	271.031
23	25.716	32.453	41.430	46.996	60.893	79.543	104.603	138.297	183.601	244.487	326.237
24	26.973	34.426	44.502	50.816	66.765	88.497	118.155	158.659	213.978	289.494	392.484
25	28.243	36.459	47.727	54.865	73.106	98.347	133.334	181.871	249.214	342.603	471.981

Present Worth of an Annuity Factor

How much 1 received or paid annually for X years is worth today

Year	1%	3%	5%	6%	8%	10%	12%	14%	16%	18%	20%
1	0.990	0.971	0.952	0.943	0.926	0.909	0.893	0.877	0.862	0.847	0.833
2	1.970	1.913	1.859	1.833	1.783	1.736	1.690	1.647	1.605	1.566	1.528
3	2.941	2.829	2.723	2.673	2.577	2.487	2.402	2.322	2.246	2.174	2.106
4	3.902	3.717	3.546	3.465	3.312	3.170	3.037	2.914	2.798	2.690	2.589
5	4.853	4.580	4.329	4.212	3.993	3.791	3.605	3.433	3.274	3.127	2.991
6	5.795	5.417	5.076	4.917	4.623	4.355	4.111	3.889	3.685	3.498	3.326
7	6.728	6.230	5.786	5.582	5.206	4.868	4.564	4.288	4.039	3.812	3.605
8	7.652	7.020	6.463	6.210	5.747	5.335	4.968	4.639	4.344	4.078	3.837
9	8.566	7.786	7.108	6.802	6.247	5.759	5.328	4.946	4.607	4.303	4.031
10	9.471	8.530	7.722	7.360	6.710	6.145	5.650	5.216	4.833	4.494	4.192
11	10.368	9.253	8.306	7.887	7.139	6.495	5.938	5.453	5.029	4.656	4.327
12	11.255	9.954	8.863	8.384	7.536	6.814	6.194	5.660	5.197	4.793	4.439
13	12.134	10.635	9.394	8.853	7.904	7.103	6.424	5.842	5.342	4.910	4.533
14	13.004	11.296	9.899	9.295	8.244	7.367	6.628	6.002	5.468	5.008	4.611
15	13.865	11.938	10.380	9.712	8.559	7.606	6.811	6.142	5.575	5.092	4.675
16	14.718	12.561	10.838	10.106	8.851	7.824	6.974	6.265	5.668	5.162	4.730
17	15.562	13.166	11.274	10.477	9.122	8.022	7.120	6.373	5.749	5.222	4.775
18	16.398	13.754	11.690	10.828	9.372	8.201	7.250	6.467	5.818	5.273	4.812
19	17.226	14.324	12.085	11.158	9.604	8.365	7.366	6.550	5.877	5.316	4.843
20	18.046	14.877	12.462	11.470	9.818	8.514	7.469	6.623	5.929	5.353	4.870
21	18.857	15.415	12.821	11.764	10.017	8.649	7.562	6.687	5.973	5.384	4.891
22	19.660	15.937	13.163	12.042	10.201	8.772	7.645	6.743	6.011	5.410	4.909
23	20.456	16.444	13.489	12.303	10.371	8.883	7.718	6.792	6.044	5.432	4.925
24	21.243	16.936	13.799	12.550	10.529	8.985	7.784	6.835	6.073	5.451	4.937
25	22.023	17.413	14.094	12.783	10.675	9.077	7.843	6.873	6.097	5.467	4.948

Sinking Fund Factor
Level deposit required each year to reach 1 by given year

Year	1%	3%	5%	6%	8%	10%	12%	14%	16%	18%	20%
1	1.000	1.000	1.000	1.000	1.000	1.000	1.000	1.000	1.000	1.000	1.000
2	0.498	0.493	0.488	0.485	0.481	0.476	0.472	0.467	0.463	0.459	0.455
3	0.330	0.324	0.317	0.314	0.308	0.302	0.296	0.291	0.285	0.280	0.275
4	0.246	0.239	0.232	0.229	0.222	0.215	0.209	0.203	0.197	0.192	0.186
5	0.196	0.188	0.181	0.177	0.170	0.164	0.157	0.151	0.145	0.140	0.134
6	0.163	0.155	0.147	0.143	0.136	0.130	0.123	0.117	0.111	0.106	0.101
7	0.139	0.131	0.123	0.119	0.112	0.105	0.099	0.093	0.088	0.082	0.077
8	0.121	0.112	0.105	0.101	0.094	0.087	0.081	0.076	0.070	0.065	0.061
9	0.107	0.098	0.091	0.087	0.080	0.074	0.068	0.062	0.057	0.052	0.048
10	0.096	0.087	0.080	0.076	0.069	0.063	0.057	0.052	0.047	0.043	0.039
11	0.086	0.078	0.070	0.067	0.060	0.054	0.048	0.043	0.039	0.035	0.031
12	0.079	0.070	0.063	0.059	0.053	0.047	0.041	0.037	0.032	0.029	0.025
13	0.072	0.064	0.056	0.053	0.047	0.041	0.036	0.031	0.027	0.024	0.021
14	0.067	0.059	0.051	0.048	0.041	0.036	0.031	0.027	0.023	0.020	0.017
15	0.062	0.054	0.046	0.043	0.037	0.031	0.027	0.023	0.019	0.016	0.014
16	0.058	0.050	0.042	0.039	0.033	0.028	0.023	0.020	0.016	0.014	0.011
17	0.054	0.046	0.039	0.035	0.030	0.025	0.020	0.017	0.014	0.011	0.009
18	0.051	0.043	0.036	0.032	0.027	0.022	0.018	0.015	0.012	0.010	0.008
19	0.048	0.040	0.033	0.030	0.024	0.020	0.016	0.013	0.010	0.008	0.006
20	0.045	0.037	0.030	0.027	0.022	0.017	0.014	0.011	0.009	0.007	0.005
21	0.043	0.035	0.028	0.025	0.020	0.016	0.012	0.010	0.007	0.006	0.004
22	0.041	0.033	0.026	0.023	0.018	0.014	0.011	0.008	0.006	0.005	0.004
23	0.039	0.031	0.024	0.021	0.016	0.013	0.010	0.007	0.005	0.004	0.003
24	0.037	0.029	0.022	0.020	0.015	0.011	0.008	0.006	0.005	0.003	0.003
25	0.035	0.027	0.021	0.018	0.014	0.010	0.007	0.005	0.004	0.003	0.002

Capital Recovery Factor

Annual payment that will repay a $1 loan in X years, with compound interest on the unpaid balance

Year	1%	3%	5%	6%	8%	10%	12%	14%	16%	18%	20%
1	1.010	1.030	1.050	1.060	1.080	1.100	1.120	1.140	1.160	1.180	1.200
2	0.508	0.523	0.538	0.545	0.561	0.576	0.592	0.607	0.623	0.639	0.655
3	0.340	0.354	0.367	0.374	0.388	0.402	0.416	0.431	0.445	0.460	0.475
4	0.256	0.269	0.282	0.289	0.302	0.315	0.329	0.343	0.357	0.372	0.386
5	0.206	0.218	0.231	0.237	0.250	0.264	0.277	0.291	0.305	0.320	0.334
6	0.173	0.185	0.197	0.203	0.216	0.230	0.243	0.257	0.271	0.286	0.301
7	0.149	0.161	0.173	0.179	0.192	0.205	0.219	0.233	0.248	0.262	0.277
8	0.131	0.142	0.155	0.161	0.174	0.187	0.201	0.216	0.230	0.245	0.261
9	0.117	0.128	0.141	0.147	0.160	0.174	0.188	0.202	0.217	0.232	0.248
10	0.106	0.117	0.130	0.136	0.149	0.163	0.177	0.192	0.207	0.223	0.239
11	0.096	0.108	0.120	0.127	0.140	0.154	0.168	0.183	0.199	0.215	0.231
12	0.089	0.100	0.113	0.119	0.133	0.147	0.161	0.177	0.192	0.209	0.225
13	0.082	0.094	0.106	0.113	0.127	0.141	0.156	0.171	0.187	0.204	0.221
14	0.077	0.089	0.101	0.108	0.121	0.136	0.151	0.167	0.183	0.200	0.217
15	0.072	0.084	0.096	0.103	0.117	0.131	0.147	0.163	0.179	0.196	0.214
16	0.068	0.080	0.092	0.099	0.113	0.128	0.143	0.160	0.176	0.194	0.211
17	0.064	0.076	0.089	0.095	0.110	0.125	0.140	0.157	0.174	0.191	0.209
18	0.061	0.073	0.086	0.092	0.107	0.122	0.138	0.155	0.172	0.190	0.208
19	0.058	0.070	0.083	0.090	0.104	0.120	0.136	0.153	0.170	0.188	0.206
20	0.055	0.067	0.080	0.087	0.102	0.117	0.134	0.151	0.169	0.187	0.205
21	0.053	0.065	0.078	0.085	0.100	0.116	0.132	0.150	0.167	0.186	0.204
22	0.051	0.063	0.076	0.083	0.098	0.114	0.131	0.148	0.166	0.185	0.204
23	0.049	0.061	0.074	0.081	0.096	0.113	0.130	0.147	0.165	0.184	0.203
24	0.047	0.059	0.072	0.080	0.095	0.111	0.128	0.146	0.165	0.183	0.203
25	0.045	0.057	0.071	0.078	0.094	0.110	0.127	0.145	0.164	0.183	0.202

Areas and Volumes

Area of a square:	Area = length × width, or length of one side squared
Area of a rectangle:	Area = length × width
Area of a triangle:	Area = $\frac{\text{base} \times \text{perpendicular height}}{2}$
Area of a circle:	Area = square of the radius × pi (3.1416)
Area of an ellipse:	Area = long diameter × short diameter × 0.7854
Area of a pentagon (5 sides):	Area = square of length of one side × 1.720
Area of a hexagon (6 sides):	Area = square of length of one side × 2.598
Area of an octagon (8 sides):	Area = square of length of one side × 4.828
Area (surface) of a cube:	Area = square of length of one side × 6
Area (surface) of a sphere:	Area = square of the diameter × pi (3.1416)
Area (surface) of a cylinder:	Area = diameter × height × pi (3.1416)
Circumference of a circle:	Circumference = diameter × pi (3.1416)
Area of an annulus:	Area = (square of O.D. − square of I.D.) × 0.7854
Volume of a cube:	Volume = length of one side cubed, or x^3
Volume of a sphere:	Volume = cube of the radius (r^3) × pi (3.1416) × $\frac{4}{3}$
Volume of a cylinder:	Volume = base radius squared (r^2) × pi (3.1416) × height
Volume of a pyramid:	Volume = area of the base × height × $\frac{1}{3}$
Volume of a cone:	Volume = base radius squared (r^2) × pi (3.1416) × ⅓ height
Volume of rectangular solid:	Volume = length × width × height

Triangles

Solution of Right Triangles:

Of the three sides and two acute angles, two of the parts (one being a side) must be known:

$$c^2 = a^2 + b^2$$

$$\text{angle } A + \text{angle } B = 90 \text{ degrees}$$

$$\sin A = a/c = \cos B$$

$$\cos A = b/c = \sin B$$

$$\tan A = a/b = \cot B$$

$$\cot A = b/a = \tan B$$

Solution of Oblique Triangles:

$$\text{angle } A + \text{angle } B + \text{angle } C = 180 \text{ degrees}$$

$$a/\sin A = b/\sin B = c/\sin C \qquad \text{(law of sines)}$$

$$a^2 = b^2 + c^2 - 2bc \cos A \qquad \text{(law of cosines)}$$

$$\frac{(a+b)}{(a-b)} = \frac{\tan 1/2\,(A+B)}{\tan 1/2\,(A-B)} \qquad \text{(law of tangents)}$$

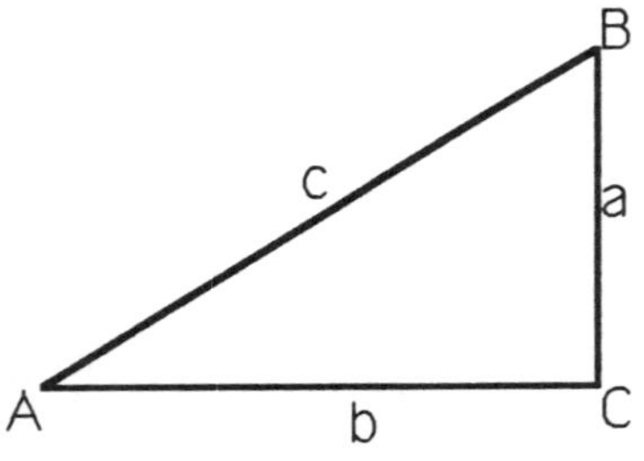

Right Triangle

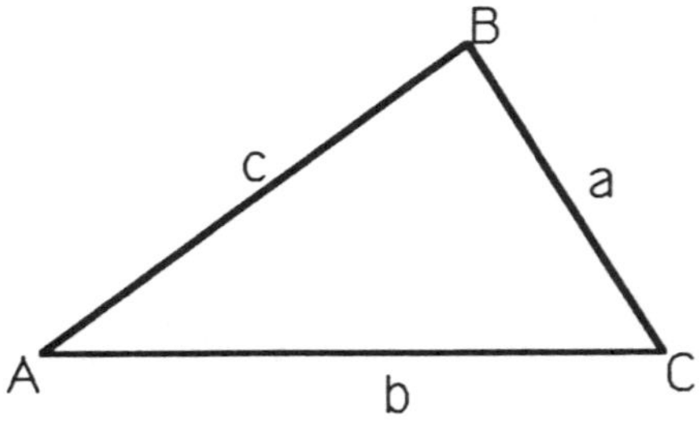

Oblique Triangle

Area and Circumference of Circles

Diam.	Area	Circum.	Diam.	Area	Circum.	Diam.	Area	Circum.
0	0	0	7	38.4845	21.9911	17	226.980	53.4071
1/8	0.0123	0.3927	1/8	39.8712	22.3838	1/2	240.528	54.9779
1/4	0.0491	0.7854	1/4	41.2825	22.7765	18	254.469	56.5487
3/8	0.1104	1.1781	3/8	42.7183	23.1692	1/2	268.803	58.1195
1/2	0.1963	1.5708	1/2	44.1786	23.5619	19	283.529	59.6903
5/8	0.3068	1.9635	5/8	45.6635	23.9546	1/2	298.648	61.2611
3/4	0.4418	2.3562	3/4	47.1730	24.3473	20	314.159	62.8319
7/8	0.6013	2.7489	7/8	48.7070	24.7400	1/2	330.064	64.4026
1	0.7854	3.1416	8	50.2655	25.1327	21	346.361	65.9734
1/8	0.9940	3.5343	1/8	51.8486	25.5254	1/2	363.050	67.5442
1/4	1.2272	3.9270	1/4	53.4562	25.9181	22	380.133	69.1150
3/8	1.4849	4.3197	3/8	55.0883	26.3108	1/2	397.608	70.6858
1/2	1.7671	4.7124	1/2	56.7450	26.7035	23	415.476	72.2566
5/8	2.0739	5.1051	5/8	58.4263	27.0962	1/2	433.736	73.8274
3/4	2.4053	5.4978	3/4	60.1320	27.4889	24	452.389	75.3982
7/8	2.7612	5.8905	7/8	61.8624	27.8816	1/2	471.435	76.9690
2	3.1416	6.2832	9	63.6173	28.2743	25	490.874	78.5398
1/8	3.5466	6.6759	1/8	65.3967	28.6670	1/2	510.705	80.1106
1/4	3.9761	7.0686	1/4	67.2006	29.0597	26	530.929	81.6814
3/8	4.4301	7.4613	3/8	69.0291	29.4524	1/2	551.546	83.2522
1/2	4.9087	7.8540	1/2	70.8822	29.8451	27	572.555	84.8230
5/8	5.4119	8.2467	5/8	72.7598	30.2378	1/2	593.957	86.3938
3/4	5.9396	8.6394	3/4	74.6619	30.6305	28	615.752	87.9646
7/8	6.4918	9.0321	7/8	76.5886	31.0232	1/2	637.940	89.5354
3	7.0686	9.4248	10	78.5398	31.4159	29	660.520	91.1062
1/8	7.6699	9.8175	1/8	80.5156	31.8086	1/2	683.493	92.6770
1/4	8.2958	10.2102	1/4	82.5159	32.2013	30	706.858	94.2478
3/8	8.9462	10.6029	3/8	84.5407	32.5940	31	754.768	97.3894
1/2	9.6211	10.9956	1/2	86.5901	32.9867	32	804.248	100.5310
5/8	10.3206	11.3883	5/8	88.6641	33.3794	33	855.299	103.6726
3/4	11.0447	11.7810	3/4	90.7626	33.7721	34	907.920	106.8142
7/8	11.7932	12.1737	7/8	92.8856	34.1648	35	962.113	109.9557
4	12.5664	12.5664	11	95.0332	34.5575	36	1017.876	113.0973
1/8	13.3640	12.9591	1/8	97.2053	34.9502	37	1075.210	116.2389
1/4	14.1863	13.3518	1/4	99.4020	35.3429	38	1134.115	119.3805
3/8	15.0330	13.7445	3/8	101.6232	35.7356	39	1194.591	122.5221
1/2	15.9043	14.1372	1/2	103.8689	36.1283	40	1256.637	125.6637
5/8	16.8002	14.5299	5/8	106.1392	36.5210	45	1590.431	141.3717
3/4	17.7205	14.9226	3/4	108.4340	36.9137	50	1963.495	157.0796
7/8	18.6655	15.3153	7/8	110.7534	37.3064	55	2375.829	172.7876
5	19.6350	15.7080	12	113.0973	37.6991	60	2827.433	188.4956
1/8	20.6290	16.1007	1/8	115.4658	38.0918	65	3318.307	204.2035
1/4	21.6475	16.4934	1/4	117.8588	38.4845	70	3848.451	219.9115
3/8	22.6906	16.8861	3/8	120.2764	38.8772	75	4417.865	235.6194
1/2	23.7583	17.2788	1/2	122.7185	39.2699	80	5026.548	251.3274
5/8	24.8505	17.6715	5/8	125.1851	39.6626	85	5674.502	267.0354
3/4	25.9672	18.0642	3/4	127.6763	40.0553	90	6361.725	282.7433
7/8	27.1085	18.4569	7/8	130.1920	40.4480	95	7088.218	298.4513
6	28.2743	18.8496	13	132.7323	40.8407	100	7853.982	314.1593
1/8	29.4647	19.2423	1/2	143.1388	42.4115	105	8659.015	329.8672
1/4	30.6796	19.6350	14	153.9380	43.9823	110	9503.318	345.5752
3/8	31.9191	20.0277	1/2	165.1300	45.5531	115	10386.891	361.2832
1/2	33.1831	20.4204	15	176.7146	47.1239	120	11309.734	376.9911
5/8	34.4716	20.8131	1/2	188.6919	48.6947	130	13273.229	408.4070
3/4	35.7847	21.2058	16	201.0619	50.2655	140	15393.804	439.8230
7/8	37.1223	21.5984	1/2	213.8246	51.8363	150	17671.459	471.2389

Natural Trigonometric Functions

Angle	Sine	Cosine	Tangent	Cotangent	Secant	Cosecant	
0	0.0000	1.0000	0.0000	—	1.0000	—	90
1	0.0175	0.9998	0.0175	57.2900	1.0002	57.2987	89
2	0.0349	0.9994	0.0349	28.6363	1.0006	28.6537	88
3	0.0523	0.9986	0.0524	19.0811	1.0014	19.1073	87
4	0.0698	0.9976	0.0699	14.3007	1.0024	14.3356	86
5	0.0872	0.9962	0.0875	11.4301	1.0038	11.4737	85
6	0.1045	0.9945	0.1051	9.5144	1.0055	9.5668	84
7	0.1219	0.9925	0.1228	8.1443	1.0075	8.2055	83
8	0.1392	0.9903	0.1405	7.1154	1.0098	7.1853	82
9	0.1564	0.9877	0.1584	6.3138	1.0125	6.3925	81
10	0.1736	0.9848	0.1763	5.6713	1.0154	5.7588	80
11	0.1908	0.9816	0.1944	5.1446	1.0187	5.2408	79
12	0.2079	0.9781	0.2126	4.7046	1.0223	4.8097	78
13	0.2250	0.9744	0.2309	4.3315	1.0263	4.4454	77
14	0.2419	0.9703	0.2493	4.0108	1.0306	4.1336	76
15	0.2588	0.9659	0.2679	3.7321	1.0353	3.8637	75
16	0.2756	0.9613	0.2867	3.4874	1.0403	3.6280	74
17	0.2924	0.9563	0.3057	3.2709	1.0457	3.4203	73
18	0.3090	0.9511	0.3249	3.0777	1.0515	3.2361	72
19	0.3256	0.9455	0.3443	2.9042	1.0576	3.0716	71
20	0.3420	0.9397	0.3640	2.7475	1.0642	2.9238	70
21	0.3584	0.9336	0.3839	2.6051	1.0711	2.7904	69
22	0.3746	0.9272	0.4040	2.4751	1.0785	2.6695	68
23	0.3907	0.9205	0.4245	2.3559	1.0864	2.5593	67
24	0.4067	0.9135	0.4452	2.2460	1.0946	2.4586	66
25	0.4226	0.9063	0.4663	2.1445	1.1034	2.3662	65
26	0.4384	0.8988	0.4877	2.0503	1.1126	2.2812	64
27	0.4540	0.8910	0.5095	1.9626	1.1223	2.2027	63
28	0.4695	0.8829	0.5317	1.8807	1.1326	2.1301	62
29	0.4848	0.8746	0.5543	1.8040	1.1434	2.0627	61
30	0.5000	0.8660	0.5774	1.7321	1.1547	2.0000	60
31	0.5150	0.8572	0.6009	1.6643	1.1666	1.9416	59
32	0.5299	0.8480	0.6249	1.6003	1.1792	1.8871	58
33	0.5446	0.8387	0.6494	1.5399	1.1924	1.8361	57
34	0.5592	0.8290	0.6745	1.4826	1.2062	1.7883	56
35	0.5736	0.8192	0.7002	1.4281	1.2208	1.7434	55
36	0.5878	0.8090	0.7265	1.3764	1.2361	1.7013	54
37	0.6018	0.7986	0.7536	1.3270	1.2521	1.6616	53
38	0.6157	0.7880	0.7813	1.2799	1.2690	1.6243	52
39	0.6293	0.7771	0.8098	1.2349	1.2868	1.5890	51
40	0.6428	0.7660	0.8391	1.1918	1.3054	1.5557	50
41	0.6561	0.7547	0.8693	1.1504	1.3250	1.5243	49
42	0.6691	0.7431	0.9004	1.1106	1.3456	1.4945	48
43	0.6820	0.7314	0.9325	1.0724	1.3673	1.4663	47
44	0.6947	0.7193	0.9657	1.0355	1.3902	1.4396	46
45	0.7071	0.7071	1.0000	1.0000	1.4142	1.4142	45
	Cosine	Sine	Cotangent	Tangent	Cosecant	Secant	Angle

Numerical Table

Number	Square	Cube	Square Root	Cube Root	Reciprocal
1	1	1	1.0000000	1.0000000	1.000000000
2	4	8	1.4142136	1.2598919	0.500000000
3	9	27	1.7320508	1.4421968	0.333333333
4	16	64	2.0000000	1.5873277	0.250000000
5	25	125	2.2360680	1.7098842	0.200000000
6	36	216	2.4494897	1.8170121	0.166666667
7	49	343	2.6457513	1.9128071	0.142857143
8	64	512	2.8284271	1.9998614	0.125000000
9	81	729	3.0000000	2.0799315	0.111111111
10	100	1,000	3.1622777	2.1542693	0.100000000
11	121	1,331	3.3166248	2.2238023	0.090909091
12	144	1,728	3.4641016	2.2892389	0.083333333
13	169	2,197	3.6055513	2.3511337	0.076923077
14	196	2,744	3.7416574	2.4099303	0.071428571
15	225	3,375	3.8729833	2.4659895	0.066666667
16	256	4,096	4.0000000	2.5196092	0.062500000
17	289	4,913	4.1231056	2.5710388	0.058823529
18	324	5,832	4.2426407	2.6204889	0.055555556
19	361	6,859	4.3588989	2.6681398	0.052631579
20	400	8,000	4.4721360	2.7141466	0.050000000
21	441	9,261	4.5825757	2.7586442	0.047619048
22	484	10,648	4.6904158	2.8017506	0.045454545
23	529	12,167	4.7958315	2.8435698	0.043478261
24	576	13,824	4.8989795	2.8841936	0.041666667
25	625	15,625	5.0000000	2.9237040	0.040000000
26	676	17,576	5.0990195	2.9621743	0.038461538
27	729	19,683	5.1961524	2.9996704	0.037037037
28	784	21,952	5.2915026	3.0362517	0.035714286
29	841	24,389	5.3851648	3.0719720	0.034482759
30	900	27,000	5.4772256	3.1068802	0.033333333
31	961	29,791	5.5677644	3.1410211	0.032258065
32	1,024	32,768	5.6568542	3.1744354	0.031250000
33	1,089	35,937	5.7445626	3.2071605	0.030303030
34	1,156	39,304	5.8309519	3.2392310	0.029411765
35	1,225	42,875	5.9160798	3.2706787	0.028571429
36	1,296	46,656	6.0000000	3.3015329	0.027777778
37	1,369	50,653	6.0827625	3.3318208	0.027027027
38	1,444	54,872	6.1644140	3.3615678	0.026315789
39	1,521	59,319	6.2449980	3.3907973	0.025641026
40	1,600	64,000	6.3245553	3.4195314	0.025000000
41	1,681	68,921	6.4031242	3.4477904	0.024390244
42	1,764	74,088	6.4807407	3.4755936	0.023809524
43	1,849	79,507	6.5574385	3.5029589	0.023255814
44	1,936	85,184	6.6332496	3.5299030	0.022727273
45	2,025	91,125	6.7082039	3.5564420	0.022222222
46	2,116	97,336	6.7823300	3.5825906	0.021739130
47	2,209	103,823	6.8556546	3.6083630	0.021276596
48	2,304	110,592	6.9282032	3.6337723	0.020833333
49	2,401	117,649	7.0000000	3.6588310	0.020408163
50	2,500	125,000	7.0710678	3.6835511	0.020000000

Numerical Table (Cont'd)

Number	Square	Cube	Square Root	Cube Root	Reciprocal
51	2,601	132,651	7.1414284	3.7079438	0.019607843
52	2,704	140,608	7.2111026	3.7320196	0.019230769
53	2,809	148,877	7.2801099	3.7557887	0.018867925
54	2,916	157,464	7.3484692	3.7792606	0.018518519
55	3,025	166,375	7.4161985	3.8024445	0.018181818
56	3,136	175,616	7.4833148	3.8253491	0.017857143
57	3,249	185,193	7.5498344	3.8479825	0.017543860
58	3,364	195,112	7.6157731	3.8703528	0.017241379
59	3,481	205,379	7.6811457	3.8924673	0.016949153
60	3,600	216,000	7.7459667	3.9143334	0.016666667
61	3,721	226,981	7.8102497	3.9359578	0.016393443
62	3,844	238,328	7.8740079	3.9573472	0.016129032
63	3,969	250,047	7.9372539	3.9785077	0.015873016
64	4,096	262,144	8.0000000	3.9994455	0.015625000
65	4,225	274,625	8.0622577	4.0201663	0.015384615
66	4,356	287,496	8.1240384	4.0406757	0.015151515
67	4,489	300,763	8.1853528	4.0609789	0.014925373
68	4,624	314,432	8.2462113	4.0810811	0.014705882
69	4,761	328,509	8.3066239	4.1009871	0.014492754
70	4,900	343,000	8.3666003	4.1207017	0.014285714
71	5,041	357,911	8.4261498	4.1402294	0.014084507
72	5,184	373,248	8.4852814	4.1595746	0.013888889
73	5,329	389,017	8.5440037	4.1787415	0.013698630
74	5,476	405,224	8.6023253	4.1977342	0.013513514
75	5,625	421,875	8.6602540	4.2165565	0.013333333
76	5,776	438,976	8.7177979	4.2352122	0.013157895
77	5,929	456,533	8.7749644	4.2537049	0.012987013
78	6,084	474,552	8.8317609	4.2720382	0.012820513
79	6,241	493,039	8.8881944	4.2902155	0.012658228
80	6,400	512,000	8.9442719	4.3082400	0.012500000
81	6,561	531,441	9.0000000	4.3261150	0.012345679
82	6,724	551,368	9.0553851	4.3438434	0.012195122
83	6,889	571,787	9.1104336	4.3614282	0.012048193
84	7,056	592,704	9.1651514	4.3788724	0.011904762
85	7,225	614,125	9.2195445	4.3961786	0.011764706
86	7,396	636,056	9.2736185	4.4133496	0.011627907
87	7,569	658,503	9.3273791	4.4303880	0.011494253
88	7,744	681,472	9.3808315	4.4472964	0.011363636
89	7,921	704,969	9.4339811	4.4640771	0.011235955
90	8,100	729,000	9.4868330	4.4807326	0.011111111
91	8,281	753,571	9.5393920	4.4972652	0.010989011
92	8,464	778,688	9.5916630	4.5136771	0.010869565
93	8,649	804,357	9.6436508	4.5299704	0.010752688
94	8,836	830,584	9.6953597	4.5461474	0.010638298
95	9,025	857,375	9.7467943	4.5622101	0.010526316
96	9,216	884,736	9.7979590	4.5781604	0.010416667
97	9,409	912,673	9.8488578	4.5940003	0.010309278
98	9,604	941,192	9.8994949	4.6097317	0.010204082
99	9,801	970,299	9.9498744	4.6253565	0.010101010
100	10,000	1,000,000	10.0000000	4.6408764	0.010000000

Numerical Table (Cont'd)

Number	Square	Cube	Square Root	Cube Root	Reciprocal
101	10,201	1,030,301	10.0498756	4.6562931	0.009900990
102	10,404	1,061,208	10.0995049	4.6716085	0.009803922
103	10,609	1,092,727	10.1488916	4.6868240	0.009708738
104	10,816	1,124,864	10.1980390	4.7019414	0.009615385
105	11,025	1,157,625	10.2469508	4.7169622	0.009523810
106	11,236	1,191,016	10.2956301	4.7318879	0.009433962
107	11,449	1,225,043	10.3440804	4.7467200	0.009345794
108	11,664	1,259,712	10.3923048	4.7614600	0.009259259
109	11,881	1,295,029	10.4403065	4.7761092	0.009174312
110	12,100	1,331,000	10.4880885	4.7906692	0.009090909
111	12,321	1,367,631	10.5356538	4.8051411	0.009009009
112	12,544	1,404,928	10.5830052	4.8195264	0.008928571
113	12,769	1,442,897	10.6301458	4.8338264	0.008849558
114	12,996	1,481,544	10.6770783	4.8480421	0.008771930
115	13,225	1,520,875	10.7238053	4.8621750	0.008695652
116	13,456	1,560,896	10.7703296	4.8762262	0.008620690
117	13,689	1,601,613	10.8166538	4.8901969	0.008547009
118	13,924	1,643,032	10.8627805	4.9040882	0.008474576
119	14,161	1,685,159	10.9087121	4.9179012	0.008403361
120	14,400	1,728,000	10.9544512	4.9316371	0.008333333
121	14,641	1,771,561	11.0000000	4.9452968	0.008264463
122	14,884	1,815,848	11.0453610	4.9588815	0.008196721
123	15,129	1,860,867	11.0905365	4.9723922	0.008130081
124	15,376	1,906,624	11.1355287	4.9858298	0.008064516
125	15,625	1,953,125	11.1803399	4.9991953	0.008000000
126	15,876	2,000,376	11.2249722	5.0124898	0.007936508
127	16,129	2,048,383	11.2694277	5.0257141	0.007874016
128	16,384	2,097,152	11.3137085	5.0388692	0.007812500
129	16,641	2,146,689	11.3578167	5.0519559	0.007751938
130	16,900	2,197,000	11.4017543	5.0649752	0.007692308
131	17,161	2,248,091	11.4455231	5.0779278	0.007633588
132	17,424	2,299,968	11.4891253	5.0908147	0.007575758
133	17,689	2,352,637	11.5325626	5.1036367	0.007518797
134	17,956	2,406,104	11.5758369	5.1163946	0.007462687
135	18,225	2,460,375	11.6189500	5.1290891	0.007407407
136	18,496	2,515,456	11.6619038	5.1417211	0.007352941
137	18,769	2,571,353	11.7046999	5.1542914	0.007299270
138	19,044	2,628,072	11.7473401	5.1668006	0.007246377
139	19,321	2,685,619	11.7898261	5.1792495	0.007194245
140	19,600	2,744,000	11.8321596	5.1916389	0.007142857
141	19,881	2,803,221	11.8743421	5.2039694	0.007092199
142	20,164	2,863,288	11.9163753	5.2162417	0.007042254
143	20,449	2,924,207	11.9582607	5.2284565	0.006993007
144	20,736	2,985,984	12.0000000	5.2406146	0.006944444
145	21,025	3,048,625	12.0415946	5.2527164	0.006896552
146	21,316	3,112,136	12.0830460	5.2647628	0.006849315
147	21,609	3,176,523	12.1243557	5.2767542	0.006802721
148	21,904	3,241,792	12.1655251	5.2886914	0.006756757
149	22,201	3,307,949	12.2065556	5.3005750	0.006711409
150	22,500	3,375,000	12.2474487	5.3124055	0.006666667

Numerical Table (Cont'd)

Number	Square	Cube	Square Root	Cube Root	Reciprocal
151	22,801	3,442,951	12.2882057	5.3241835	0.006622517
152	23,104	3,511,808	12.3288280	5.3359097	0.006578947
153	23,409	3,581,577	12.3693169	5.3475845	0.006535948
154	23,716	3,652,264	12.4096736	5.3592085	0.006493506
155	24,025	3,723,875	12.4498996	5.3707824	0.006451613
156	24,336	3,796,416	12.4899960	5.3823065	0.006410256
157	24,649	3,869,893	12.5299641	5.3937816	0.006369427
158	24,964	3,944,312	12.5698051	5.4052080	0.006329114
159	25,281	4,019,679	12.6095202	5.4165862	0.006289308
160	25,600	4,096,000	12.6491106	5.4279169	0.006250000
161	25,921	4,173,281	12.6885775	5.4392005	0.006211180
162	26,244	4,251,528	12.7279221	5.4504374	0.006172840
163	26,569	4,330,747	12.7671453	5.4616282	0.006134969
164	26,896	4,410,944	12.8062485	5.4727732	0.006097561
165	27,225	4,492,125	12.8452326	5.4838731	0.006060606
166	27,556	4,574,296	12.8840987	5.4949282	0.006024096
167	27,889	4,657,463	12.9228480	5.5059391	0.005988024
168	28,224	4,741,632	12.9614814	5.5169060	0.005952381
169	28,561	4,826,809	13.0000000	5.5278295	0.005917160
170	28,900	4,913,000	13.0384048	5.5387100	0.005882353
171	29,241	5,000,211	13.0766968	5.5495479	0.005847953
172	29,584	5,088,448	13.1148770	5.5603436	0.005813953
173	29,929	5,177,717	13.1529464	5.5710976	0.005780347
174	30,276	5,268,024	13.1909060	5.5818102	0.005747126
175	30,625	5,359,375	13.2287566	5.5924818	0.005714286
176	30,976	5,451,776	13.2664992	5.6031129	0.005681818
177	31,329	5,545,233	13.3041347	5.6137037	0.005649718
178	31,684	5,639,752	13.3416641	5.6242548	0.005617978
179	32,041	5,735,339	13.3790882	5.6347664	0.005586592
180	32,400	5,832,000	13.4164079	5.6452389	0.005555556
181	32,761	5,929,741	13.4536240	5.6556727	0.005524862
182	33,124	6,028,568	13.4907376	5.6660681	0.005494505
183	33,489	6,128,487	13.5277493	5.6764256	0.005464481
184	33,856	6,229,504	13.5646600	5.6867453	0.005434783
185	34,225	6,331,625	13.6014705	5.6970278	0.005405405
186	34,596	6,434,856	13.6381817	5.7072732	0.005376344
187	34,969	6,539,203	13.6747943	5.7174820	0.005347594
188	35,344	6,644,672	13.7113092	5.7276545	0.005319149
189	35,721	6,751,269	13.7477271	5.7377909	0.005291005
190	36,100	6,859,000	13.7840488	5.7478917	0.005263158
191	36,481	6,967,871	13.8202750	5.7579571	0.005235602
192	36,864	7,077,888	13.8564065	5.7679874	0.005208333
193	37,249	7,189,057	13.8924440	5.7779829	0.005181347
194	37,636	7,301,384	13.9283883	5.7879439	0.005154639
195	38,025	7,414,875	13.9642400	5.7978708	0.005128205
196	38,416	7,529,536	14.0000000	5.8077638	0.005102041
197	38,809	7,645,373	14.0356688	5.8176233	0.005076142
198	39,204	7,762,392	14.0712473	5.8274494	0.005050505
199	39,601	7,880,599	14.1067360	5.8372424	0.005025126
200	40,000	8,000,000	14.1421356	5.8470027	0.005000000

Logarithms (Base 10)

N	0	1	2	3	4	5	6	7	8	9
10	0000	0043	0086	0128	0170	0212	0253	0294	0334	0374
11	0414	0453	0492	0531	0569	0607	0645	0682	0719	0755
12	0792	0828	0864	0899	0934	0969	1004	1038	1072	1106
13	1139	1173	1206	1239	1271	1303	1335	1367	1399	1430
14	1461	1492	1523	1553	1584	1614	1644	1673	1703	1732
15	1761	1790	1818	1847	1875	1903	1931	1959	1987	2014
16	2041	2068	2095	2122	2148	2175	2201	2227	2253	2279
17	2304	2330	2355	2380	2405	2430	2455	2480	2504	2529
18	2553	2577	2601	2625	2648	2672	2695	2718	2742	2765
19	2788	2810	2833	2856	2878	2900	2923	2945	2967	2989
20	3010	3032	3054	3075	3096	3118	3139	3160	3181	3201
21	3222	3243	3263	3284	3304	3324	3345	3365	3385	3404
22	3424	3444	3464	3483	3502	3522	3541	3560	3579	3598
23	3617	3636	3655	3674	3692	3711	3729	3747	3766	3784
24	3802	3820	3838	3856	3874	3892	3909	3927	3945	3962
25	3979	3997	4014	4031	4048	4065	4082	4099	4116	4133
26	4150	4166	4183	4200	4216	4232	4249	4265	4281	4298
27	4314	4330	4346	4362	4378	4393	4409	4425	4440	4456
28	4472	4487	4502	4518	4533	4548	4564	4579	4594	4609
29	4624	4639	4654	4669	4683	4698	4713	4728	4742	4757
30	4771	4786	4800	4814	4829	4843	4857	4871	4886	4900
31	4914	4928	4942	4955	4969	4983	4997	5011	5024	5038
32	5051	5065	5079	5092	5105	5119	5132	5145	5159	5172
33	5185	5198	5211	5224	5237	5250	5263	5276	5289	5302
34	5315	5328	5340	5353	5366	5378	5391	5403	5416	5428
35	5441	5453	5465	5478	5490	5502	5514	5527	5539	5551
36	5563	5575	5587	5599	5611	5623	5635	5647	5658	5670
37	5682	5694	5705	5717	5729	5740	5752	5763	5775	5786
38	5798	5809	5821	5832	5843	5855	5866	5877	5888	5899
39	5911	5922	5933	5944	5955	5966	5977	5988	5999	6010
40	6021	6031	6042	6053	6064	6075	6085	6096	6107	6117
41	6128	6138	6149	6160	6170	6180	6191	6201	6212	6222
42	6232	6243	6253	6263	6274	6284	6294	6304	6314	6325
43	6335	6345	6355	6365	6375	6385	6395	6405	6415	6425
44	6435	6444	6454	6464	6474	6484	6493	6503	6513	6522
45	6532	6542	6551	6561	6571	6580	6590	6599	6609	6618
46	6628	6637	6646	6656	6665	6675	6684	6693	6702	6712
47	6721	6730	6739	6749	6758	6767	6776	6785	6794	6803
48	6812	6821	6830	6839	6848	6857	6866	6875	6884	6893
49	6902	6911	6920	6928	6937	6946	6955	6964	6972	6981
50	6990	6998	7007	7016	7024	7033	7042	7050	7059	7067
51	7076	7084	7093	7101	7110	7118	7126	7135	7143	7152
52	7160	7168	7177	7185	7193	7202	7210	7218	7226	7235
53	7243	7251	7259	7267	7275	7284	7292	7300	7308	7316
54	7324	7332	7340	7348	7356	7364	7372	7380	7388	7396
N	0	1	2	3	4	5	6	7	8	9

Log π = 0.49715   Log (Base e) = 2.30259 × Log (Base 10)

Logarithms (Base 10) (Cont'd)

N	0	1	2	3	4	5	6	7	8	9
55	7404	7412	7419	7427	7435	7443	7451	7459	7466	7474
56	7482	7490	7497	7505	7513	7520	7528	7536	7543	7551
57	7559	7566	7574	7582	7589	7597	7604	7612	7619	7627
58	7634	7642	7649	7657	7664	7672	7679	7686	7694	7701
59	7709	7716	7723	7731	7738	7745	7752	7760	7767	7774
60	7782	7789	7796	7803	7810	7818	7825	7832	7839	7846
61	7853	7860	7868	7875	7882	7889	7896	7903	7910	7917
62	7924	7931	7938	7945	7952	7959	7966	7973	7980	7987
63	7993	8000	8007	8014	8021	8028	8035	8041	8048	8055
64	8062	8069	8075	8082	8089	8096	8102	8109	8116	8122
65	8129	8136	8142	8149	8156	8162	8169	8176	8182	8189
66	8195	8202	8209	8215	8222	8228	8235	8241	8248	8254
67	8261	8267	8274	8280	8287	8293	8299	8306	8312	8319
68	8325	8331	8338	8344	8351	8357	8363	8370	8376	8382
69	8388	8395	8401	8407	8414	8420	8426	8432	8439	8445
70	8451	8457	8463	8470	8476	8482	8488	8494	8500	8506
71	8513	8519	8525	8531	8537	8543	8549	8555	8561	8567
72	8573	8579	8585	8591	8597	8603	8609	8615	8621	8627
73	8633	8639	8645	8651	8657	8663	8669	8675	8681	8686
74	8692	8698	8704	8710	8716	8722	8727	8733	8739	8745
75	8751	8756	8762	8768	8774	8779	8785	8791	8797	8802
76	8808	8814	8820	8825	8831	8837	8842	8848	8854	8859
77	8865	8871	8876	8882	8887	8893	8899	8904	8910	8915
78	8921	8927	8932	8938	8943	8949	8954	8960	8965	8971
79	8976	8982	8987	8993	8998	9004	9009	9015	9020	9025
80	9031	9036	9042	9047	9053	9058	9063	9069	9074	9079
81	9085	9090	9096	9101	9106	9112	9117	9122	9128	9133
82	9138	9143	9149	9154	9159	9165	9170	9175	9180	9186
83	9191	9196	9201	9206	9212	9217	9222	9227	9232	9238
84	9243	9248	9253	9258	9263	9269	9274	9279	9284	9289
85	9294	9299	9304	9309	9315	9320	9325	9330	9335	9340
86	9345	9350	9355	9360	9365	9370	9375	9380	9385	9390
87	9395	9400	9405	9410	9415	9420	9425	9430	9435	9440
88	9445	9450	9455	9460	9465	9469	9474	9479	9484	9489
89	9494	9499	9504	9509	9513	9518	9523	9528	9533	9538
90	9542	9547	9552	9557	9562	9566	9571	9576	9581	9586
91	9590	9595	9600	9605	9609	9614	9619	9624	9628	9633
92	9638	9643	9647	9652	9657	9661	9666	9671	9675	9680
93	9685	9689	9694	9699	9703	9708	9713	9717	9722	9727
94	9731	9736	9741	9745	9750	9754	9759	9763	9768	9773
95	9777	9782	9786	9791	9795	9800	9805	9809	9814	9818
96	9823	9827	9832	9836	9841	9845	9850	9854	9859	9863
97	9868	9872	9877	9881	9886	9890	9894	9899	9903	9908
98	9912	9917	9921	9926	9930	9934	9939	9943	9948	9952
99	9956	9961	9965	9969	9974	9978	9983	9987	9991	9996
N	0	1	2	3	4	5	6	7	8	9

Natural (Napierian) Logarithms

N	0.0	0.1	0.2	0.3	0.4	0.5	0.6	0.7	0.8	0.9
1	0.0000	0.0953	0.1823	0.2624	0.3365	0.4055	0.4700	0.5306	0.5878	0.6419
2	0.6931	0.7419	0.7885	0.8329	0.8755	0.9163	0.9555	0.9933	1.0296	1.0647
3	1.0986	1.1314	1.1632	1.1939	1.2238	1.2528	1.2809	1.3083	1.3350	1.3610
4	1.3863	1.4110	1.4351	1.4586	1.4816	1.5041	1.5261	1.5476	1.5686	1.5892
5	1.6094	1.6292	1.6487	1.6677	1.6864	1.7047	1.7228	1.7405	1.7579	1.7750
6	1.7918	1.8083	1.8245	1.8405	1.8563	1.8718	1.8871	1.9021	1.9169	1.9315
7	1.9459	1.9601	1.9741	1.9879	2.0015	2.0149	2.0281	2.0412	2.0541	2.0669
8	2.0794	2.0919	2.1041	2.1163	2.1282	2.1401	2.1518	2.1633	2.1748	2.1861
9	2.1972	2.2083	2.2192	2.2300	2.2407	2.2513	2.2618	2.2721	2.2824	2.2925
10	2.3026	2.3125	2.3224	2.3321	2.3418	2.3514	2.3609	2.3702	2.3795	2.3888

To obtain the natural logarithm of a number that is 1/10 or 10 times that given in the table, subtract from or add to the given logarithm (in the table) the logarithm of 10 (2.3026).

Two-Thirds Power of Numbers

n	.00	.01	.02	.03	.04	.05	.06	.07	.08	.09
0.0	0.000	0.046	0.074	0.097	0.117	0.136	0.153	0.170	0.186	0.201
0.1	0.215	0.230	0.243	0.257	0.270	0.282	0.295	0.307	0.319	0.330
0.2	0.342	0.353	0.364	0.375	0.386	0.397	0.407	0.418	0.428	0.438
0.3	0.448	0.458	0.468	0.478	0.487	0.497	0.506	0.515	0.525	0.534
0.4	0.543	0.552	0.561	0.570	0.578	0.587	0.596	0.604	0.613	0.622
0.5	0.630	0.638	0.647	0.655	0.663	0.671	0.679	0.687	0.695	0.703
0.6	0.711	0.719	0.727	0.735	0.743	0.750	0.758	0.766	0.773	0.781
0.7	0.788	0.796	0.803	0.811	0.818	0.825	0.833	0.840	0.847	0.855
0.8	0.862	0.869	0.876	0.883	0.890	0.897	0.904	0.911	0.918	0.925
0.9	0.932	0.939	0.946	0.953	0.960	0.966	0.973	0.980	0.987	0.993
1.0	1.000	1.007	1.013	1.020	1.026	1.033	1.040	1.046	1.053	1.059
1.1	1.066	1.072	1.078	1.085	1.091	1.098	1.104	1.110	1.117	1.123
1.2	1.129	1.136	1.142	1.148	1.154	1.160	1.167	1.173	1.179	1.185
1.3	1.191	1.197	1.203	1.209	1.215	1.222	1.228	1.234	1.240	1.246
1.4	1.251	1.257	1.263	1.269	1.275	1.281	1.287	1.293	1.299	1.305
1.5	1.310	1.316	1.322	1.328	1.334	1.339	1.345	1.351	1.357	1.362
1.6	1.368	1.374	1.379	1.385	1.391	1.396	1.402	1.408	1.413	1.419
1.7	1.424	1.430	1.436	1.441	1.447	1.452	1.458	1.463	1.469	1.474
1.8	1.480	1.485	1.491	1.496	1.502	1.507	1.512	1.518	1.523	1.529
1.9	1.534	1.539	1.545	1.550	1.556	1.561	1.566	1.572	1.577	1.582
2.0	1.587	1.593	1.598	1.603	1.609	1.614	1.619	1.624	1.629	1.635
2.1	1.640	1.645	1.650	1.656	1.661	1.666	1.671	1.676	1.681	1.686
2.2	1.692	1.697	1.702	1.707	1.712	1.717	1.722	1.727	1.732	1.737
2.3	1.742	1.748	1.753	1.758	1.763	1.768	1.773	1.778	1.783	1.788
2.4	1.793	1.798	1.803	1.808	1.812	1.817	1.822	1.827	1.832	1.837
2.5	1.842	1.847	1.852	1.857	1.862	1.867	1.871	1.876	1.881	1.886
2.6	1.891	1.896	1.901	1.905	1.910	1.915	1.920	1.925	1.929	1.934
2.7	1.939	1.944	1.949	1.953	1.958	1.963	1.968	1.972	1.977	1.982
2.8	1.987	1.991	1.996	2.001	2.006	2.010	2.015	2.020	2.024	2.029
2.9	2.034	2.038	2.043	2.048	2.052	2.057	2.062	2.066	2.071	2.076
3.0	2.080	2.085	2.089	2.094	2.099	2.103	2.108	2.112	2.117	2.122
3.1	2.126	2.131	2.135	2.140	2.144	2.149	2.153	2.158	2.163	2.167
3.2	2.172	2.176	2.181	2.185	2.190	2.194	2.199	2.203	2.208	2.212
3.3	2.217	2.221	2.226	2.230	2.235	2.239	2.243	2.248	2.252	2.257
3.4	2.261	2.266	2.270	2.274	2.279	2.283	2.288	2.292	2.297	2.301
3.5	2.305	2.310	2.314	2.318	2.323	2.327	2.332	2.336	2.340	2.345
3.6	2.349	2.353	2.358	2.362	2.366	2.371	2.375	2.379	2.384	2.388
3.7	2.392	2.397	2.401	2.405	2.410	2.414	2.418	2.422	2.427	2.431
3.8	2.435	2.440	2.444	2.448	2.452	2.457	2.461	2.465	2.469	2.474
3.9	2.478	2.482	2.486	2.490	2.495	2.499	2.503	2.507	2.512	2.516
4.0	2.520	2.524	2.528	2.533	2.537	2.541	2.545	2.549	2.553	2.558
4.1	2.562	2.566	2.570	2.574	2.578	2.583	2.587	2.591	2.595	2.599
4.2	2.603	2.607	2.612	2.616	2.620	2.624	2.628	2.632	2.636	2.640
4.3	2.644	2.649	2.653	2.657	2.661	2.665	2.669	2.673	2.677	2.681
4.4	2.685	2.689	2.693	2.697	2.702	2.706	2.710	2.714	2.718	2.722
4.5	2.726	2.730	2.734	2.738	2.742	2.746	2.750	2.754	2.758	2.762
4.6	2.766	2.770	2.774	2.778	2.782	2.786	2.790	2.794	2.798	2.802
4.7	2.806	2.810	2.814	2.818	2.822	2.826	2.830	2.834	2.838	2.842
4.8	2.846	2.850	2.854	2.858	2.861	2.865	2.869	2.873	2.877	2.881
4.9	2.885	2.889	2.893	2.897	2.901	2.905	2.909	2.912	2.916	2.920
5.0	2.924	2.928	2.932	2.936	2.940	2.944	2.948	2.951	2.955	2.959

Three-Halves Power of Numbers

n	.00	.01	.02	.03	.04	.05	.06	.07	.08	.09
0.0	0.000	0.001	0.003	0.005	0.008	0.011	0.015	0.019	0.023	0.027
0.1	0.032	0.036	0.042	0.047	0.052	0.058	0.064	0.070	0.076	0.083
0.2	0.089	0.096	0.103	0.110	0.118	0.125	0.133	0.140	0.148	0.156
0.3	0.164	0.173	0.181	0.190	0.198	0.207	0.216	0.225	0.234	0.244
0.4	0.253	0.263	0.272	0.282	0.292	0.302	0.312	0.322	0.333	0.343
0.5	0.354	0.364	0.375	0.386	0.397	0.408	0.419	0.430	0.442	0.453
0.6	0.465	0.476	0.488	0.500	0.512	0.524	0.536	0.548	0.561	0.573
0.7	0.586	0.598	0.611	0.624	0.637	0.650	0.663	0.676	0.689	0.702
0.8	0.716	0.729	0.743	0.756	0.770	0.784	0.798	0.811	0.826	0.840
0.9	0.854	0.868	0.882	0.897	0.911	0.926	0.941	0.955	0.970	0.985
1.0	1.000	1.015	1.030	1.045	1.061	1.076	1.091	1.107	1.122	1.138
1.1	1.154	1.169	1.185	1.201	1.217	1.233	1.249	1.266	1.282	1.298
1.2	1.315	1.331	1.348	1.364	1.381	1.398	1.414	1.431	1.448	1.465
1.3	1.482	1.499	1.517	1.534	1.551	1.569	1.586	1.604	1.621	1.639
1.4	1.657	1.674	1.692	1.710	1.728	1.746	1.764	1.782	1.800	1.819
1.5	1.837	1.856	1.874	1.893	1.911	1.930	1.948	1.967	1.986	2.005
1.6	2.024	2.043	2.062	2.081	2.100	2.119	2.139	2.158	2.178	2.197
1.7	2.217	2.236	2.256	2.275	2.295	2.315	2.335	2.355	2.375	2.395
1.8	2.415	2.435	2.455	2.476	2.496	2.516	2.537	2.557	2.578	2.598
1.9	2.619	2.640	2.660	2.681	2.702	2.723	2.744	2.765	2.786	2.807
2.0	2.828	2.850	2.871	2.892	2.914	2.935	2.957	2.978	3.000	3.021
2.1	3.043	3.065	3.087	3.109	3.131	3.153	3.17	3.20	3.22	3.24
2.2	3.26	3.29	3.31	3.33	3.35	3.38	3.40	3.42	3.44	3.47
2.3	3.49	3.51	3.53	3.56	3.58	3.60	3.63	3.65	3.67	3.69
2.4	3.72	3.74	3.76	3.79	3.81	3.83	3.86	3.88	3.91	3.93
2.5	3.95	3.98	4.00	4.02	4.05	4.07	4.10	4.12	4.14	4.17
2.6	4.19	4.22	4.24	4.27	4.29	4.31	4.34	4.36	4.39	4.41
2.7	4.44	4.46	4.49	4.51	4.54	4.56	4.59	4.61	4.64	4.66
2.8	4.69	4.71	4.74	4.76	4.79	4.81	4.84	4.86	4.89	4.91
2.9	4.94	4.96	4.99	5.02	5.04	5.07	5.09	5.12	5.14	5.17
3.0	5.20	5.22	5.25	5.27	5.30	5.33	5.35	5.38	5.41	5.43
3.1	5.46	5.48	5.51	5.54	5.56	5.59	5.62	5.64	5.67	5.70
3.2	5.72	5.75	5.78	5.81	5.83	5.86	5.89	5.91	5.94	5.97
3.3	5.99	6.02	6.05	6.08	6.10	6.13	6.16	6.19	6.21	6.24
3.4	6.27	6.30	6.32	6.35	6.38	6.41	6.44	6.46	6.49	6.52
3.5	6.55	6.58	6.60	6.63	6.66	6.69	6.72	6.75	6.77	6.80
3.6	6.83	6.86	6.89	6.92	6.94	6.97	7.00	7.03	7.06	7.09
3.7	7.12	7.15	7.17	7.20	7.23	7.26	7.29	7.32	7.35	7.38
3.8	7.41	7.44	7.47	7.50	7.52	7.55	7.58	7.61	7.64	7.67
3.9	7.70	7.73	7.76	7.79	7.82	7.85	7.88	7.91	7.94	7.97
4.0	8.00	8.03	8.06	8.09	8.12	8.15	8.18	8.21	8.24	8.27
4.1	8.30	8.33	8.36	8.39	8.42	8.45	8.48	8.52	8.55	8.58
4.2	8.61	8.64	8.67	8.70	8.73	8.76	8.79	8.82	8.85	8.89
4.3	8.92	8.95	8.98	9.01	9.04	9.07	9.10	9.14	9.17	9.20
4.4	9.23	9.26	9.29	9.32	9.36	9.39	9.42	9.45	9.48	9.51
4.5	9.55	9.58	9.61	9.64	9.67	9.71	9.74	9.77	9.80	9.83
4.6	9.87	9.90	9.93	9.96	9.99	10.03	10.06	10.09	10.12	10.16
4.7	10.19	10.22	10.25	10.29	10.32	10.35	10.39	10.42	10.45	10.48
4.8	10.52	10.55	10.58	10.62	10.65	10.68	10.71	10.75	10.78	10.81
4.9	10.85	10.88	10.91	10.95	10.98	11.01	11.05	11.08	11.11	11.15
5.0	11.18	11.21	11.25	11.28	11.31	11.35	11.38	11.42	11.45	11.48

Five-Halves Powers of Numbers

n	.00	.01	.02	.03	.04	.05	.06	.07	.08	.09
0.0	0.0000	0.0000	0.0001	0.0002	0.0003	0.0006	0.0009	0.0013	0.0018	0.0024
0.1	0.0032	0.0040	0.0050	0.0061	0.0073	0.0087	0.0102	0.0119	0.0137	0.0157
0.2	0.0179	0.0202	0.0227	0.0254	0.0282	0.0313	0.0345	0.0379	0.0415	0.0453
0.3	0.0493	0.0535	0.0579	0.0626	0.0674	0.0725	0.0778	0.0833	0.0890	0.0950
0.4	0.1012	0.1076	0.1143	0.1212	0.1284	0.1358	0.1435	0.1514	0.1596	0.1681
0.5	0.1768	0.1857	0.1950	0.2045	0.2143	0.2243	0.2347	0.2453	0.2562	0.2674
0.6	0.2789	0.2906	0.3027	0.3150	0.3277	0.3406	0.3539	0.3674	0.3813	0.3955
0.7	0.4100	0.4248	0.4399	0.4553	0.4711	0.4871	0.5035	0.5203	0.5373	0.5547
0.8	0.5724	0.5905	0.6089	0.6276	0.6467	0.6661	0.6859	0.7060	0.7265	0.7473
0.9	0.7684	0.7900	0.8118	0.8341	0.8567	0.8796	0.9030	0.9267	0.9507	0.9752
1.0	1.0000	1.0252	1.0508	1.0767	1.1030	1.1297	1.1568	1.1843	1.2122	1.2404
1.1	1.2691	1.2981	1.3275	1.3574	1.3876	1.4182	1.4493	1.4807	1.5125	1.5448
1.2	1.5774	1.6105	1.6440	1.6779	1.7122	1.7469	1.7821	1.8176	1.8536	1.8901
1.3	1.9269	1.9642	2.0019	2.0400	2.0786	2.1176	2.1570	2.1969	2.2372	2.2779
1.4	2.3191	2.3607	2.4028	2.4453	2.4883	2.5317	2.5756	2.6200	2.6647	2.7100
1.5	2.7557	2.8018	2.8485	2.8955	2.9431	2.9911	3.0396	3.0885	3.1379	3.1878
1.6	3.2382	3.2890	3.3403	3.3921	3.4444	3.4971	3.5503	3.6041	3.6582	3.7129
1.7	3.7681	3.8238	3.8799	3.9365	3.9937	4.0513	4.1094	4.1681	4.2272	4.2868
1.8	4.3469	4.4075	4.4687	4.5303	4.5925	4.6551	4.7183	4.7819	4.8461	4.9108
1.9	4.9760	5.0418	5.1080	5.1748	5.2421	5.3099	5.3782	5.4471	5.5165	5.5864
2.0	5.6569	5.7278	5.7993	5.8714	5.9440	6.0171	6.0907	6.1649	6.2396	6.3149
2.1	6.3907	6.4670	6.5439	6.6214	6.6994	6.7779	6.8570	6.9366	7.0168	7.0976
2.2	7.1789	7.2607	7.3432	7.4261	7.5097	7.5938	7.6784	7.7636	7.8494	7.9358
2.3	8.0227	8.1102	8.1982	8.2869	8.3761	8.4658	8.5562	8.6471	8.7386	8.8307
2.4	8.9234	9.0166	9.1104	9.2048	9.2998	9.3954	9.4916	9.5883	9.6857	9.7836
2.5	9.8821	9.9812	10.081	10.181	10.282	10.384	10.486	10.588	10.692	10.796
2.6	10.900	11.005	11.111	11.217	11.324	11.432	11.540	11.649	11.758	11.868
2.7	11.979	12.090	12.202	12.314	12.427	12.541	12.655	12.770	12.886	13.002
2.8	13.119	13.236	13.354	13.473	13.592	13.712	13.833	13.954	14.076	14.199
2.9	14.322	14.445	14.570	14.695	14.821	14.947	15.074	15.202	15.330	15.459
3.0	15.588	15.719	15.850	15.981	16.113	16.246	16.380	16.514	16.649	16.784
3.1	16.920	17.057	17.194	17.332	17.471	17.611	17.751	17.892	18.033	18.175
3.2	18.318	18.461	18.605	18.750	18.896	19.042	19.189	19.336	19.484	19.633
3.3	19.783	19.933	20.084	20.235	20.388	20.541	20.694	20.848	21.004	21.159
3.4	21.316	21.473	21.630	21.789	21.948	22.108	22.268	22.430	22.592	22.754
3.5	22.918	23.082	23.246	23.412	23.578	23.745	23.913	24.081	24.250	24.419
3.6	24.590	24.761	24.933	25.105	25.279	25.453	25.627	25.803	25.979	26.156
3.7	26.333	26.512	26.691	26.870	27.051	27.232	27.414	27.596	27.780	27.964
3.8	28.149	28.334	28.521	28.708	28.895	29.084	29.273	29.463	29.654	29.845
3.9	30.037	30.230	30.424	30.618	30.813	31.009	31.206	31.403	31.601	31.800
4.0	32.000	32.200	32.402	32.603	32.806	33.009	33.214	33.418	33.624	33.830
4.1	34.038	34.246	34.454	34.664	34.874	35.085	35.297	35.509	35.722	35.936
4.2	36.151	36.367	36.583	36.800	37.018	37.237	37.456	37.676	37.897	38.119
4.3	38.342	38.565	38.789	39.014	39.240	39.466	39.693	39.921	40.150	40.380
4.4	40.610	40.841	41.073	41.306	41.539	41.773	42.008	42.244	42.481	42.718
4.5	42.957	43.196	43.436	43.676	43.918	44.160	44.403	44.647	44.891	45.137
4.6	45.383	45.630	45.878	46.127	46.376	46.626	46.878	47.129	47.382	47.636
4.7	47.890	48.145	48.401	48.658	48.915	49.174	49.433	49.693	49.954	50.216
4.8	50.478	50.741	51.006	51.271	51.536	51.803	52.070	52.339	52.608	52.878
4.9	53.148	53.420	53.692	53.966	54.240	54.515	54.790	55.067	55.344	55.623
5.0	55.902	56.182	56.462	56.744	57.026	57.310	57.594	57.879	58.165	58.451

Length of Circular Arc for Radius of One

Angle (Deg)	Factor	Angle (Deg)	Factor	Angle (Deg)	Factor
0	0.00000	60	1.04720	120	2.09440
5	0.08727	65	1.13446	125	2.18166
10	0.17453	70	1.22173	130	2.26893
15	0.26180	75	1.30900	135	2.35619
20	0.34907	80	1.39626	140	2.44346
25	0.43633	85	1.48353	145	2.53073
30	0.52360	90	1.57080	150	2.61799
35	0.61087	95	1.65806	155	2.70526
40	0.69813	100	1.74533	160	2.79253
45	0.78540	105	1.83260	165	2.87979
50	0.87266	110	1.91986	170	2.96706
55	0.95993	115	2.00713	175	3.05433
				180	3.14159

$$\text{Arc length} = \frac{\text{central angle (deg)}}{180} \times 3.1416 \times \text{radius}$$

Chapter 2

Drilling

Introduction to Chapter 2

As stated elsewhere, this handbook is not a textbook on well drilling methods or techniques. This chapter is intended to present some of the reference materials frequently used or required by well-drilling contractors. For information on subjects relating to the actual drilling and completion techniques, the reader should consult one or more of the following reference materials:

American Water Works Association, (1984). AWWA Standard for water wells, A 100-84. Denver, Colo.

Bennison, E.W., (1947). Ground water, its development, uses, and conservation. Edward E. Johnson Co., St. Paul, Minn.

Campbell, M.D. and Lehr, J.H., (1973). Water well technology. McGraw-Hill, New York, N.Y.

Driscoll, F.G., (1986). Groundwater and wells. Johnson Division, St. Paul, Minn.

Gibson, U.P. and Singer, R.D., (1969). Small wells manual. U.S. Agency for International Development, Washington, D.C.

Gordon, R.W., (1958). Water well drilling with cable tools. Bucyrus-Erie Co., South Milwaukee, Wisconsin.

National Water Well Association, (1971). Water well drillers beginning training manual. Columbus, Ohio.

Speed Star Division of Koehring Co., (1967). Well drilling manual. Enid, Oklahoma.

U.S. Department of the Army, (1965). Well drilling. Technical Manual 5-297, Washington, D.C.

U.S. Department of Interior, Bureau of Reclamation (rev. 1981). Ground water manual. Denver, Colo.

(plus)

Various trade journals and manufacturers' literature.

Capacity of Holes

Diameter Inches	U.S. Gallons Per Foot	Cubic Ft. Per Foot	Sax Cement Per Foot*	Diameter Inches	U.S. Gallons Per Foot	Cubic Ft. Per Foot	Sax Cement Per Foot*
2	0.163	0.022	0.020	10	4.080	0.545	0.496
2½	0.255	0.034	0.031	10¼	4.287	0.573	0.521
3	0.367	0.049	0.045	10½	4.498	0.601	0.547
3½	0.500	0.067	0.061	10¾	4.715	0.630	0.573
4	0.653	0.087	0.079	11	4.937	0.660	0.600
4¼	0.737	0.099	0.090	11¼	5.164	0.690	0.628
4½	0.826	0.110	0.100	11½	5.396	0.721	0.656
4¾	0.921	0.123	0.112	11¾	5.633	0.753	0.685
5	1.020	0.136	0.124	12	5.875	0.785	0.714
5¼	1.125	0.150	0.137	12¼	6.123	0.819	0.744
5½	1.234	0.165	0.150	12½	6.375	0.852	0.775
5¾	1.349	0.180	0.164	12¾	6.633	0.887	0.806
6	1.469	0.196	0.179	13	6.895	0.922	0.838
6¼	1.594	0.213	0.194	13½	7.436	0.994	0.904
6½	1.724	0.230	0.210	14	7.997	1.069	0.972
6¾	1.859	0.249	0.226	14½	8.578	1.147	1.043
7	1.999	0.267	0.243	15	9.180	1.227	1.116
7¼	2.145	0.287	0.261	15½	9.802	1.310	1.191
7½	2.295	0.307	0.279	16	10.445	1.396	1.269
7¾	2.451	0.328	0.298	16½	11.108	1.485	1.350
8	2.611	0.349	0.317	17	11.791	1.576	1.433
8¼	2.777	0.371	0.338	18	13.219	1.767	1.607
8½	2.948	0.394	0.358	19	14.729	1.969	1.790
8¾	3.124	0.418	0.380	20	16.320	2.182	1.983
9	3.305	0.442	0.402	22	19.747	2.640	2.400
9¼	3.491	0.467	0.424	24	23.501	3.142	2.856
9½	3.682	0.492	0.448	26	27.581	3.687	3.352
9¾	3.879	0.519	0.471	28	31.987	4.276	3.888

*Cement calculations based on an average grout mixture of 1.1 cubic feet per sack of cement.

Volume Between Casing and Hole

Casing O.D. Inches	Hole Diam. Inches	U.S. Gallons Per Foot	Cubic Ft. Per Foot	Sax Cement Per Foot*
4½	5	0.1938	0.0259	0.0236
	6	0.6426	0.0859	0.0781
	7	1.1730	0.1568	0.1426
	8	1.7850	0.2386	0.2169
	10	3.2538	0.4350	0.3955
5½	6	0.2346	0.0314	0.0285
	7	0.7650	0.1023	0.0930
	8	1.3770	0.1841	0.1674
	9	2.0706	0.2768	0.2517
	10	2.8458	0.3805	0.3459
6⅝	7	0.2085	0.0279	0.0253
	8	0.8205	0.1097	0.0997
	9	1.5141	0.2024	0.1840
	10	2.2893	0.3061	0.2782
	12	4.0845	0.5461	0.4964
8⅝	9	0.2697	0.0361	0.0328
	10	1.0449	0.1397	0.1270
	11	1.9017	0.2542	0.2311
	12	2.8401	0.3797	0.3452
	14	4.9617	0.6633	0.6030
10¾	12	1.1603	0.1551	0.1410
	13	2.1803	0.2915	0.2650
	14	3.2819	0.4388	0.3989
	16	5.7299	0.7660	0.6964
	18	8.5043	1.1369	1.0336
12¾	14	1.3643	0.1824	0.1658
	15	2.5475	0.3406	0.3096
	16	3.8123	0.5097	0.4633
	18	6.5867	0.8806	0.8005
	20	9.6875	1.2951	1.1774
13⅝	15	1.6059	0.2147	0.1952
	16	2.8707	0.3838	0.3489
	17	4.2171	0.5638	0.5125
	18	5.6451	0.7547	0.6861
	20	8.7459	1.1692	1.0629
14	15	1.1832	0.1582	0.1438
	16	2.4480	0.3273	0.2975
	17	3.7944	0.5073	0.4612
	18	5.2224	0.6982	0.6347
	20	8.3232	1.1127	1.0116
16	18	2.7744	0.3709	0.3372
	19	4.2840	0.5727	0.5207
	20	5.8752	0.7855	0.7140
	22	9.3024	1.2436	1.1306
	24	13.0560	1.7455	1.5868
20	22	3.4272	0.4582	0.4165
	23	5.2632	0.7036	0.6397
	24	7.1808	0.9600	0.8727
	26	11.2608	1.5055	1.3686
	28	15.6672	2.0945	1.9041

*Cement calculations based on an average grout mixture of 1.1 cubic feet per 94-lb sack of cement.

Capacity of Suction Pits or Rectangular Tanks
(in Cubic Feet per Inch of Depth)

Width	Length 6′0″	6′4″	6′8″	7′0″	7′4″	7′8″	8′0″	8′4″	8′8″	9′0″	9′4″	9′8″	10′0″
4′0″	2.00	2.11	2.22	2.33	2.44	2.56	2.67	2.78	2.89	3.00	3.11	3.22	3.33
4′4″	2.17	2.29	2.41	2.53	2.65	2.77	2.89	3.01	3.13	3.25	3.37	3.49	3.61
4′8″	2.33	2.46	2.59	2.72	2.85	2.98	3.11	3.24	3.37	3.50	3.63	3.76	3.89
5′0″	2.50	2.64	2.78	2.92	3.05	3.19	3.33	3.47	3.61	3.75	3.89	4.03	4.17
5′4″	2.67	2.81	2.96	3.11	3.26	3.41	3.56	3.70	3.85	4.00	4.15	4.30	4.44
5′8″	2.83	2.99	3.15	3.31	3.46	3.62	3.78	3.94	4.09	4.25	4.41	4.57	4.72
6′0″	3.00	3.17	3.33	3.50	3.67	3.83	4.00	4.17	4.33	4.50	4.67	4.83	5.00
6′4″	3.17	3.34	3.52	3.69	3.87	4.05	4.22	4.40	4.57	4.75	4.93	5.10	5.28
6′8″	3.33	3.52	3.70	3.89	4.07	4.26	4.44	4.63	4.82	5.00	5.19	5.37	5.56
7′0″	3.50	3.69	3.89	4.08	4.28	4.47	4.67	4.86	5.06	5.25	5.44	5.64	5.83
7′4″	3.67	3.87	4.07	4.28	4.48	4.69	4.89	5.09	5.30	5.50	5.70	5.91	6.11
7′8″	3.83	4.04	4.26	4.47	4.68	4.90	5.11	5.32	5.54	5.75	5.96	6.18	6.39
8′0″	4.00	4.22	4.44	4.67	4.89	5.11	5.33	5.56	5.78	6.00	6.22	6.44	6.67

One cubic foot = 7.48 U.S. gallons
One cubic foot = 0.178 barrels (42 gal)
One barrel = 42 U.S. gallons
One barrel = 5.615 cubic feet

Displacement of Mud Pumps

(Gallons per Stroke at 100% Volumetric Efficiency)

Liner Size (Inches)	Single-Acting Triplex Pumps Stroke Length – (Inches)										Double-Acting Duplex Pumps Stroke Length – (Inches)									
	7	8	9	10	11	12	14	16	18	20	7	8	9	10	11	12	14	16	18	20
3.00	0.643	0.735	0.826	0.918	1.01	1.10	1.29	1.47	1.65	1.84	0.857	0.979	1.10	1.22	1.35	1.47	1.71	1.96	2.20	2.45
3.25	0.754	0.862	0.970	1.078	1.19	1.29	1.51	1.72	1.94	2.16	1.01	1.15	1.29	1.44	1.58	1.72	2.01	2.30	2.59	2.87
3.50	0.875	1.00	1.12	1.25	1.37	1.50	1.75	2.00	2.25	2.50	1.17	1.33	1.50	1.67	1.83	2.00	2.33	2.67	3.00	3.33
3.75	1.00	1.15	1.29	1.43	1.58	1.72	2.01	2.30	2.58	2.87	1.34	1.53	1.72	1.91	2.10	2.30	2.68	3.06	3.44	3.83
4.00	1.14	1.31	1.47	1.63	1.80	1.96	2.29	2.61	2.94	3.26	1.52	1.74	1.96	2.18	2.39	2.61	3.05	3.48	3.92	4.35
4.25	1.29	1.47	1.66	1.84	2.03	2.21	2.58	2.95	3.32	3.69	1.72	1.97	2.21	2.46	2.70	2.95	3.44	3.93	4.42	4.91
4.50	1.45	1.65	1.86	2.07	2.27	2.48	2.89	3.31	3.72	4.13	1.93	2.20	2.48	2.75	3.03	3.31	3.86	4.41	4.96	5.51
4.75	1.61	1.84	2.07	2.30	2.53	2.76	3.22	3.68	4.14	4.60	2.15	2.46	2.76	3.07	3.38	3.68	4.30	4.91	5.52	6.14
5.00	1.79	2.04	2.30	2.55	2.81	3.06	3.57	4.08	4.59	5.10	2.38	2.72	3.06	3.40	3.74	4.08	4.76	5.44	6.12	6.80
5.25	1.97	2.25	2.53	2.81	3.09	3.37	3.94	4.50	5.06	5.62	2.62	3.00	3.37	3.75	4.12	4.50	5.25	6.00	6.75	7.50
5.50	2.16	2.47	2.78	3.09	3.39	3.70	4.32	4.94	5.56	6.17	2.88	3.29	3.70	4.11	4.53	4.94	5.76	6.58	7.41	8.23
5.75	2.36	2.70	3.04	3.37	3.71	4.05	4.72	5.40	6.07	6.75	3.15	3.60	4.05	4.50	4.95	5.40	6.30	7.20	8.10	9.00
6.00	2.57	2.94	3.31	3.67	4.04	4.41	5.14	5.88	6.61	7.35	3.43	3.92	4.41	4.90	5.39	5.88	6.86	7.84	8.81	9.79
6.25	2.79	3.19	3.59	3.99	4.38	4.78	5.58	6.38	7.17	7.97	3.72	4.25	4.78	5.31	5.85	6.38	7.44	8.50	9.56	10.63
6.50	3.02	3.45	3.88	4.31	4.74	5.17	6.03	6.90	7.76	8.62	4.02	4.60	5.17	5.75	6.32	6.90	8.05	9.20	10.35	11.49
6.75	3.25	3.72	4.18	4.65	5.11	5.58	6.51	7.44	8.37	9.30	4.34	4.96	5.58	6.20	6.82	7.44	8.68	9.92	11.16	12.40
7.00	3.50	4.00	4.50	5.00	5.50	6.00	7.00	8.00	9.00	10.00	4.67	5.33	6.00	6.67	7.33	8.00	9.33	10.66	12.00	13.33
7.25	3.75	4.29	4.83	5.36	5.90	6.44	7.51	8.58	9.65	10.73	5.01	5.72	6.44	7.15	7.87	8.58	10.01	11.44	12.87	14.30
7.50	4.02	4.59	5.16	5.74	6.31	6.89	8.03	9.18	10.33	11.48	5.36	6.12	6.89	7.65	8.42	9.18	10.71	12.24	13.77	15.30
7.75	4.29	4.90	5.52	6.13	6.74	7.35	8.58	9.80	11.03	12.26	5.72	6.54	7.35	8.17	8.99	9.80	11.44	13.07	14.71	16.34
8.00	4.57	5.22	5.88	6.53	7.18	7.84	9.14	10.45	11.75	13.06	6.09	6.96	7.84	8.71	9.58	10.45	12.19	13.93	15.67	17.41
8.25	4.86	5.56	6.25	6.94	7.64	8.33	9.72	11.11	12.50	13.89	6.48	7.41	8.33	9.26	10.18	11.11	12.96	14.81	16.67	18.52
8.50	5.16	5.90	6.63	7.37	8.11	8.85	10.32	11.79	13.27	14.74	6.88	7.86	8.85	9.83	10.81	11.79	13.76	15.73	17.69	19.66

(NOTE: To obtain displacement in barrels per stroke, divide numbers in table by 42.)

Ascending Mud Velocities — Feet/Minute

Diameter (Inches)		Circulation Rate (U.S. Gallons/Minute)									
Drill Pipe	Hole	100	200	300	400	500	600	700	800	900	1000
3½	6	103	206	310							
	6¼	91	183	274	366						
	6¾	74	147	221	294	368					
	7⅝	53	107	160	214	267	320	374			
	8⅜	42	85	127	169	212	254	296	339	381	
	8½	41	82	123	163	204	245	286	327	368	
	9	36	71	107	143	178	214	250	285	321	357
	9⅝	30	61	91	122	152	183	213	244	274	305
4	6¼	106	213	319							
	6¾	83	166	249	332						
	7⅝	58	116	174	233	291	349				
	8⅜	45	91	136	181	226	272	317	362		
	8½	44	87	131	174	218	261	305	349	392	
	9	38	75	113	151	189	226	264	302	339	377
	9⅝	32	64	96	128	160	192	224	256	288	320
	9⅞	30	60	90	120	150	180	210	241	271	301
4½	7⅞	59	117	176	235	293	352				
	8⅜	49	98	147	197	246	295	344	393		
	8½	47	94	141	189	236	283	330	377		
	9	40	81	121	161	202	242	282	323	363	
	9⅝	34	68	102	135	169	203	237	271	305	339
	10⅝	26	53	79	106	132	159	185	212	238	265
	11	24	49	73	97	122	146	170	195	219	243
	12¼	19	38	57	76	94	113	132	151	170	189
5 9/16	10⅝	30	60	90	120	150	179	209	239	269	299
	11	27	54	82	109	136	163	191	218	245	272
	12¼	21	41	62	82	103	123	144	165	185	206
	13	18	36	53	71	89	107	124	142	160	178
	13¾	16	31	47	62	78	93	109	124	140	155
	15	13	25	38	51	63	76	88	101	114	126
	17	9	19	28	38	47	57	66	76	85	95
	18	8	17	25	33	42	50	59	67	75	84
6⅝	11	32	64	95	127	159	191	223	254	286	318
	12¼	23	46	69	92	115	139	162	185	208	231
	13	20	39	59	78	98	118	137	157	176	196
	13¾	17	34	51	68	84	101	118	135	152	169
	15	14	27	41	54	68	81	95	108	122	135
	17	10	20	30	40	50	60	70	80	90	100
	20	7	14	21	28	34	41	48	55	62	69
	22	6	11	17	22	28	33	39	45	50	56

Mud Weight Conversions

Lbs/Gal	Lbs/Cu Ft	PSI/Ft Depth	Sp. Gravity	Lbs/Gal	Lbs/Cu Ft	PSI/Ft Depth	Sp. Gravity
8.0	59.8	0.415	0.96	12.0	89.8	0.623	1.44
8.2	61.3	0.426	0.98	12.2	91.3	0.633	1.46
8.34	62.4	0.433	1.00	12.4	92.8	0.644	1.49
8.6	64.3	0.446	1.03	12.6	94.3	0.654	1.51
8.8	65.8	0.457	1.06	12.8	95.8	0.664	1.53
9.0	67.3	0.467	1.08	13.0	97.3	0.675	1.56
9.2	68.8	0.478	1.10	13.2	98.7	0.685	1.58
9.4	70.3	0.488	1.13	13.4	100.2	0.696	1.61
9.6	71.8	0.498	1.15	13.6	101.7	0.706	1.63
9.8	73.3	0.509	1.17	13.8	103.2	0.716	1.65
10.0	74.8	0.519	1.20	14.0	104.7	0.727	1.68
10.2	76.3	0.529	1.22	14.2	106.2	0.737	1.70
10.4	77.8	0.540	1.25	14.4	107.7	0.748	1.73
10.6	79.3	0.550	1.27	14.6	109.2	0.758	1.75
10.8	80.8	0.561	1.29	14.8	110.7	0.768	1.77
11.0	82.3	0.571	1.32	15.0	112.2	0.779	1.80
11.2	83.8	0.581	1.34	15.2	113.7	0.789	1.82
11.4	85.3	0.592	1.37	15.4	115.2	0.799	1.85
11.6	86.8	0.602	1.39	15.6	116.7	0.810	1.87
11.8	88.3	0.613	1.41	15.8	118.2	0.820	1.89

Solids Content Effect on Mud Weight

Solids Content (% by volume)	Mud Weight (lbs/gallon)	Solids Content (% by volume)	Mud Weight (lbs/gallon)
0	8.33	10	9.7
1	8.5	11	9.8
2	8.6	12	10.0
3	8.7	13	10.1
4	8.9	14	10.3
5	9.0	15	10.4
6	9.2	16	10.5
7	9.3	17	10.7
8	9.4	18	10.8
9	9.6	19	10.9
		20	11.1

(Based on specific gravity of 2.65 for solids)

Salt-Water Drilling Fluid

Percent Salt (NaCl)	Specific Gravity	Resulting Fluid Density		Pounds of Salt (Sodium Chloride) To Be Added to Fresh Water	
		(Lbs/Gal)	(Lbs/Cu Ft)	(Per 100 Gal)	(Per 42-gal Bbl)
0	1.000	8.33	62.4	0	0
2	1.013	8.44	63.2	16.9	7.1
4	1.027	8.55	64.1	34.3	14.4
6	1.041	8.67	65.0	52.1	21.9
8	1.056	8.80	65.9	70.5	29.6
10	1.071	8.92	66.8	89.4	37.5
12	1.086	9.05	67.8	109	45.8
14	1.101	9.17	68.7	129	54.2
16	1.116	9.30	69.6	149	62.6
18	1.132	9.43	70.6	170	71.4
20	1.148	9.56	71.6	192	80.6
22	1.164	9.70	72.6	214	89.9
24	1.180	9.83	73.6	236	99.1
26	1.197	9.97	74.7	260	109

Weight of Common Drilling Fluid Additives

Item	Specific Gravity	Weight		
		Lbs/Cu Ft	Grams/Cu Cm	Kilograms/Cu M
Water	1.00	62.4	1.00	1,000
Clay	2.50	156	2.50	2,500
Barite	4.30	268	4.30	4,300
Salt (NaCl)	2.16	135	2.16	2,160

Drilling Fluid Weight Adjustment
(With Water or Barite)

Initial Drill Fluid Weight (lbs/gal)	Final Drill Fluid Weight (lbs/gal)											
	9.5	10.0	10.5	11.0	11.5	12.0	12.5	13.0	13.5	14.0	14.5	15.0
9.0	68	139	212	289	368	451	538	628	722	821	924	1033
9.5		69	142	217	295	376	461	550	642	739	840	947
10.0	43		71	144	221	301	384	471	562	657	756	861
10.5	85	30		72	147	226	307	393	482	575	672	775
11.0	128	60	23		74	150	230	314	401	493	588	688
11.5	171	90	46	19		75	154	236	321	411	504	602
12.0	214	120	69	37	16		77	157	241	328	420	516
12.5	256	150	92	56	32	14		79	161	246	336	430
13.0	299	180	115	75	47	27	12		80	164	252	344
13.5	342	210	138	94	63	41	24	11		82	168	258
14.0	385	240	161	112	79	54	36	21	10		84	172
14.5	427	269	184	131	95	68	48	32	19	9		86
15.0	470	299	207	150	110	82	60	43	29	18	8	

The lower left (shaded) part of table gives number of gallons of water to add to 100 gallons of drilling fluid to reduce weight — from initial weight (left column) to desired final weight (top column).

The upper right (unshaded) part of table gives pounds of barite to add to 100 gallons of drilling fluid to increase weight —from initial weight (left column) to desired final weight (top column).

Neat Cement Slurry Table

Slurry Density			Mixing Water		Slurry Volume
Sp.Gravity	Lbs/US Gal	Lbs/Cu Ft	US Gal/Sack*	Cu Ft/Sack*	Cu ft/Sack*
2.01	16.8	125.5	4.0	0.53	1.01
1.98	16.6	123.8	4.2	0.56	1.04
1.96	16.4	122.3	4.4	0.59	1.07
1.94	16.2	120.8	4.6	0.61	1.09
1.91	16.0	119.4	4.8	0.64	1.12
1.89	15.8	118.1	5.0	0.67	1.15
1.87	15.6	116.8	5.2	0.70	1.18
1.85	15.5	115.6	5.4	0.72	1.20
1.83	15.3	114.5	5.6	0.75	1.23
1.82	15.2	113.4	5.8	0.78	1.26
1.80	15.0	112.3	6.0	0.80	1.28
1.78	14.9	111.3	6.2	0.83	1.31
1.77	14.7	110.3	6.4	0.86	1.34
1.75	14.6	109.4	6.6	0.88	1.36
1.74	14.5	108.4	6.8	0.91	1.39
1.72	14.4	107.6	7.0	0.94	1.42
1.71	14.3	106.7	7.2	0.96	1.44
1.70	14.2	105.9	7.4	0.99	1.47
1.69	14.1	105.1	7.6	1.02	1.50
1.67	14.0	104.4	7.8	1.04	1.52
1.66	13.9	103.7	8.0	1.07	1.55
1.65	13.8	103.0	8.2	1.10	1.58
1.64	13.7	102.3	8.4	1.12	1.60
1.63	13.6	101.6	8.6	1.15	1.63
1.62	13.5	101.0	8.8	1.18	1.66
1.61	13.4	100.4	9.0	1.20	1.68
1.60	13.3	99.8	9.2	1.23	1.71
1.59	13.3	99.2	9.4	1.26	1.74
1.58	13.2	98.7	9.6	1.28	1.76
1.57	13.1	98.1	9.8	1.31	1.79
1.56	13.0	97.6	10.0	1.34	1.82

* - One sack of cement = 94 lbs (Sp.Gr. 3.15)

Bentonite/Cement Slurry Table

Bentonite (as percent of cement wt)	Mixing Water		Slurry Volume	Slurry Density		
	US Gal/Sack*	Cu Ft/Sack*	Cu Ft/Sack*	Sp. Gravity	Lbs/Cu Ft	Lbs/US Gal
2	5.0	0.67	1.16	1.90	118.6	15.9
	5.5	0.74	1.24	1.85	115.4	15.4
	6.0	0.80	1.30	1.81	112.9	15.1
	6.5	0.87	1.37	1.76	109.8	14.7
	7.0	0.94	1.44	1.73	108.0	14.4
4	7.0	0.94	1.44	1.74	108.6	14.5
	7.5	1.00	1.51	1.71	106.7	14.3
	8.0	1.07	1.58	1.67	104.2	13.9
	8.5	1.14	1.64	1.64	102.3	13.7
	9.0	1.20	1.71	1.62	101.1	13.5
6	8.0	1.07	1.58	1.68	104.8	14.0
	8.5	1.14	1.65	1.65	103.0	13.8
	9.0	1.20	1.72	1.63	101.7	13.6
	9.5	1.27	1.78	1.61	100.5	13.4
	10.0	1.34	1.85	1.59	99.2	13.3

*One sack of cement = 94 pounds; Sp. gravity of cement = 3.15, of bentonite = 2.5

Friction Loss of Air In Pipe

Nominal Pipe Size (inches)	Gage Pressure (psi)	Flow of Free Air (Cubic Feet per Minute) Through 100-Foot Length of Pipe								
		200	400	600	800	1,000	1,200	1,500	2,000	2,500
		Loss of Pressure Due To Friction (psi)								
2	60	0.40	1.53	3.53	6.40	9.96	15.00	24.10		
	80	0.31	1.20	2.75	4.90	7.62	11.40	18.30		
	100	0.26	0.99	2.23	4.00	6.20	9.05	14.50		
	125	0.21	0.81	1.83	3.23	5.04	7.45	11.70		
2½	60	0.16	0.61	1.37	2.37	3.70	5.45	8.54	15.80	
	80	0.13	0.48	1.08	1.87	2.89	4.21	6.62	12.00	
	100	0.11	0.40	0.89	1.54	2.37	3.45	5.39	9.66	
	125	0.09	0.33	0.73	1.25	1.94	2.78	4.38	7.80	
3	60	0.06	0.20	0.45	0.78	1.20	1.70	2.70	4.85	7.80
	80	0.04	0.16	0.35	0.61	0.94	1.37	2.15	3.77	6.00
	100	0.04	0.13	0.29	0.50	0.78	1.12	1.73	3.09	4.85
	125	0.03	0.11	0.24	0.41	0.64	0.92	1.43	2.52	4.00
4	60		0.05	0.11	0.19	0.30	0.43	0.66	1.16	1.82
	80		0.04	0.09	0.15	0.23	0.34	0.52	0.91	1.43
	100		0.03	0.07	0.13	0.19	0.28	0.43	0.75	1.16
	125		0.03	0.06	0.10	0.16	0.23	0.35	0.61	0.95

Friction Loss of Air In Hose

Hose I.D. (inches)	Gage Pressure (psi)	Flow of Free Air (Cubic Feet per Minute) Through 50-Foot Length of Smooth Hose: 20	30	40	50	60	70	80	90	100	110	120	130	140	150
		Loss of Pressure Due To Friction (psi)													
½	50	1.8	5.0	10.1	18.1										
	60	1.3	4.0	8.4	14.8	23.4									
	70	1.0	3.4	7.0	12.4	20.0	28.4								
	80	0.9	2.8	6.0	10.8	17.4	25.2	34.6							
	90	0.8	2.4	5.4	9.5	14.8	22.0	30.5	41.0						
	100	0.7	2.3	4.8	8.4	13.3	19.3	27.2	36.6						
¾	50	0.4	0.8	1.5	2.4	3.5	4.4	6.5	8.5	11.4	14.2				
	60	0.3	0.6	1.2	1.9	2.8	3.8	5.2	6.8	8.6	11.2	14.1			
	70	0.2	0.5	0.9	1.5	2.3	3.2	4.2	5.5	7.0	8.8	11.0			
	80	0.2	0.5	0.8	1.3	1.9	2.8	3.6	4.7	5.8	7.2	8.8	10.6		
	90	0.2	0.4	0.7	1.1	1.6	2.3	3.1	4.0	5.0	6.2	7.5	9.0	10.8	
	100	0.2	0.4	0.6	1.0	1.4	2.0	2.7	3.5	4.4	5.4	6.6	7.9	9.4	11.1
1	50		0.2	0.3	0.5	0.8	1.1	1.5	2.0	2.6	3.5	4.8	7.0		
	60		0.2	0.3	0.4	0.6	0.8	1.2	1.5	2.0	2.5	3.3	4.2	5.5	7.2
	70		0.1	0.2	0.4	0.5	0.7	1.0	1.3	1.6	2.0	2.5	3.1	3.8	4.7
	80			0.2	0.3	0.5	0.7	0.8	1.1	1.4	1.7	2.0	2.4	2.7	3.5
	90			0.2	0.3	0.4	0.6	0.7	0.9	1.2	1.4	1.7	2.0	2.4	2.8
	100			0.2	0.2	0.4	0.5	0.6	0.8	1.0	1.2	1.5	1.8	2.1	2.4
1¼	50				0.2	0.2	0.3	0.4	0.5	0.7	1.0				
	60				0.1	0.2	0.3	0.3	0.5	0.6	0.8	1.0	1.2	1.5	
	70					0.2	0.2	0.3	0.4	0.4	0.6	0.7	0.8	1.0	1.3
	80					0.1	0.2	0.2	0.3	0.4	0.5	0.6	0.7	0.8	1.0
	90						0.2	0.2	0.3	0.3	0.4	0.5	0.6	0.7	0.8
	100						0.1	0.2	0.2	0.3	0.4	0.4	0.5	0.6	0.7

Based on 50-ft lengths of smooth hose, coupled at both ends;loss in different lengths is proportional (multiply by 2 for 100 ft, etc)

Delivered Air Volume for Desired Annulus Velocities

Table gives air flow requirement (cfm) to produce air velocity shown, for various hole and drill pipe

Hole Diam. (Inches)	Pipe Diam. (Inches)	Delivered Air Volume Required (CFM)				
		3000 Ft/Min	4000 Ft/Min	5000 Ft/Min	6000 Ft/Min	7000 Ft/Min
4¼	2⅜	203	271	339	407	474
	2⅞	160	214	267	321	374
	3½	95	127	159	190	222
4½	2⅜	239	319	398	478	558
	2⅞	196	261	327	392	458
	3½	131	175	218	262	305
	4	70	93	116	139	162
4¾	2⅜	277	369	461	554	646
	2⅞	234	312	390	468	546
	3½	169	225	281	337	394
	4	107	143	179	215	251
5⅛	2⅞	295	393	491	589	687
	3½	229	306	382	459	535
	4	168	224	280	336	392
5⅝	2⅞	382	510	637	765	892
	3½	317	423	529	635	740
	4	256	341	427	512	597
6¼	3½	439	585	731	877	1024
	4	377	503	629	755	881
	5	230	307	383	460	537
6¾	3½	545	727	908	1090	1272
	4	484	645	806	967	1129
	5	336	449	561	673	785
	5½	251	334	418	501	585
7⅜	3½	690	919	1149	1379	1609
	4	628	838	1047	1256	1466
	5	481	641	802	962	1122
	5½	395	527	658	790	922
7⅞	3½	814	1086	1357	1629	1900
	4	753	1004	1255	1506	1757
	5½	520	693	866	1040	1213
	6¼	376	501	626	751	876
	6⅝	297	395	494	593	692
9	4	1064	1418	1773	2127	2482
	5½	830	1107	1384	1661	1938
	6⅝	607	810	1012	1214	1417
	7	524	698	873	1047	1222
	7¾	343	457	571	685	799
9⅞	7	794	1058	1323	1588	1852
	7¾	613	817	1021	1226	1430
	8⅝	378	505	631	757	883
10⅝	7	1045	1394	1742	2091	2439
	7¾	864	1153	1441	1729	2017
	8⅝	630	840	1050	1260	1470
	9	522	696	870	1044	1218
12¼	8⅝	1238	1651	2064	2476	2889
	9	1130	1507	1883	2260	2637
	10	819	1092	1365	1638	1911

Weight of Drill Collars (Pounds/Foot)

Collar O.D.	Diameter Of Collar Bore (Inches)										
(Inches)	1½	1¾	2	2¼	2½	2 13/16	3	3¼	3½	3¾	4
3⅜	24.4	22.2	19.7	16.9							
3½	26.7	24.5	22.0	19.2							
3¾	31.5	29.4	26.9	24.0	20.9						
3⅞	34.1	31.9	29.4	26.6	23.4						
4	36.7	34.5	32.0	29.2	26.0						
4⅛	39.4	37.3	34.8	31.9	28.7						
4¼	42.2	40.1	37.5	34.7	31.5						
4½	48.1	45.9	43.4	40.6	37.4						
4¾	54.2	52.1	49.6	46.7	43.6						
5	60.7	58.6	56.1	53.2	50.1						
5¼	67.6	65.4	62.9	60.1	56.9	52.5	49.6				
5½	74.8	72.6	70.1	67.3	64.1	59.6	56.7				
5¾	82.3	80.1	77.6	74.8	71.6	67.2	64.2	60.1	55.6	50.7	45.6
6	90.1	87.9	85.4	82.6	79.4	75.0	72.1	67.9	63.4	58.6	53.4
6¼	98.3	96.1	93.6	90.8	87.6	83.2	80.3	76.1	71.6	66.8	61.6
6½	106.8	104.6	102.1	99.3	96.1	91.7	88.8	84.6	80.1	75.3	70.1
6¾	115.6	113.5	111.0	108.1	105.0	100.5	97.6	93.5	88.9	84.1	78.9
7	124.8	122.7	120.2	117.3	114.1	109.7	106.8	102.6	98.1	93.3	88.1
7¼	134.3	132.2	129.7	126.8	123.7	119.2	116.3	112.1	107.6	102.8	97.6
7½	144.2	142.0	139.5	136.7	133.5	129.1	126.2	122.0	117.5	112.6	107.5
7¾	154.4	152.2	149.7	146.9	143.7	139.2	136.3	132.2	127.7	122.8	117.6
8	164.9	162.7	160.2	157.4	154.2	149.8	146.9	142.7	138.2	133.3	128.2

Based on weight of steel of 0.2833 pounds per cubic inch.

Capacity of Drill Collars – (U.S. gal/Ft)			
Bore Diameter (Inches	Capacity (Gal/Ft)	Bore Diameter (Inches)	Capacity (Gal/Ft)
1½	0.09	3	0.37
1¾	0.12	3¼	0.43
2	0.16	3½	0.50
2¼	0.21	3¾	0.57
2½	0.26	4	0.65
2.813	0.32		

Displacement of Drill Collars (U.S. Gal/Foot)											
Collar O.D.	Diameter of Collar Bore (Inches)										
(Inches)	1½	1¾	2	2¼	2½	2 13/16	3	3¼	3½	3¾	4
3⅜	0.37	0.34	0.30	0.26							
3½	0.41	0.37	0.34	0.29							
3¾	0.48	0.45	0.41	0.37	0.32						
3⅞	0.52	0.49	0.45	0.41	0.36						
4	0.56	0.53	0.49	0.45	0.40						
4⅛	0.60	0.57	0.53	0.49	0.44						
4¼	0.65	0.61	0.57	0.53	0.48						
4½	0.73	0.70	0.66	0.62	0.57						
4¾	0.83	0.80	0.76	0.71	0.67						
5	0.93	0.90	0.86	0.81	0.76						
5¼	1.03	1.00	0.96	0.92	0.87	0.80	0.76				
5½	1.14	1.11	1.07	1.03	0.98	0.91	0.87				
5¾	1.26	1.22	1.19	1.14	1.09	1.03	0.98	0.92	0.85	0.78	0.70
6	1.38	1.34	1.31	1.26	1.21	1.15	1.10	1.04	0.97	0.90	0.82
6¼	1.50	1.47	1.43	1.39	1.34	1.27	1.23	1.16	1.09	1.02	0.94
6½	1.63	1.60	1.56	1.52	1.47	1.40	1.36	1.29	1.22	1.15	1.07
6¾	1.77	1.73	1.70	1.65	1.60	1.54	1.49	1.43	1.36	1.29	1.21
7	1.91	1.87	1.84	1.79	1.74	1.68	1.63	1.57	1.50	1.43	1.35
7¼	2.05	2.02	1.98	1.94	1.89	1.82	1.78	1.71	1.64	1.57	1.49
7½	2.20	2.17	2.13	2.09	2.04	1.97	1.93	1.86	1.80	1.72	1.64
7¾	2.36	2.33	2.29	2.24	2.20	2.13	2.08	2.02	1.95	1.88	1.80
8	2.52	2.49	2.45	2.40	2.36	2.29	2.24	2.18	2.11	2.04	1.96

A.P.I. Drill Pipe

Outside Diam. (Inches)	Wall Thickness (Inches)	Inside Diam. (Inches)	Nominal Wt. (Lbs/Ft)	Grade	Contents (U.S. Gal/Ft)	Displacement (U.S. Gal/Ft)
2⅜	0.190	1.995	4.9	D,E	0.16	0.07
	0.280	1.815	6.7	D,E	0.13	0.10
2⅞	0.217	2.441	6.9	D,E	0.24	0.09
	0.362	2.151	10.4	D,E	0.19	0.15
3½	0.254	2.992	9.5	D,E	0.37	0.13
	0.368	2.764	13.3	D,E	0.31	0.19
	0.449	2.602	15.5	D,E	0.28	0.22
4	0.262	3.476	11.9	D,E	0.49	0.16
	0.330	3.340	14.0	D,E	0.46	0.20
4½	0.271	3.958	13.8	D,E	0.64	0.19
	0.337	3.826	16.6	D,E	0.60	0.23
	0.373	3.754	18.1	D,E	0.57	0.25
	0.430	3.640	20.0	D,E	0.54	0.29
5	0.296	4.408	15.0	D,E	0.79	0.23
	0.362	4.276	19.5	D,E	0.75	0.27
5½	0.361	4.778	21.9	D,E	0.93	0.30
	0.415	4.670	24.7	D,E	0.89	0.34
6⅝	0.330	5.965	25.2	D,E	1.45	0.34

Yield Strength: Grade D = 55,000 psi; Grade E = 75,000 psi

Rotary Tricone Roller Bits					
Diameter (Inches)	API Pin Size (Inches)	Diameter (Inches)	API Pin Size (Inches)	Diameter (Inches)	API Pin Size (Inches)
2⅞	N Rod	6	3½	11	6⅝
2 15/16	N Rod	6¼	3½	12¼	6⅝
3⅛	N Rod	6¾	3½	13¾	6⅝
3½	N Rod or 2⅜	7⅜	3½	14¾	6⅝
3¾	2⅜	7⅝	4½	15	6⅝ or 7⅝
3⅞	2⅜	7⅞	4½	16	6⅝ or 7⅝
4¼	2⅜	8½	4½	17½	6⅝ or 7⅝
4½	2⅜	8¾	4½	18½	6⅝ or 7⅝
4¾	2⅞	9	4½	20	6⅝ or 7⅝
5	2⅞	9⅝	6⅝	22	6⅝ or 7⅝
5⅛	2⅞	9⅞	6⅝	24	6⅝ or 7⅝
5⅝	3½	10⅝	6⅝	26	6⅝ or 7⅝

Diamond Core Drilling – Nominal Dimensions

(Diameters-Inches; Weights-Lbs/Ft; Volumes-U.S. Gals/100 Ft)

Coring Bits

Size	Hole Diam.	Core Diam.	Hole Volume
EX,EWX,EWG	1.500	0.875	9.2
AX,AWX,AWG	1.875	1.125	14.3
BX,BWX,BWG	2.375	1.625	23.0
NX,NWX,NWG	3.000	2.125	36.7

Drill Rods & Couplings

Size	Rod O.D.	Rod I.D.	Coupling I.D.	Weight	Volume
E	1.313	0.813	0.438	2.8	2.7
EW	1.375	0.875	0.500	3.1	3.1
A	1.625	1.125	0.563	3.7	5.2
AW	1.750	1.438	0.625	4.2	8.4
B	1.875	1.250	0.625	5.0	6.4
BW	2.125	1.813	0.750	5.3	13.4
N	2.375	1.625	1.000	5.2	10.8
NW	2.625	2.313	1.375	5.5	21.8

Casing Bits and Shoes

Size	Bit O.D.	Bit I.D	Shoe O.D.	Shoe I.D.
EX,EW	1.875	1.405	1.875	1.500
AX,AW	2.345	1.780	2.345	1.905
BX,BW	2.965	2.215	2.965	2.375
NX,NW	3.615	2.840	3.615	2.995

Flush Joint Casing

Size	O.D.	I.D.	Weight	Volume
EW	1.813	1.500	2.8	9.2
AW	2.250	1.906	3.8	14.8
BW	2.875	2.375	7.0	23.0
NW	3.500	3.000	8.7	36.7

Flush Coupled Casing

Size	O.D.	I.D.	Coupling I.D.	Casing Wt.	Volume
EX	1.813	1.625	1.500	1.7	10.8
AX	2.250	2.000	1.906	2.8	16.3
BX	2.875	2.469	2.375	5.8	24.9
NX	3.500	3.063	3.000	7.7	38.3

Standard Dimensions for Cable Tools

(All Dimensions in Inches)

Nominal API Taper Joint Size	Drilling Tool Joints					Rope Sockets				
	Threads Per Inch	Taper Pin Length	Wrench Square Size	Pin Collar Diameter	Box Collar Diameter	Socket Diameter	Wire Rope Socket Neck	Manila Rope Socket Neck	Swivel O.D.	Swivel Length
1⅛ × 1¾	8	2½	1¾	2½	2⅝	2⅝	1⅝	*	1 3/16	5
1½ × 2¼	8	3	2¼	3⅛	3¼	3¼	2	*	1 7/16	6
1⅝ × 2⅝	7	3½	2¾	3⅝	3¾	3¾	2½	2¾	1¾	7
2 × 3	7	4	3¼	4¼	4⅜	4⅜	2¾	3¼	2 1/16	8
2¼ × 3¼	7	4	3½	4½	4¾	4¾	3¼	3¼	2 1/16	8
2½ × 3½	7	4¼	4	5	5¼	5¼	3½	3½	2 7/16	10
2¾ × 3¾	7	4½	4	5¼	5½	5½	3½	3½	2 9/16	10
3 × 4	7	4¾	4½	6	6¼	6¼	4	4	2 9/16	10
3¼ × 4¼	7	5	5	6¼	6½	6½	4½	4½	3	10
3½ × 4½	7	5½	5	6½	6¾	6¾	4½	4½	3	10
4 × 5	7	5½	5	7	7⅜	7⅜	4½	4½	3½	11
4 × 5½	7	6½	5½	7½	7⅞	7⅞	5½	5½	3½	11
4¼ × 6	7	7	6	8¼	8⅝	8⅝	5½	5½	3½	11

Approximate Dimensions – Cable Tool Drilling Bits

Hole Size (inches)	Length (feet)	Weight (pounds)	API Pin Sizes (inches)
3	4	50	1 3/4
3 1/2	4 1/2	75	1 3/4
4	3 1/2	75	1 3/4 - 2 5/8
	4 1/2	100	
4 1/2	4	100	2 5/8
	5	125	
	6	150	
5 - 5 3/16	3	100	2 1/4 - 2 5/8
	4	135	
	5	165	
	5	190	2 5/8 - 3
	6	220	
	7	250	
5 5/8	4	175	2 1/4 - 2 5/8
	4 1/2	200	3 - 3 3/4
	5	225	
6	3 1/2	150	2 1/4 - 2 5/8
	4	175	
	4	200	3 - 3 3/4
	5	250	
	6	300	
6 1/4	5	270	3 - 3 3/4
	6	320	
6 1/4 - 6 5/8	5	300	3 1/2 - 3 3/4
	6	350	
	6	400	
	6 1/2	425	
	7	450	
	7 1/2	475	
	8	500	
8 - 8 1/4	3 1/2	250	3 - 3 1/4
	4	300	
	4 1/2	350	
	5	400	
	5	400	3 1/2 - 3 3/4
	5 1/2	450	
	6	500	
	7	600	
8 - 8 1/4	4 1/2	400	4 1/4 - 4 1/2
	5	450	
	5 1/2	500	
	6	550	
	6 1/2	600	
	7	650	
	7 1/2	700	
	8	750	
10	3 1/2	350	3 - 3 1/4
	4	400	
	4 1/2	450	
	5	500	
	4	450	3 1/2 - 3 3/4
	4 1/2	500	
	5	575	
	4	500	4 1/4 - 4 1/2
	4 1/2	625	
	5	750	
	7	875	4 1/2 - 5
	7 1/2	950	
	8	1000	
12 - 12 1/2	3	400	3 1/2 - 3 3/4
	3 1/2	500	
	4	600	
	4 1/2	700	
	5	800	
	3 1/2	600	4 1/4 - 4 1/2
	4 1/2	800	5
	5 1/2	1000	
	6 1/2	1200	
	7	1300	
	7 1/2	1400	
	8	1500	
14	4	600	4 1/4 - 4 1/2
	4 1/2	800	5
	5	1000	
	5 1/2	1200	
	6	1400	
16	3 1/2	1000	4 1/4 - 4 1/2
	4 1/2	1200	5
	5 1/2	1400	
	6 1/2	1600	
	7 1/2	1800	

Capacity of Rope Drums

The approximate capacity of rope drums can be calculated from the following formula:

$$L = (A + B) \times A \times C \times K \text{(see sketch below)}$$

Where: A = Flange depth, in inches (reduced by 1-2 inches for clearance)
B = Drum diameter, in inches
C = Drum width, in inches
L = Rope length, in feet
K = Constant (from table below)

Rope Diam. (inches)	"K"	Rope Diam. (inches)	"K"
1/4	4.3	3/4	0.48
5/16	2.8	7/8	0.35
3/8	1.9	1	0.27
7/16	1.4	1 1/8	0.21
1/2	1.1	1 1/4	0.17
9/16	0.85	1 3/8	0.14
5/8	0.70	1 1/2	0.12

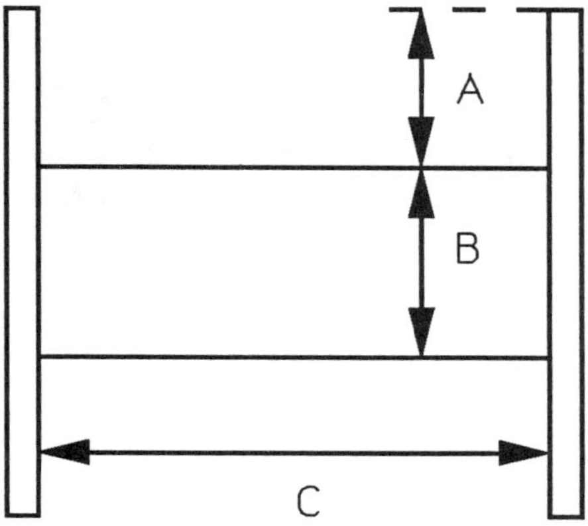

Safe Loads on Chains

Diam. of Metal (inches)	Approx. Weight (Lbs/100 Ft.)	Approximate Safe Load (Tons)		
		Wrought Iron	Soft Steel	Special Steel
3/8	160	1.35	1.65	3.20
7/16	210	1.73	2.18	4.15
1/2	280	2.25	2.63	5.25
5/8	430	3.47	4.23	7.60
3/4	630	5.07	6.00	10.50
7/8	840	7.00	8.25	14.33
1	1100	9.30	10.60	18.20

Approximate Safe Loads on Steel Hooks

Diam. of Metal (Inches)	Inside Diameter of Eye (Inches)	Width of Opening (Inches)	Safe Load (Tons)
5/8	3/4	1	0.5
11/16	7/8	1 1/16	0.6
3/4	1	1 1/8	0.7
7/8	1 1/8	1 1/4	1.2
1	1 1/4	1 3/8	1.7
1 1/8	1 3/8	1 1/2	2.1
1 1/4	1 1/2	1 11/16	2.5
1 3/8	1 5/8	1 7/8	3.0
1 1/2	1 3/4	2 1/16	4.0
1 5/8	2	2 1/4	4.7
1 7/8	2 1/8	2 1/2	5.5
2 1/4	2 3/4	3	6.8
2 5/8	3 1/8	3 3/8	8.5
3	3 1/2	4	12.0

Fiber Ropes (3-Strand) – Approximate Safe Working Loads

Diameter (Inches)	Manila		Nylon (dry*)		Cotton	
	(Pounds)	(Kilograms)	(Pounds)	(Kilograms)	(Pounds)	(Kilograms)
3/8	135	61	540	245	65	30
1/2	250	114	970	441	120	55
5/8	360	164	1,480	673	180	82
3/4	540	245	2,080	945	260	118
7/8	700	318	2,820	1,282	340	155
1	950	432	3,660	1,664	450	205
1 1/8	1,150	523	—	—	550	250
1 1/4	1,350	614	—	—	690	314
1 1/2	1,900	864	—	—	950	432
1 3/4	2,450	1,114	—	—	1,250	568
2	3,100	1,409	—	—	1,550	705
2 1/2	4,650	2,114	—	—	2,320	1,055

* – Deduct 20% from values in table when wet

Wire Rope (6 × 19) – Approximate Breaking Strength in Tons

Diameter (Inches)	3/8	1/2	5/8	3/4	7/8	1	1 1/8	1 1/4	1 1/2
Wrought Iron	2.1	3.6	5.5	7.9	10.6	13.7	17.2	21.0	29.7
Traction Steel	4.0	6.8	10.4	14.8	20.0	26.0	32.7	40.6	56.6
Cast Steel	4.5	7.7	11.8	16.8	22.8	29.5	37.0	46.0	65.0
Mild Plow Steel	5.0	8.5	13.1	18.7	25.4	33.0	41.5	51.0	72.5
Plow Steel	5.5	9.4	14.4	20.6	28.0	36.5	46.0	56.5	80.5
Improved Plow Steel	6.3	10.8	16.6	23.7	32.2	42.0	53.0	65.0	92.5
Weight (Lbs/Ft)	0.23	0.40	0.63	0.90	1.23	1.60	2.03	2.50	3.60

Working load should not exceed one-fifth of breaking strength; to determine proper working load divide the breaking strength (from table above) by 5.

Sheave Diameters for Wire Rope		
Type of Rope (Strands × Wires)	**Recommended Sheave Or Drum Diameter**	**Minimum Sheave Or Drum Diameter**
6 × 7	72 × rope diameter	42 × rope diameter
6 × 19	45 × rope diameter	30 × rope diameter
6 × 37	27 × rope diameter	18 × rope diameter
8 × 19	31 × rope diameter	21 × rope diameter

Well Development

Upon completion of drilling, most wells require development to reach maximum efficiency. This is particularly true of wells producing from unconsolidated aquifer materials and those in which an artificial filter pack has been placed around well screens. In addition, many wells may require periodic re-development to restore production capacity that has been lost as a result of such factors as incrustation of screens, clogging of screens by bacterial deposits, or migration of fine particles into a gravel pack (filter).

The principal purpose of development is to remove the fine materials adjacent to the well bore, to increase permeability of the aquifer (and filter pack), to remove any mud cake or compacted zone, and to minimize or eliminate sand pumping. The following discussion summarizes some development methods and procedures; the reader is referred to the standard manuals on well drilling and trade journal publications for additional detail.

Development methods commonly used include:

- Overpumping; or surging with pump
- Surge blocks
- Compressed air
- Acids and chemicals
- Hydraulic jetting
- Hydraulic fracturing

Overpumping/Pump Surging:

Development by overpumping consists of pumping a well at a rate of two or more times the designed discharge, usually for a prolonged period, to remove fine sands or clays. Pump surging (sometimes called "rawhiding") consists of pumping a well at a high rate for a short period (usually 5 to 10 minutes), shutting down the pump to allow water in the column to fall and backwash the screen, and then repeating the process until the discharge is clear. Surging with a pump requires that there not be a non-reverse ratchet on the shaft, so that the water in the pump column can drain rapidly back into the well. These methods, although simple and inexpensive to perform, are seldom satisfactory in providing good development as they tend to develop only a short section of the length of a well screen.

Surge Blocks:

The use of a surge block is a common and effective means of well development. The block is a circular plate (or short cylinder) attached to the lower end of a drill stem or drill pipe which can be raised and lowered to produce a strong surging action in the well. The block is commonly solid, but can be equipped with valves which close on the up-stroke. The block is usually made of discs of belting bolted between steel plates, or of wood. The outside diameter of the block should be only slightly less than the inside diameter of the screen (⅛- to ¼-inch). Before development, the hole should be bailed clean to the bottom. Surging should start above the screen to reduce the possibility of "sand-locking" the surge block. Initial surging should be with a long stroke and at a slow rate; after surging above the screen the hole should be cleaned and surging started at the lower end of the screen — gradually working upward until the entire screen has been developed. The process is then repeated, beginning at the bottom of the screen, but with a faster stroke. The hole should be cleaned periodically, and records kept of the duration of surging and amount of material brought into the well at each cleaning. It is important that the surge-block assembly have sufficient weight for the particular diameter of screen; suggested minimum weights are shown in the accompanying table.

Compressed Air:

Compressed air can be used to provide both pumping and surging action for well development. In the simplest form, the well is pumped rapidly by air-lifting using drill pipe or tubing — then allowed to recover, and the process repeated. This is similar to surging or "rawhiding" with a turbine pump. The effectiveness is often enhanced, particularly with long screens, by using a double-packer tool that confines the pumping to a short length of the screen. Equipment can also be designed to provide a stronger backwash effect in addition to the pumping. As with surge blocks, the well should be periodically cleaned and records kept of the amount of material brought in during development.

Compressed air can also be used to horizontally "jet" a well screen or well bore, similar to hydraulic jetting. Occasionally this may result in undesirable air-entrainment ("air-lock") in the aquifer.

Acids and Chemicals:

Various acids and/or chemicals can be very effective in well development under certain conditions, both upon completion of drilling and in re-development. Acids and chemicals can be used singly, or in combination with other methods such as surge blocks or hydraulic jetting. Acids can be very effective in limestone and dolomite aquifers by enlarging fractures adjacent to the well bore and thereby significantly reducing drawdown at a given pumping rate. Acids can also be effective in removing incrusting deposits on well screens. Various chemical additives can be effective in breaking down clays or muds.

Until recent years, the most commonly used acid was hydrochloric (muriatic) acid (HCl). Sulfamic acid (H_2NSO_3H) is now being used increasingly in well development or rehabilitation. Both are effective, but sulfamic acid — though more expensive — is easier and safer to handle. Hydrochloric acid is available commercially in strengths of 18, 20, and 22 degrees Baumé; the 18 degree strength (27.92%) is usually used in treating wells. Sulfamic acid is used "full strength" by dissolving as much of the dry acid in water as possible. Physical data on these two acids are given in accompanying tables. The usual procedure is to introduce a volume of acid several times that of the section of hole or screen to be treated, followed by frequent agitation, and then removed by pumping after 6 to 8 hours; repeat treatment may be required. Both types of acid may have "inhibitors" added to reduce adverse effects on metals in the pump, casing, or screen.

Chemicals frequently used in well development are polyphosphates and surfactants ("wetting agents"). The commonly used polyphosphates include sodium tripolyphosphate ($Na_5P_3O_{10}$), sodium pyrophosphate ($Na_4P_2O_7$), tetra sodium pyrophosphate (NaP_2O_7), and sodium hexametaphosphate $(NaPO_3)_6$. The polyphosphates act to deflocculate and disperse clays and drilling muds. They are typically introduced as a solution of from 5 to 15 lbs. polyphosphate in 100 gals. of water in conjunction with surge-block or hydraulic jetting development. The addition of wetting agents aids in allowing the polyphosphates to penetrate clays.

Manufacturers or distributors of commercially available acids and chemicals for well development should be consulted for detailed specifications and recommendations.

Hydraulic Jetting:

Hydraulic jetting is a very effective way to develop sand and gravel aquifers and artificial filter packs. A jetting tool is lowered on a drop pipe and water pumped down at high pressure to produce nozzle velocities of 100 ft/sec or more, with a minimum desirable velocity of about 150 ft/sec. The tool should have 2, 3, or 4 nozzles, spaced equally around the circumference and directed horizontally toward the inside of the well screen or perforated casing. The nozzles should extend to within about 1/2-inch from the inside of the screen or casing. In operation, the jetting tool is lowered into the lower part of the screen, rotated slowly (1 rpm) and raised slowly throughout the length to be developed. The process is then repeated until develoment is complete. The tool *must* be rotated continuously when jetting to avoid damage to casing or screen. Pumping the well at a low rate (such as by air lift) assists in bringing fine sands into the well. Adding polyphosphates to the water introduced can also be helpful. It is essential that the water used in jetting be clear and free from turbidity. Approximate jet nozzle velocities attained at various discharge rates and pressures, as well as suggested drop-pipe sizes to minimize friction loss, are given in the accompanying tables.

Hydraulic Fracturing:

Hydraulic fracturing has been successfully used to increase production from consolidated aquifers (limestone, quartzite, granite, etc.) where water enters the well through narrow joints or fractures in the rock. The zone to be treated is isolated with packers above and below, and water pumped in at extremely high pressure to enlarge the natural openings. Sometimes fine sand is added to the water being pumped to "prop up" the fractures that are enlarged. Commercial equipment for hydraulic fracturing is available, or it can be performed by many of the oil-field service companies.

Recommended Surge-Block Development of Screens

Screen Diam. (Inches)	Surge-Block Assembly (Minimum Wt – lbs)	Drill Stem	
		(Diam. – Inches)	(Weight – Lbs/Ft)
5	1,000	2	10.7
6	1,200	2½	16.7
8	1,500	3	24.0
10	1,800	3½	32.7
12	2,200	4	42.7
14	2,500	4½	54.1
16	2,800	5	66.8
18	3,150	5½	80.8
20	3,400	6	96.1
24	4,000		

Suggested Surge Block Diameters:

Belting Discs: Equal to screen inside diameter
Wooden Discs: ½-inch less than screen inside diameter
Steel Plate: 1-inch less than screen inside diameter

Suggested Surging Rates:

Initial: 20–25 strokes per minute
Intermediate: 30-35 strokes per minute
Final: Rapid

Hydrochloric Acid Data

Degrees Baumé	Specific Gravity	Percent HCl	Weight (Lbs/U.S. Gal)	HCl (Lbs/U.S. Gal)
22	1.18	35.21	9.83	3.54
20	1.16	31.45	9.66	3.10
18	1.14	27.92	9.50	2.72
10	1.07	14.83	8.91	1.37
7	1.05	10.17	8.75	0.88
5	1.04	7.15	8.66	0.61
3	1.02	4.25	8.50	0.36

Sulfamic Acid Data

Solubility in Water

Temperature		Solubility
Degrees C	Degrees F	(Lbs/U.S. Gal)
0	32	1.38
5	41	1.45
10	50	1.54
15	59	1.66
20	68	1.79

pH of Acid Solutions

Solution Concentration (Percent by Weight)	Solution pH
1	1.20
2	0.95
3	0.80
4	0.71
5	0.63
6	0.57
8	0.50
10	0.46

Hydraulic Jetting – Approximate Velocities and Flow Rates (Per Nozzle)

Nozzle	100 Psi		150 Psi		200 Psi		250 Psi		300 Psi	
Orifice Size (Inches)	Velocity (Ft/Sec)	Flow (US Gpm)	Velocity (Ft/Sec)	Flow (US Gpm)	Velocity (Ft/Sec)	Flow (US Gpm)	Velocity (Ft/Sec)	Flow (US Gpm)	Velocity (Ft/Sec)	Flow (US Gpm)
3/16	98	8	119	10	138	12	154	13	169	15
1/4	98	15	119	18	138	21	154	24	169	26
5/16	98	23	119	28	138	33	154	37	169	40
3/8	98	34	119	41	138	47	154	53	169	58
7/16	98	46	119	56	138	64	154	72	169	79
1/2	98	60	119	73	138	84	154	94	169	103

Nozzle coefficient of discharge assumed to be 0.8

Hydraulic Jetting – Recommended Drop-Pipe Size (Inches)

Total Flow Rate (Gpm)	Depth of Setting (Feet)						
	100	150	200	250	300	400	500
25	1½	1½	1½	1½	1½	1½	1½
50	1½	1½	1½	1½	2	2	2
75	1½	2	2	2	2½	2½	2½
100	2	2	2½	2½	2½	2½	2½
125	2	2½	2½	2½	3	3	3
150	2½	2½	3	3	3	3	3
175	2½	3	3	3	3	3	4
200	2½	3	3	3	3	3	4
250	3	3	3	4	4	4	4
300	3	3	4	4	4	4	4
350	3	4	4	4	4	4	4
400	4	4	4	4	4	5	5

Chapter 3

Pipe and Casing

Steel Pipe Dimensions

Nominal Diam. (In)	Outside Diam. (In)	Wall Thickness (In)	Inside Diam. (In)	Plain End Weight (Lbs/Ft)	Threaded Coupling O.D. (In)	Capacity (Gal/Ft)	Surface Area (SqFt/Ft)	Tensile Strength (Tons)	Remarks
0.125	0.405	0.068	0.269	0.24	0.563	0.003	0.106	2	Sch 40 (Std)
(1/8)		0.095	0.215	0.31		0.002		3	Sch 80
0.25	0.540	0.088	0.364	0.42	0.719	0.005	0.141	4	Sch 40 (Std)
(1/4)		0.119	0.302	0.54		0.004		5	Sch 80
0.375	0.675	0.091	0.493	0.57	0.875	0.010	0.177	5	Sch 40 (Std)
(3/8)		0.126	0.423	0.74		0.007		7	Sch 80
0.5	0.840	0.109	0.622	0.85	1.063	0.016	0.220	8	Sch 40 (Std)
(1/2)		0.147	0.546	1.09		0.012		10	Sch 80
		0.188	0.464	1.31		0.009		12	Sch 160
0.75	1.050	0.113	0.824	1.13	1.313	0.028	0.275	10	Sch 40 (Std)
(3/4)		0.154	0.742	1.47		0.022		13	Sch 80
		0.219	0.612	1.94		0.015		17	Sch 160
1	1.315	0.133	1.049	1.68	1.576	0.045	0.344	15	Sch 40 (Std)
		0.179	0.957	2.17		0.037		19	Sch 80
		0.250	0.815	2.84		0.027		25	Sch 160
		0.358	0.599	3.66		0.015		32	XX Strong
1.25	1.660	0.140	1.380	2.27	2.054	0.078	0.435	20	Sch 40 (Std)
(1 1/4)		0.191	1.278	3.00		0.067		26	Sch 80
		0.250	1.160	3.76		0.055		33	Sch 160
		0.382	0.896	5.21		0.033		46	XX Strong
1.5	1.900	0.145	1.610	2.72	2.200	0.106	0.497	24	Sch 40 (Std)
(1 1/2)		0.200	1.500	3.63		0.092		32	Sch 80
		0.281	1.338	4.86		0.073		43	Sch 160
		0.400	1.100	6.41		0.049		57	XX Strong
2	2.375	0.154	2.067	3.65	2.875	0.174	0.622	32	Sch 40 (Std)
		0.218	1.939	5.02		0.153		44	Sch 80
		0.344	1.687	7.46		0.116		66	Sch 160
		0.436	1.503	9.03		0.092		80	XX Strong
2.5	2.875	0.203	2.469	5.79	3.375	0.249	0.753	51	Sch 40 (Std)
(2 1/2)		0.276	2.323	7.66		0.220		68	Sch 80
		0.375	2.125	10.0		0.184		88	Sch 160
		0.552	1.771	13.7		0.128		121	XX Strong
3	3.500	0.125	3.250	4.51	4.000	0.431	0.916	40	
		0.156	3.188	5.57		0.415		49	
		0.188	3.124	6.65		0.398		59	
		0.216	3.068	7.58		0.384		67	Sch 40 (Std)
		0.250	3.000	8.68		0.367		77	
		0.281	2.938	9.66		0.352		85	
		0.300	2.900	10.3		0.343		90	Sch 80
		0.438	2.624	14.3		0.281		126	Sch 160
		0.600	2.300	18.6		0.216		164	XX Strong
3.5	4.000	0.125	3.750	5.17	4.625	0.574	1.05	46	
(3 1/2)		0.256	3.688	6.40		0.555		57	
		0.188	3.624	7.65		0.536		68	
		0.226	3.548	9.11		0.514		80	Sch 40 (Std)
		0.250	3.500	10.0		0.500		88	
		0.281	3.438	11.2		0.482		98	
		0.318	3.364	12.5		0.462		110	Sch 80
		0.636	2.728	22.9		0.304		202	XX Strong

Steel Pipe Dimensions (Cont'd)

Nominal Diam. (In)	Outside Diam. (In)	Wall Thickness (In)	Inside Diam. (In)	Plain End Weight (Lbs/Ft)	Threaded Coupling O.D. (In)	Capacity (Gal/Ft)	Surface Area (SqFt/Ft)	Tensile Strength (Tons)	Remarks
4	4.500	0.125	4.250	5.84	5.200	0.737	1.18	52	
		0.156	4.188	7.24		0.716		64	
		0.188	4.124	8.66		0.694		76	
		0.219	4.062	10.0		0.673		88	
		0.237	4.026	10.8		0.661		95	Sch 40 (Std)
		0.250	4.000	11.3		0.653		100	
		0.281	3.938	12.7		0.633		112	
		0.312	3.876	14.0		0.613		123	
		0.337	3.826	15.0		0.597		132	Sch 80
		0.438	3.624	19.0		0.536		168	Sch 120
		0.531	3.438	22.5		0.482		199	Sch 160
		0.674	3.152	27.5		0.405		243	XX Strong
5	5.563	0.156	5.251	9.01	6.296	1.12	1.46	79	
		0.188	5.187	10.8		1.10		95	
		0.219	5.125	12.5		1.07		110	
		0.244	5.075	13.9		1.05		122	
		0.258	5.047	14.6		1.04		129	Sch 40 (Std)
		0.275	5.013	15.5		1.03		137	
		0.281	5.001	15.9		1.02		140	
		0.312	4.939	17.5		1.00		154	
		0.344	4.875	19.2		0.97		169	
		0.375	4.813	20.8		0.95		183	Sch 80
		0.500	4.563	27.0		0.85		239	Sch 120
		0.625	4.313	33.0		0.76		291	Sch 160
		0.750	4.063	38.6		0.67		340	XX Strong
6	6.625	0.188	6.249	12.9	7.390	1.59	1.73	114	
		0.219	6.187	15.0		1.56		132	
		0.250	6.125	17.0		1.53		150	
		0.280	6.065	19.0		1.50		167	Sch 40 (Std)
		0.312	6.001	21.0		1.47		186	
		0.344	5.937	23.1		1.44		204	
		0.375	5.875	25.0		1.41		221	
		0.432	5.761	28.6		1.35		252	Sch 80
		0.562	5.501	36.4		1.23		321	Sch 120
		0.719	5.187	45.4		1.10		400	Sch 160
		0.864	4.897	53.2		0.98		469	XX Strong
8	8.625	0.188	8.249	16.9	9.625	2.78	2.26	149	
		0.203	8.219	18.3		2.76		161	
		0.219	8.187	19.7		2.73		174	
		0.250	8.125	22.4		2.69		197	Sch 20
		0.277	8.071	24.7		2.66		218	Sch 30
		0.312	8.001	27.7		2.61		244	
		0.322	7.981	28.6		2.60		252	Sch 40 (Std)
		0.344	7.937	30.4		2.57		268	
		0.375	7.875	33.0		2.53		292	
		0.406	7.813	35.6		2.49		314	Sch 60
		0.438	7.749	38.3		2.45		338	
		0.500	7.625	43.4		2.37		383	Sch 80
		0.594	7.437	50.9		2.26		450	Sch 100
		0.719	7.187	60.7		2.11		536	Sch 120
		0.812	7.001	67.8		2.00		598	Sch 140
		0.906	6.813	74.7		1.89		659	Sch 160
10	10.750	0.188	10.374	21.2	11.750	4.39	2.81	187	
		0.203	10.344	22.9		4.37		202	
		0.219	10.312	24.6		4.34		217	
		0.250	10.250	28.0		4.29		247	Sch 20
		0.279	10.192	31.2		4.24		275	
		0.307	10.136	34.2		4.19		302	Sch 30

Steel Pipe Dimensions (Cont'd)

Nominal Diam. (In)	Outside Diam. (In)	Wall Thickness (In)	Inside Diam. (In)	Plain End Weight (Lbs/Ft)	Threaded Coupling O.D. (In)	Capacity (Gal/Ft)	Surface Area (SqFt/Ft)	Tensile Strength (Tons)	Remarks
10	10.750	0.344	10.062	38.2	11.750	4.13	2.81	337	
		0.365	10.020	40.5		4.10		357	Sch 40 (Std)
		0.438	9.874	48.2		3.98		426	
		0.500	9.750	54.7		3.88		483	Sch 60
		0.594	9.562	64.4		3.73		569	Sch 80
		0.719	9.312	77.0		3.54		680	Sch 100
		0.844	9.062	89.3		3.35		788	Sch 120
		1.000	8.750	104.1		3.12		919	Sch 140
		1.125	8.500	115.6		2.95		1,021	Sch 160
12	12.750	0.203	12.344	27.2	14.000	6.22	3.34	240	
		0.219	12.312	29.3		6.18		259	
		0.250	12.250	33.4		6.12		295	Sch 20
		0.281	12.188	37.4		6.06		330	
		0.312	12.126	41.4		6.00		366	
		0.330	12.090	43.8		5.96		386	Sch 30
		0.344	12.062	45.6		5.94		402	
		0.375	12.000	49.6		5.88		437	(Std)
		0.406	11.938	53.5		5.81		472	Sch 40
		0.438	11.874	57.6		5.75		508	
		0.500	11.750	65.4		5.63		577	
		0.562	11.626	73.2		5.51		646	Sch 60
		0.688	11.374	88.6		5.28		782	Sch 80
		0.844	11.062	107.3		4.99		947	Sch 100
		1.000	10.750	125.5		4.71		1,107	Sch 120
		1.125	10.500	139.7		4.50		1,233	Sch 140
		1.312	10.126	160.3		4.18		1,414	Sch 160
14	14.000	0.210	13.580	30.9	15.000	7.52	3.67	273	Sch 10
		0.219	13.562	32.2		7.50		284	
		0.250	13.500	36.7		7.44		324	
		0.281	13.438	41.2		7.37		363	
		0.312	13.376	45.6		7.30		403	Sch 20
		0.344	13.312	50.2		7.23		443	
		0.375	13.250	54.6		7.16		482	Sch 30 (Std)
		0.438	13.124	63.4		7.03		560	Sch 40
		0.469	13.062	67.8		6.96		598	
		0.500	13.000	72.1		6.90		636	
		0.594	12.812	85.0		6.70		751	Sch 60
		0.750	12.500	106.1		6.38		937	Sch 80
		0.938	12.124	130.9		6.00		1,155	Sch 100
		1.094	11.812	150.8		5.69		1,331	Sch 120
		1.250	11.500	170.2		5.40		1,502	Sch 140
		1.406	11.188	189.1		5.11		1,669	Sch 160
16	16.000	0.219	15.562	36.9	17.000	9.88	4.19	326	
		0.250	15.500	42.1		9.80		371	Sch 10
		0.281	15.438	47.2		9.72		416	
		0.312	15.376	52.3		9.65		461	Sch 20
		0.344	15.312	57.5		9.57		508	
		0.375	15.250	62.6		9.49		552	Sch 30 (Std)
		0.438	15.124	72.8		9.33		642	
		0.469	15.062	77.8		9.26		687	
		0.500	15.000	82.8		9.18		730	Sch 40
		0.656	14.688	107.5		8.80		949	Sch 60
		0.844	14.312	136.6		8.36		1,206	Sch 80
		1.031	13.938	164.8		7.93		1,455	Sch 100
		1.219	13.562	192.4		7.50		1,698	Sch 120
		1.438	13.124	223.6		7.03		1,974	Sch 140
		1.594	12.812	245.3		6.70		2,164	Sch 160

Steel Pipe Dimensions (Cont'd)

Nominal Diam. (In)	Outside Diam. (In)	Wall Thickness (In)	Inside Diam. (In)	Plain End Weight (Lbs/Ft)	Threaded Coupling O.D. (In)	Capacity (Gal/Ft)	Surface Area (SqFt/Ft)	Tensile Strength (Tons)	Remarks
18	18.000	0.250	17.500	47.4	19.000	12.50	4.71	418	Sch 10
		0.281	17.438	53.2		12.41		469	
		0.312	17.376	58.9		12.32		520	Sch 20
		0.344	17.312	64.9		12.23		572	
		0.375	17.250	70.6		12.14		623	(Std)
		0.406	17.188	76.3		12.05		673	
		0.438	17.124	82.2		11.96		725	Sch 30
		0.469	17.062	87.8		11.88		775	
		0.500	17.000	93.5		11.79		825	
		0.562	16.876	104.7		11.62		924	Sch 40
		0.750	16.500	138.2		11.11		1,219	Sch 60
		0.938	16.124	170.9		10.61		1,508	Sch 80
		1.156	15.688	208.0		10.04		1,835	Sch 100
		1.375	15.250	244.1		9.49		2,154	Sch 120
		1.562	14.876	274.2		9.03		2,420	Sch 140
		1.781	14.438	308.5		8.50		2,722	Sch 160
20	20.000	0.250	19.500	52.7	21.000	15.51	5.24	465	Sch 10
		0.281	19.438	59.2		15.42		522	
		0.312	19.376	65.6		15.32		579	
		0.344	19.312	72.2		15.22		637	
		0.375	19.250	78.6		15.12		694	Sch 20 (Std)
		0.406	19.188	85.0		15.02		750	
		0.438	19.124	91.5		14.92		808	
		0.469	19.062	97.8		14.83		863	
		0.500	19.000	104.1		14.73		919	Sch 30
		0.594	18.812	123.1		14.44		1,086	Sch 40
		0.812	18.376	166.4		13.78		1,468	Sch 60
		1.031	17.938	208.9		13.13		1,843	Sch 80
		1.281	17.438	256.1		12.41		2,260	Sch 100
		1.500	17.000	296.4		11.79		2,615	Sch 120
		1.750	16.500	341.1		11.11		3,010	Sch 140
		1.969	16.062	379.2		10.53		3,346	Sch 160
22	22.000	0.250	21.500	58.1		18.86	5.76	512	Sch 10
		0.281	21.438	65.2		18.75		575	
		0.312	21.376	72.3		18.64		638	
		0.344	21.312	79.6		18.53		702	
		0.375	21.250	86.6		18.42		764	Sch 20 (Std)
		0.406	21.188	93.6		18.32		826	
		0.438	21.124	100.9		18.21		890	
		0.469	21.062	107.8		18.10		952	
		0.500	21.000	114.8		17.99		1,013	Sch 30
		0.625	20.750	142.7		17.57		1,259	
		0.875	20.250	197.4		16.73		1,742	Sch 60
		1.125	19.750	250.8		15.91		2,213	Sch 80
		1.375	19.250	302.9		15.12		2,673	Sch 100
		1.625	18.750	353.6		14.34		3,120	Sch 120
		1.875	18.250	403.0		13.59		3,556	Sch 140
		2.125	17.750	451.1		12.85		3,981	Sch 160
24	24.000	0.250	23.500	63.4		22.53	6.28	560	Sch 10
		0.281	23.438	71.2		22.41		628	
		0.312	23.376	78.9		22.29		697	
		0.344	23.312	86.9		22.17		767	
		0.375	23.250	94.6		22.05		835	Sch 20 (Std)
		0.406	23.188	102.3		21.94		903	
		0.438	23.124	110.2		21.82		973	
		0.469	23.062	117.9		21.70		1,040	
		0.500	23.000	125.5		21.58		1,107	

Steel Pipe Dimensions (Cont'd)

Nominal Diam. (In)	Outside Diam. (In)	Wall Thickness (In)	Inside Diam. (In)	Plain End Weight (Lbs/Ft)	Threaded Coupling O.D. (In)	Capacity (Gal/Ft)	Surface Area (SqFt/Ft)	Tensile Strength (Tons)	Remarks
24	24.000	0.562	22.876	140.7		21.35	6.28	1,241	Sch 30
		0.688	22.624	171.3		20.88		1,512	Sch 40
		0.969	22.062	238.4		19.86		2,103	Sch 60
		1.219	21.562	296.6		18.97		2,617	Sch 80
		1.531	20.938	367.4		17.89		3,242	Sch 100
		1.812	20.376	429.4		16.94		3,789	Sch 120
		2.062	19.876	483.1		16.12		4,263	Sch 140
		2.344	19.312	542.1		15.22		4,784	Sch 160
26	26.000	0.250	25.500	68.8		26.53	6.81	607	
		0.281	25.438	77.2		26.40		681	
		0.312	25.376	85.6		26.27		755	Sch 10
		0.344	25.312	94.3		26.14		832	
		0.375	25.250	102.6		26.01		906	(Std)
		0.406	25.188	111.0		25.88		979	
		0.438	25.124	119.6		25.75		1,055	
		0.469	25.062	127.9		25.63		1,129	
		0.500	25.000	136.2		25.50		1,202	Sch 20
		0.562	24.876	152.7		25.25		1,347	
28	28.000	0.250	27.500	74.1		30.86	7.33	654	
		0.281	27.438	83.2		30.72		734	
		0.312	27.376	92.3		30.58		814	Sch 10
		0.344	27.312	101.6		30.43		897	
		0.375	27.250	110.6		30.30		976	(Std)
		0.406	27.188	119.7		30.16		1,056	
		0.438	27.124	128.9		30.02		1,138	
		0.469	27.062	137.9		29.88		1,217	
		0.500	27.000	146.9		29.74		1,296	Sch 20
		0.625	26.750	182.7		29.19		1,613	Sch 30
30	30.000	0.250	29.500	79.4		35.51	7.85	701	
		0.281	29.438	89.2		35.36		787	
		0.312	29.376	98.9		35.21		873	Sch 10
		0.344	29.312	109.0		35.06		961	
		0.375	29.250	118.7		34.91		1,047	(Std)
		0.406	29.188	128.3		34.76		1,132	
		0.438	29.124	138.3		34.61		1,220	
		0.469	29.062	147.9		34.46		1,305	
		0.500	29.000	157.5		34.31		1,390	Sch 20
		0.625	28.750	196.1		33.72		1,730	Sch 30
32	32.000	0.250	31.500	84.8		40.48	8.38	748	
		0.281	31.438	95.2		40.32		840	
		0.312	31.376	105.6		40.17		932	Sch 10
		0.344	31.312	116.3		40.00		1,026	
		0.375	31.250	126.7		39.84		1,118	(Std)
		0.406	31.188	137.0		39.69		1,209	
		0.438	31.124	147.6		39.52		1,303	
		0.469	31.062	157.9		39.37		1,394	
		0.500	31.000	168.2		39.21		1,484	Sch 20
		0.625	30.750	209.4		38.58		1,848	Sch 30
		0.688	30.624	230.1		38.26		2,030	Sch 40
34	34.000	0.250	33.500	90.1		45.79	8.90	795	
		0.281	33.438	101.2		45.62		893	
		0.312	33.376	112.3		45.45		991	Sch 10
		0.344	33.312	123.7		45.28		1,091	
		0.375	33.250	134.7		45.11		1,188	(Std)
		0.406	33.188	145.7		44.94		1,285	
		0.438	33.124	157.0		44.77		1,385	
		0.469	33.062	168.0		44.60		1,482	

Steel Pipe Dimensions (Cont'd)

Nominal Diam. (In)	Outside Diam. (In)	Wall Thickness (In)	Inside Diam. (In)	Plain End Weight (Lbs/Ft)	Threaded Coupling O.D. (In)	Capacity (Gal/Ft)	Surface Area (SqFt/Ft)	Tensile Strength (Tons)	Remarks
34	34.000	0.500	33.000	178.9		44.43	8.90	1,579	Sch 20
		0.625	32.750	222.8		43.76		1,966	Sch 30
		0.688	32.624	244.8		43.42		2,160	Sch 40
36	36.000	0.250	35.500	95.5		51.42	9.42	842	
		0.281	35.438	107.2		51.24		946	
		0.312	35.376	118.9		51.06		1,049	Sch 10
		0.344	35.312	131.0		50.88	1,156		
		0.375	35.250	142.7		50.70		1,259	(Std)
		0.406	35.188	154.3		50.52		1,362	
		0.438	35.124	166.4		50.33		1,468	
		0.469	35.062	178.0		50.16		1,571	
		0.500	35.000	189.6		49.98		1,673	Sch 20
		0.562	34.876	212.7		49.63		1,877	
		0.625	34.750	236.1		49.27		2,084	Sch 30
		0.750	34.500	282.4		48.56		2,492	Sch 40
38	38.000	0.312	37.376	125.6		57.00	9.95	1,108	
		0.344	37.312	138.3		56.80		1,221	
		0.375	37.250	150.7		56.61		1,330	
		0.438	37.124	175.7		56.23		1,551	
		0.500	37.000	200.3		55.86		1,767	
		0.625	36.750	249.5		55.10		2,202	
40	40.000	0.312	39.376	132.2		63.26	10.47	1,167	
		0.344	39.312	145.7		63.05		1,286	
		0.375	39.250	158.7		62.85		1,400	
		0.438	39.124	185.1		62.45		1,633	
		0.500	39.000	210.9		62.06		1,861	
		0.625	38.750	262.8		61.26		2,319	
42	42.000	0.312	41.376	138.9		69.85	11.00	1,226	
		0.344	41.312	153.0		69.63		1,351	
		0.375	41.250	166.7		69.42		1,471	
		0.438	41.124	194.4		69.00		1,716	
		0.500	41.000	221.6		68.58		1,956	
		0.625	40.750	276.2		67.75		2,437	
48	48.000	0.375	47.250	190.7		91.09		1,683	
		0.500	47.000	253.7		90.13		2,238	

Large Diameter Fabricated Steel Pipe

Nominal Size (I.D. Inches)	Wall Thickness (Inches)									Approx. Capacity (U.S. Gal/Ft)
	0.250	0.312	0.375	0.438	0.500	0.625	0.750	0.875	1.000	
	Approx. Weight (Lbs/Ft)									
51	137	171	206	241	275	345	415			106
54	145	181	218	255	291	365	439			119
57	153	191	230	269	307	385	463			133
60	161	201	242	283	323	405	487	569	651	147
63	169	211	254	297	339	425	511	597	684	162
66	177	221	266	311	355	445	535	625	716	178
69	185	231	278	325	371	465	559	653	748	194
72	193	241	290	339	387	485	583	681	780	212
75	201	251	302	353	403	505	607	709	812	230
78	209	261	314	367	419	525	631	737	844	248
81	217	271	326	381	435	545	655	765	876	268
84	225	281	338	395	451	565	679	793	908	288
87	233	291	350	409	467	585	703	821	940	309
90	241	301	362	423	483	605	727	849	972	330
93	249	311	374	437	499	625	751	877	1004	353
96	257	321	386	451	515	645	775	905	1036	376

Weight of Thin-Wall Steel Pipe (Pounds Per Foot)

O.D. Inches	Wall Thickness (Gage Number)						
	14	12	10	9	8	7	3
4	3.13	4.35	5.55				
4½	3.53	4.91	6.27				
6⅝	5.23	7.28	9.32	10.34	11.34		
8⅝		9.52	12.20	13.53	14.86	16.90	22.36
10¾		11.86	15.25	16.92	18.60	21.15	28.04
12¾		14.13	18.12	20.12	22.11	25.16	33.38
14		15.52	19.92	22.11	24.31	27.66	36.71
16		17.79	22.79	25.31	27.82	31.66	42.05
18		19.99	25.66	28.50	31.33	35.67	47.39
20		22.23	28.54	31.69	34.83	39.67	52.73
24		26.69	34.28	38.08	41.85	47.08	63.41

A.P.I. Casing

Outside Diameter (In)	Wall Thickness (In)	Inside Diameter (In)	Coupling O.D. (In)	Grade	Nominal Weight (T&C) (Lbs/Ft)	Capacity (Gal/Foot)
4½	0.205	4.090	5.000	H,J,K	9.5	0.68
	0.224	4.052		J,K	10.5	0.67
	0.250	4.000		J,K	11.6	0.65
	0.250	4.000		N	11.6	0.65
	0.290	3.920		N	13.5	0.63
5	0.220	4.560	5.563	J,K	11.5	0.85
	0.253	4.494		J,K	13.0	0.82
	0.296	4.408		J,K	15.0	0.79
	0.296	4.408		N	15.0	0.79
	0.362	4.276		N	18.0	0.75
5½	0.244	5.012	6.050	H,J,K	14.0	1.02
	0.275	4.950		J,K	15.5	1.00
	0.304	4.892		J,K	17.0	0.98
	0.304	4.892		N	17.0	0.98
	0.361	4.778		N	20.0	0.93
	0.415	4.670		N	23.0	0.89
6⅝	0.288	6.049	7.390	H	20.0	1.49
	0.288	6.049		J,K	20.0	1.49
	0.352	5.921		J,K	24.0	1.43
	0.352	5.921		N	24.0	1.43
	0.417	5.791		N	28.0	1.37
	0.475	5.675		N	32.0	1.31
7	0.231	6.538	7.656	H	17.0	1.74
	0.272	6.456		H,J,K	20.0	1.70
	0.317	6.366		J,K	23.0	1.65
	0.317	6.366		N	23.0	1.65
	0.362	6.276		J,K	26.0	1.61
	0.362	6.276		N	26.0	1.61
	0.408	6.184		N	29.0	1.56
	0.453	6.094		N	32.0	1.52
	0.498	6.004		N	35.0	1.47
	0.540	5.920		N	38.0	1.43
7⅝	0.300	7.025	8.500	H	24.0	2.01
	0.328	6.969		J,K	26.4	1.98
	0.328	6.969		N	26.4	1.98
	0.375	6.875		N	29.7	1.93
	0.430	6.765		N	33.7	1.87
	0.500	6.625		N	39.0	1.79

A.P.I. Casing (Cont'd.)

Outside Diameter (In)	Wall Thickness (In)	Inside Diameter (In)	Coupling O.D. (In)	Grade	Nominal Weight (T&C) (Lbs/Ft)	Capacity (Gal/Foot)
8⅝	0.264	8.097	9.625	J,K	24.0	2.67
	0.304	8.017		H	28.0	2.62
	0.352	7.921		H	32.0	2.56
	0.352	7.921		J,K	32.0	2.56
	0.400	7.825		J,K	36.0	2.50
	0.400	7.825		N	36.0	2.50
	0.450	7.725		N	40.0	2.43
	0.500	7.625		N	44.0	2.37
	0.557	7.511		N	49.0	2.30
9⅝	0.312	9.001	10.625	H	32.3	3.31
	0.352	8.921		H	36.0	3.25
	0.352	8.921		J,K	36.0	3.25
	0.395	8.835		J,K	40.0	3.18
	0.395	8.835		N	40.0	3.18
	0.435	8.755		N	43.5	3.13
	0.472	8.681		N	47.0	3.07
	0.545	8.535		N	53.5	2.97
10¾	0.279	10.192	11.750	H	32.8	4.24
	0.350	10.050		H	40.5	4.12
	0.350	10.050		J,K	40.5	4.12
	0.400	9.950		J,K	45.5	4.04
	0.450	9.850		J,K,N	51.0	3.96
	0.495	9.760		N	55.5	3.89
11¾	0.333	11.084	12.750	H	42.0	5.01
	0.375	11.000		J,K	47.0	4.94
	0.435	10.880		J,K	54.0	4.83
	0.489	10.772		J,K,N	60.0	4.73
13⅜	0.330	12.715	14.375	H	48.0	6.60
	0.380	12.615		J,K	54.5	6.49
	0.430	12.515		J,K	61.0	6.39
	0.480	12.415		J,K	68.0	6.29
	0.514	12.347		N	72.0	6.22
16	0.375	15.250	17.000	H	65.0	9.49
	0.438	15.124		J,K	75.0	9.33
	0.495	15.010		J,K	84.0	9.19
18⅝	0.435	17.755	19.625	H,J,K	87.5	12.86
20	0.438	19.124	21.000	H,J,K	94.0	14.92
	0.500	19.000		J,K	106.5	14.73
	0.635	18.730		J,K	133.0	14.31

A.P.I. Tubing

Outside Diameter (Inches)	Wall Thickness (Inches)	Inside Diameter (Inches)	Coupling Outside Diam. (Inches)	Nominal Wt. (T&C) (Lb/Ft)	Grade	End Type	Capacity (Gal/Ft)
1.050	0.113	0.824		1.2	H,J,N	Ext. Upset	0.028
1.315	0.133	1.049		1.7	H,J,N	Non-Upset	0.045
1.315	0.133	1.049		1.8	H,J,N	Ext. Upset	0.045
1.660	0.140	1.380	2.054	2.3	H,J.N	Non-Upset	0.078
1.660	0.140	1.380	2.200	2.4	H,J,N	Ext. Upset	0.078
1.900	0.145	1.610	2.200	2.8	H,J,N	Non-Upset	0.106
1.900	0.145	1.610	2.500	2.9	H,J,N	Ext. Upset	0.106
2⅜	0.167	2.041	2.875	4.0	H,J,N	Non-Upset	0.170
2⅜	0.190	1.995	2.875	4.6	H,J,N	Non-Upset	0.162
2⅜	0.190	1.995	3.063	4.7	H,J,N	Ext. Upset	0.162
2⅜	0.254	1.867	2.875	5.8	N	Non-Upset	0.142
2⅜	0.254	1.867	3.063	6.0	N	Ext. Upset	0.142
2⅞	0.217	2.441	3.500	6.4	H.J,N	Non-Upset	0.243
2⅞	0.217	2.441	3.668	6.5	H,J,N	Ext. Upset	0.243
2⅞	0.308	2.259	3.500	8.6	N	Non-Upset	0.208
2⅞	0.308	2.259	3.668	8.7	N	Ext. Upset	0.208
3½	0.216	3.068	4.250	7.7	H,J,N	Non-Upset	0.384
3½	0.254	2.992	4.250	9.2	H,J,N	Non-Upset	0.365
3½	0.254	2.992	4.500	9.3	H,J,N	Ext. Upset	0.365
3½	0.289	2.922	4.250	10.2	H,J,N	Non-Upset	0.348
3½	0.375	2.750	4.250	12.7	N	Non-Upset	0.309
3½	0.375	2.750	4.500	13.0	N	Ext. Upset	0.309
4	0.226	3.548	4.750	9.5	H,J,N	Non-Upset	0.514
4	0.226	3.548	5.000	11.0	H,J,N	Ext. Upset	0.514
4½	0.271	3.958	5.200	12.6	H,J,N	Non-Upset	0.639
4½	0.271	3.958	5.563	12.8	H,J,N	Ext. Upset	0.639

Stretch of Suspended Steel Tubing, Drill Pipe, or Casing
(Stretch in Inches When Suspended in Various Fluids)

Pipe Length (Feet)	Air	Fresh Water	Drilling Fluid Weight (Lbs/Gal)											
			9	9.5	10	10.5	11	12	13	14	15	16	17	18
500	0.17	0.14	0.14	0.13	0.13	0.13	0.13	0.13	0.12	0.12	0.11	0.11	0.11	0.10
1000	0.68	0.56	0.55	0.54	0.53	0.52	0.52	0.50	0.49	0.47	0.46	0.44	0.43	0.41
1500	1.51	1.25	1.23	1.21	1.19	1.18	1.16	1.13	1.09	1.06	1.02	0.99	0.96	0.92
2000	2.72	2.22	2.18	2.15	2.12	2.09	2.06	2.00	1.94	1.88	1.82	1.76	1.70	1.64
2500	4.19	3.47	3.41	3.36	3.31	3.27	3.22	3.13	3.03	2.94	2.85	2.75	2.66	2.57
3000	6.12	5.00	4.91	4.84	4.77	4.71	4.64	4.50	4.37	4.23	4.10	3.96	3.83	3.70
4000	10.88	8.88	8.72	8.60	8.48	8.37	8.25	8.01	7.77	7.53	7.29	7.05	6.81	6.57
5000	17.00	13.88	13.63	13.44	13.26	13.07	12.88	12.51	12.14	11.76	11.39	11.01	10.64	10.27
6000	24.48	19.99	19.63	19.36	19.09	18.82	18.55	18.01	17.48	16.94	16.40	15.86	15.32	14.78
7000	33.32	27.21	26.72	26.35	25.99	25.62	25.25	24.52	23.79	23.05	22.32	21.59	20.85	20.12
8000	43.51	35.54	34.90	34.42	33.94	33.46	32.98	32.02	31.07	30.11	29.15	28.19	27.24	26.28
10000	67.99	55.53	54.53	53.78	53.03	52.28	51.53	50.04	48.54	47.05	45.55	44.05	42.56	41.06

Valid only for strings suspended freely, without float plug; not valid if string has been pulled on, or set beyond proper safety factor.

Stretch Table for Tubing, Drill Pipe, and Casing

(Approximate Pull Required, Above Weight of String, To Obtain 3 Inches of Stretch Per 1,000 Feet of Free Pipe)

Pipe Size (O.D. – Inches)	Nominal Weight (Lbs/Foot)	Approx. Pull Required (Pounds)
Tubing		
1.315	1.8	4,000
1.660	2.4	5,500
1.900	2.9	6,500
2.375	4.7	10,000
2.875	6.5	14,000
3.500	9.3	21,000
4.000	11.0	24,000
4.500	12.8	28,000
Drill Pipe		
2.375	6.7	15,000
2.875	10.4	23,000
3.500	13.3	29,000
4.500	16.6	37,000
5.000	19.5	43,000
Casing		
5.000	15.0	33,000
5.500	17.0	37,000
6.625	24.0	53,000
7.000	26.0	57,000
7.625	29.7	66,000
8.625	36.0	79,000
9.625	43.5	96,000
10.750	45.5	100,000

Figures valid only if pipe has not been, or is not being, stretched beyond elastic limit.

Steel Pipe Flange Dimensions

Nominal Pipe Size (Inches)	Class 150					Class 300				
	Flange O.D. (Inches)	Bolt Circle Diam. (In.)	Bolt Holes Diam. (In.)	No. of Bolts	Bolt Diam. (Inches)	Flange O.D. (Inches)	Bolt Circle Diam. (In.)	Bolt Holes Diam. (In.)	No. of Bolts	Bolt Diam. (Inches)
2	6.00	4.75	0.75	4	5/8	6.50	5.00	0.75	8	5/8
2½	7.00	5.50	0.75	4	5/8	7.50	5.88	0.88	8	3/4
3	7.50	6.00	0.75	4	5/8	8.25	6.62	0.88	8	3/4
3½	8.50	7.00	0.75	8	5/8	9.00	7.25	0.88	8	3/4
4	9.00	7.50	0.75	8	5/8	10.00	7.88	0.88	8	3/4
5	10.00	8.50	0.88	8	3/4	11.00	9.25	0.88	8	3/4
6	11.00	9.50	0.88	8	3/4	12.50	10.62	0.88	12	3/4
8	13.50	11.75	0.88	8	3/4	15.00	13.00	1.00	12	7/8
10	16.00	14.25	1	12	7/8	17.50	15.25	1.12	16	1
12	19.00	17.00	1	12	7/8	20.50	17.75	1.25	16	1⅛
14	21.00	18.75	1.12	12	1	23.00	20.25	1.25	20	1⅛
16	23.50	21.25	1.12	16	1	25.50	22.50	1.38	20	1¼
18	25.00	22.75	1.25	16	1⅛	28.00	24.75	1.38	24	1¼
20	27.50	25.00	1.25	20	1⅛	30.50	27.00	1.38	24	1¼
24	32.00	29.50	1.38	20	1¼	36.00	32.00	1.62	24	1½

Thermal Expansion of Steel Pipe

Temperature Change		Change In Length		Temperature Change		Change In Length	
Deg. F	Deg. C	(in/100 ft)	(cm/100 m)	Deg. F	Deg. C	(in/100 ft)	(cm/100 m)
10	5.56	0.076	0.633	90	50.00	0.684	5.700
20	11.11	0.152	1.267	100	55.56	0.760	6.333
30	16.67	0.228	1.900	110	61.11	0.836	6.966
40	22.22	0.304	2.533	120	66.67	0.912	7.600
50	27.78	0.380	3.167	130	72.22	0.988	8.233
60	33.33	0.456	3.800	140	77.78	1.064	8.866
70	38.89	0.532	4.433	150	83.33	1.140	9.500
80	44.44	0.608	5.066				

PVC Pipe Dimensions

Nominal Diam. (In)	Outside Diam. (In)	Wall Thickness (In)	Inside Diam. (In)	Weight* (Lbs/Ft)	Capacity (Gal/Ft)	Surface Area (SqFt/Ft)	Remarks
½	0.840	0.062	0.716	0.10	0.021	0.22	SDR 13.5
		0.109	0.622	0.16	0.016		Sch 40
		0.147	0.546	0.21	0.012		Sch 80
¾	1.050	0.062	0.926	0.13	0.035	0.27	SDR 17
		0.078	0.894	0.16	0.033		SDR 13.5
		0.113	0.824	0.22	0.028		Sch 40
		0.154	0.742	0.28	0.022		Sch 80
1	1.315	0.063	1.189	0.16	0.058	0.34	SDR 21
		0.077	1.161	0.20	0.055		SDR 17
		0.097	1.121	0.24	0.051		SDR 13.5
		0.133	1.049	0.32	0.045		Sch 40
		0.179	0.957	0.41	0.037		Sch 80
1¼	1.660	0.064	1.532	0.21	0.096	0.43	SDR 26
		0.079	1.502	0.26	0.092		SDR 21
		0.098	1.464	0.31	0.087		SDR 17
		0.123	1.414	0.39	0.082		SDR 13.5
		0.140	1.380	0.43	0.078		Sch 40
		0.191	1.278	0.57	0.067		Sch 80
1½	1.900	0.073	1.754	0.27	0.13	0.50	SDR 26
		0.090	1.720	0.33	0.12		SDR 21
		0.112	1.676	0.41	0.11		SDR 17
		0.141	1.618	0.51	0.11		SDR 13.5
		0.145	1.610	0.52	0.11		Sch 40
		0.200	1.500	0.69	0.092		Sch 80
2	2.375	0.091	2.193	0.43	0.20	0.62	SDR 26
		0.113	2.149	0.52	0.19		SDR 21
		0.140	2.095	0.64	0.18		SDR 17
		0.176	2.023	0.79	0.17		SDR 13.5
		0.154	2.067	0.70	0.17		Sch 40
		0.218	1.939	0.96	0.15		Sch 80
2½	2.875	0.110	2.655	0.62	0.29	0.75	SDR 26
		0.137	2.601	0.77	0.28		SDR 21
		0.169	2.537	0.93	0.26		SDR 17
		0.213	2.449	1.2	0.24		SDR 13.5
		0.203	2.469	1.1	0.25		Sch 40
		0.276	2.323	1.5	0.22		Sch 80

*Weight in air; weight in water about 28.5% of value shown

SDR = Standard Dimension Ratio = Outside Diam./Wall Thickness

PVC Pipe Dimensions (Cont'd)

Nominal Diam. (In)	Outside Diam. (In)	Wall Thickness (In)	Inside Diam. (In)	Weight* (Lbs/Ft)	Capacity (Gal/Ft)	Surface Area (SqFt/Ft)	Remarks
3	3.500	0.135	3.230	0.93	0.43	0.92	SDR 26
		0.167	3.166	1.1	0.41		SDR 21
		0.206	3.088	1.4	0.39		SDR 17
		0.259	2.982	1.7	0.36		SDR 13.5
		0.216	3.068	1.4	0.38		Sch 40
		0.300	2.900	2.0	0.34		Sch 80
3½	4.000	0.154	3.692	1.2	0.56	1.05	SDR 26
		0.190	3.620	1.5	0.53		SDR 21
		0.235	3.530	1.8	0.51		SDR 17
		0.226	3.548	1.7	0.51		Sch 40
4	4.500	0.173	4.154	1.5	0.70	1.18	SDR 26
		0.214	4.072	1.9	0.68		SDR 21
		0.265	3.970	2.3	0.64		SDR 17
		0.333	3.834	2.8	0.60		SDR 13.5
		0.237	4.026	2.1	0.66		Sch 40
		0.337	3.826	2.9	0.60		Sch 80
5	5.563	0.214	5.135	2.3	1.08	1.46	SDR 26
		0.265	5.033	2.9	1.03		SDR 21
		0.327	4.909	3.5	0.98		SDR 17
		0.258	5.047	2.8	1.04		Sch 40
		0.375	4.813	4.0	0.95		Sch 80
6	6.625	0.255	6.115	3.3	1.53	1.73	SDR 26
		0.316	5.993	4.1	1.47		SDR 21
		0.390	5.845	5.0	1.39		SDR 17
		0.491	5.643	6.1	1.30		SDR 13.5
		0.280	6.065	3.6	1.50		Sch 40
		0.432	5.761	5.5	1.35		Sch 80
8	8.625	0.332	7.961	5.6	2.59	2.26	SDR 26
		0.410	7.805	6.9	2.49		SDR 21
		0.508	7.609	8.4	2.36		SDR 17
		0.322	7.981	5.5	2.60		Sch 40
		0.500	7.625	8.3	2.37		Sch 80
10	10.750	0.413	9.924	8.7	4.02	2.81	SDR 26
		0.511	9.728	10.7	3.86		SDR 21
		0.365	10.020	7.8	4.10		Sch 40
		0.593	9.564	12.3	3.73		Sch 80
12	12.750	0.490	11.770	12.3	5.65	3.34	SDR 26
		0.606	11.538	15.0	5.43		SDR 21
		0.406	11.938	10.3	5.81		Sch 40
		0.687	11.376	16.9	5.28		Sch 80

*Weight in air; weight in water about 28.5% of value shown

SDR = Standard Dimension Ratio = Outside Diam./Wall Thickness

PVC Pipe Pressure Ratings – Pounds/Square Inch

Nominal Diameter Inches	Pipe Class					
	SDR 26	SDR 21	SDR 17	SDR 13.5	Schedule 40	Schedule 80
2	160	200	250	315	280	400
4	160	200	250	315	220	320
6	160	200	250	315	180	280
8	160	200	250	315	160	250
10	160	200	250	315	140	230
12	160	200	250	315	130	230

SDR = Standard Dimension Ratio (Outside diameter / wall thickness)

Pressure ratings are based on water at 73.4 degrees F (23 degrees C).
For higher temperature conditions, pressure ratings are to be lowered by multiplying by the derating factor from the following table:

Temperature		
Deg. F	Deg. C	Factor
73.4	23.0	1.00
80	26.7	0.88
90	32.2	0.75
100	37.8	0.62
110	43.3	0.50
120	48.9	0.40
130	54.4	0.30
140	60.0	0.22

Thermal Expansion of PVC Pipe

Temp. Change (Degrees F)	Length Change (In/100 Ft)
40	1.39
50	1.74
60	2.09
70	2.44
80	2.78
90	3.13
100	3.48

Hydraulic Collapse Pressure of Pipe

The collapse pressure resulting from external hydraulic forces on pipes is related to the outside diameter (D) and wall thickness (t). The approximate maximum uniform external pressure allowable, for various D/t ratios, is given in the table.

D/t	Steel Pipe*		PVC Pipe	
	(PSI)	(Ft of Water)	(PSI)	(Ft of Water)
10	—	—	—	—
15	—	—	330	762
20	—	—	135	312
25	—	—	66	152
30	985	2,275	37	85
35	730	1,686	22	51
40	550	1,271	15	35
45	415	959	11	25
50	320	739	—	—
55	255	589	—	—
60	200	462	—	—
65	165	381	—	—
70	135	312	—	—
75	113	261	—	—
80	95	219	—	—
85	82	189	—	—
90	70	162	—	—
95	60	139	—	—
100	52	120	—	—
110	40	92	—	—
120	31	72	—	—

*Yield strength of 35,000 psi.

Ductile Iron Pipe					
Nominal Size (inches)	Outside Diam. (inches)	Inside Diam. (inches)	Wall Thickness (inches)	Capacity (Gals/Ft)	Pressure Class (psi)
3	3.96	3.46	0.25	0.49	350
4	4.80	4.30	0.25	0.75	350
6	6.90	6.40	0.25	1.67	350
8	9.05	8.55	0.25	2.98	350
10	11.10	10.60	0.25	4.58	350
12	13.20	12.70	0.25	6.58	350
14	15.30	14.74	0.28	8.86	250
		14.70	0.30	8.82	300
		14.68	0.31	8.79	350
16	17.40	16.80	0.30	11.5	250
		16.76	0.32	11.5	300
		16.72	0.34	11.4	350
18	19.50	18.88	0.31	14.5	250
		18.82	0.34	14.5	300
		18.78	0.36	14.4	350
20	21.60	20.94	0.33	17.9	250
		20.88	0.36	17.8	300
		20.84	0.38	17.7	350
24	25.80	25.14	0.33	25.8	200
		25.06	0.37	25.6	250
		25.00	0.40	25.5	300
		24.94	0.43	25.4	350
30	32.00	31.32	0.34	40.0	150
		31.24	0.38	39.8	200
		31.16	0.42	39.6	250
		31.10	0.45	39.5	300
		31.02	0.49	39.3	350
36	38.30	37.54	0.38	57.5	150
		37.46	0.42	57.3	200
		37.36	0.47	56.9	250
		37.28	0.51	56.7	300
		37.18	0.56	56.4	350

Asbestos Cement (A/C) Pipe

Nominal Size (Inches)	Inside Diam. (Inches)	Outside Diam. (Inches)	Volume (U.S. Gal/Ft)	Approx. Weight* (Lbs/Foot)	Couplings		Class
					O.D. (In.)	Length (In.)	
4	4.00	5.05	0.65	7.2	6.25	6.50	100
	4.00	5.14	0.65	7.9	6.66	6.50	150
	3.95	5.32	0.64	9.2	6.88	6.50	200
6	6.00	7.16	1.47	11.3	8.40	7.00	100
	5.80	7.12	1.37	12.2	8.76	7.00	150
	5.70	7.36	1.33	15.8	9.35	7.00	200
8	8.00	9.32	2.61	16.2	10.80	7.00	100
	7.80	9.32	2.48	18.6	11.32	7.00	150
	7.60	9.46	2.36	23.2	12.00	7.00	200
10	10.00	11.46	4.08	23.8	13.30	7.00	100
	10.00	11.85	4.08	30.2	14.36	7.00	150
	9.60	11.88	3.76	35.7	15.15	7.00	200
12	12.00	13.70	5.88	33.1	15.55	8.00	100
	12.00	14.11	5.88	41.7	17.00	8.00	150
	11.44	14.11	5.34	49.3	17.90	8.00	200
14	13.60	15.36	7.55	39.3	17.60	9.00	100
	14.00	16.41	8.00	53.3	19.80	9.00	150
	13.60	16.44	7.55	64.8	20.50	9.00	200
16	15.50	17.50	9.80	47.1	19.80	9.00	100
	16.00	18.65	10.44	66.3	22.28	9.00	150
	15.50	18.74	9.80	84.6	23.20	9.00	200
18	18.00	20.30	13.22	—	22.60	10.00	100
	18.00	21.20	13.22	—	24.91	10.00	150
20	20.00	22.50	16.32	—	25.30	10.00	100
	20.00	23.54	16.32	—	27.55	10.00	150
24	24.00	27.00	23.50	—	30.00	10.00	100
	24.00	28.22	23.50	—	33.02	10.00	150

*Includes couplings

Concrete Irrigation Pipe					
Inside Diam. (Inches)	Wall Thickness (Inches)	Outside Diam. (Inches)	Volume (U.S. Gal/Ft)	Working Pressure	
				(Ft. of Water)	(PSI)
6	7/8	7 3/4	1.5	30	13.0
8	1	10	2.6	30	13.0
10	1 1/8	12 1/4	4.1	30	13.0
12	1 1/4	14 1/2	5.9	25	10.8
14	1 3/8	16 3/4	8.0	25	10.8
15	1 1/2	18	9.2	25	10.8
16	1 1/2	19	10.4	25	10.8
18	1 3/4	21 1/2	13.2	25	10.8
20	2	24	16.3	25	10.8
21	2 1/8	25 1/4	18.0	25	10.8
24	2 1/4	28 1/2	23.5	25	10.8

Chapter 4

Hydraulics and Water Data

Pressure Conversion – Head (Ft. of Water) and Pressure (Psi)

Ft	Ft or Psi	Psi	Ft	Ft or Psi	Psi	Ft	Ft or Psi	Psi	Ft	Ft or Psi	Psi
2.31	1	0.43	117.78	51	22.08	233.24	101	43.74	348.71	151	65.39
4.62	2	0.87	120.09	52	22.52	235.55	102	44.17	351.02	152	65.82
6.93	3	1.30	122.40	53	22.95	237.86	103	44.60	353.33	153	66.25
9.24	4	1.73	124.70	54	23.38	240.17	104	45.03	355.64	154	66.69
11.55	5	2.17	127.01	55	23.82	242.48	105	45.47	357.95	155	67.12
13.86	6	2.60	129.32	56	24.25	244.79	106	45.90	360.26	156	67.55
16.17	7	3.03	131.63	57	24.68	247.10	107	46.33	362.57	157	67.98
18.47	8	3.46	133.94	58	25.12	249.41	108	46.77	364.88	158	68.42
20.78	9	3.90	136.25	59	25.55	251.72	109	47.20	367.19	159	68.85
23.09	10	4.33	138.56	60	25.98	254.03	110	47.63	369.49	160	69.28
25.40	11	4.76	140.87	61	26.41	256.34	111	48.07	371.80	161	69.72
27.71	12	5.20	143.18	62	26.85	258.65	112	48.50	374.11	162	70.15
30.02	13	5.63	145.49	63	27.28	260.96	113	48.93	376.42	163	70.58
32.33	14	6.06	147.80	64	27.71	263.26	114	49.36	378.73	164	71.02
34.64	15	6.50	150.11	65	28.15	265.57	115	49.80	381.04	165	71.45
36.95	16	6.93	152.42	66	28.58	267.88	116	50.23	383.35	166	71.88
39.26	17	7.36	154.73	67	29.01	270.19	117	50.66	385.66	167	72.31
41.57	18	7.79	157.04	68	29.45	272.50	118	51.10	387.97	168	72.75
43.88	19	8.23	159.34	69	29.88	274.81	119	51.53	390.28	169	73.18
46.19	20	8.66	161.65	70	30.31	277.12	120	51.96	392.59	170	73.61
48.50	21	9.09	163.96	71	30.74	279.43	121	52.40	394.90	171	74.05
50.81	22	9.53	166.27	72	31.18	281.74	122	52.83	397.21	172	74.48
53.11	23	9.96	168.58	73	31.61	284.05	123	53.26	399.52	173	74.91
55.42	24	10.39	170.89	74	32.04	286.36	124	53.69	401.83	174	75.35
57.73	25	10.83	173.20	75	32.48	288.67	125	54.13	404.13	175	75.78
60.04	26	11.26	175.51	76	32.91	290.98	126	54.56	406.44	176	76.21
62.35	27	11.69	177.82	77	33.34	293.29	127	54.99	408.75	177	76.64
64.66	28	12.12	180.13	78	33.78	295.60	128	55.43	411.06	178	77.08
66.97	29	12.56	182.44	79	34.21	297.90	129	55.86	413.37	179	77.51
69.28	30	12.99	184.75	80	34.64	300.21	130	56.29	415.68	180	77.94
71.59	31	13.42	187.06	81	35.07	302.52	131	56.73	417.99	181	78.38
73.90	32	13.86	189.37	82	35.51	304.83	132	57.16	420.30	182	78.81
76.21	33	14.29	191.68	83	35.94	307.14	133	57.59	422.61	183	79.24
78.52	34	14.72	193.98	84	36.37	309.45	134	58.02	424.92	184	79.68
80.83	35	15.16	196.29	85	36.81	311.76	135	58.46	427.23	185	80.11
83.14	36	15.59	198.60	86	37.24	314.07	136	58.89	429.54	186	80.54
85.45	37	16.02	200.91	87	37.67	316.38	137	59.32	431.85	187	80.97
87.75	38	16.45	203.22	88	38.11	318.69	138	59.76	434.16	188	81.41
90.06	39	16.89	205.53	89	38.54	321.00	139	60.19	436.47	189	81.84
92.37	40	17.32	207.84	90	38.97	323.31	140	60.62	438.77	190	82.27
94.68	41	17.75	210.15	91	39.40	325.62	141	61.06	441.08	191	82.71
96.99	42	18.19	212.46	92	39.84	327.93	142	61.49	443.39	192	83.14
99.30	43	18.62	214.77	93	40.27	330.24	143	61.92	445.70	193	83.57
101.61	44	19.05	217.08	94	40.70	332.54	144	62.35	448.01	194	84.01
103.92	45	19.49	219.39	95	41.14	334.85	145	62.79	450.32	195	84.44
106.23	46	19.92	221.70	96	41.57	337.16	146	63.22	452.63	196	84.87
108.54	47	20.35	224.01	97	42.00	339.47	147	63.65	454.94	197	85.30
110.85	48	20.78	226.32	98	42.44	341.78	148	64.09	457.25	198	85.74
113.16	49	21.22	228.62	99	42.87	344.09	149	64.52	459.56	199	86.17
115.47	50	21.65	230.93	100	43.30	346.40	150	64.95	461.87	200	86.60

Pressure Conversion – Head (Ft. of Water) and Pressure (Psi)

Ft	Ft or Psi	Psi	Ft	Ft or Psi	Psi	Ft	Ft or Psi	Psi	Ft	Ft or Psi	Psi
464.18	201	87.04	579.64	251	108.69	697.42	302	130.77	928.35	402	174.07
466.49	202	87.47	581.95	252	109.12	702.04	304	131.64	932.97	404	174.94
468.80	203	87.90	584.26	253	109.55	706.66	306	132.50	937.59	406	175.81
471.11	204	88.34	586.57	254	109.99	711.28	308	133.37	942.21	408	176.67
473.41	205	88.77	588.88	255	110.42	715.90	310	134.24	946.83	410	177.54
475.72	206	89.20	591.19	256	110.85	720.51	312	135.10	951.45	412	178.40
478.03	207	89.64	593.50	257	111.29	725.13	314	135.97	956.07	414	179.27
480.34	208	90.07	595.81	258	111.72	729.75	316	136.83	960.69	416	180.14
482.65	209	90.50	598.12	259	112.15	734.37	318	137.70	965.30	418	181.00
484.96	210	90.93	600.43	260	112.59	738.99	320	138.57	969.92	420	181.87
487.27	211	91.37	602.74	261	113.02	743.61	322	139.43	974.54	422	182.73
489.58	212	91.80	605.05	262	113.45	748.23	324	140.30	979.16	424	183.60
491.89	213	92.23	607.36	263	113.88	752.84	326	141.16	983.78	426	184.47
494.20	214	92.67	609.67	264	114.32	757.46	328	142.03	988.40	428	185.33
496.51	215	93.10	611.98	265	114.75	762.08	330	142.90	993.02	430	186.20
498.82	216	93.53	614.28	266	115.18	766.70	332	143.76	997.63	432	187.06
501.13	217	93.97	616.59	267	115.62	771.32	334	144.63	1002.25	434	187.93
503.44	218	94.40	618.90	268	116.05	775.94	336	145.49	1006.87	436	188.80
505.75	219	94.83	621.21	269	116.48	780.56	338	146.36	1011.49	438	189.66
508.05	220	95.26	623.52	270	116.92	785.18	340	147.23	1016.11	440	190.53
510.36	221	95.70	625.83	271	117.35	789.79	342	148.09	1020.73	442	191.39
512.67	222	96.13	628.14	272	117.78	794.41	344	148.96	1025.35	444	192.26
514.98	223	96.56	630.45	273	118.21	799.03	346	149.82	1029.97	446	193.13
517.29	224	97.00	632.76	274	118.65	803.65	348	150.69	1034.58	448	193.99
519.60	225	97.43	635.07	275	119.08	808.27	350	151.56	1039.20	450	194.86
521.91	226	97.86	637.38	276	119.51	812.89	352	152.42	1043.82	452	195.73
524.22	227	98.30	639.69	277	119.95	817.51	354	153.29	1048.44	454	196.59
526.53	228	98.73	642.00	278	120.38	822.13	356	154.16	1053.06	456	197.46
528.84	229	99.16	644.31	279	120.81	826.74	358	155.02	1057.68	458	198.32
531.15	230	99.59	646.62	280	121.25	831.36	360	155.89	1062.30	460	199.19
533.46	231	100.03	648.92	281	121.68	835.98	362	156.75	1066.92	462	200.06
535.77	232	100.46	651.23	282	122.11	840.60	364	157.62	1071.53	464	200.92
538.08	233	100.89	653.54	283	122.54	845.22	366	158.49	1076.15	466	201.79
540.39	234	101.33	655.85	284	122.98	849.84	368	159.35	1080.77	468	202.65
542.69	235	101.76	658.16	285	123.41	854.46	370	160.22	1085.39	470	203.52
545.00	236	102.19	660.47	286	123.84	859.07	372	161.08	1090.01	472	204.39
547.31	237	102.63	662.78	287	124.28	863.69	374	161.95	1094.63	474	205.25
549.62	238	103.06	665.09	288	124.71	868.31	376	162.82	1099.25	476	206.12
551.93	239	103.49	667.40	289	125.14	872.93	378	163.68	1103.86	478	206.98
554.24	240	103.92	669.71	290	125.58	877.55	380	164.55	1108.48	480	207.85
556.55	241	104.36	672.02	291	126.01	882.17	382	165.41	1113.10	482	208.72
558.86	242	104.79	674.33	292	126.44	886.79	384	166.28	1117.72	484	209.58
561.17	243	105.22	676.64	293	126.87	891.41	386	167.15	1122.34	486	210.45
563.48	244	105.66	678.95	294	127.31	896.02	388	168.01	1126.96	488	211.31
565.79	245	106.09	681.26	295	127.74	900.64	390	168.88	1131.58	490	212.18
568.10	246	106.52	683.56	296	128.17	905.26	392	169.74	1136.20	492	213.05
570.41	247	106.96	685.87	297	128.61	909.88	394	170.61	1140.81	494	213.91
572.72	248	107.39	688.18	298	129.04	914.50	396	171.48	1145.43	496	214.78
575.03	249	107.82	690.49	299	129.47	919.12	398	172.34	1150.05	498	215.64
577.34	250	108.26	692.80	300	129.91	923.74	400	173.21	1154.67	500	216.51

Flow Rate Conversions

U.S. Gallons Per Minute	U.S. Gallons Per Hour	U.S. Gallons Per Day	Cubic Feet Per Second	Liters Per Second	Cubic Meters Per Hour
10	600	14,400	0.0223	0.6309	2.271
20	1,200	28,800	0.0446	1.2618	4.542
30	1,800	43,200	0.0668	1.8926	6.813
40	2,400	57,600	0.0891	2.5235	9.084
50	3,000	72,000	0.1114	3.1544	11.36
60	3,600	86,400	0.1337	3.7853	13.63
70	4,200	100,800	0.1560	4.4162	15.90
80	4,800	115,200	0.1782	5.0471	18.17
90	5,400	129,600	0.2005	5.6779	20.44
100	6,000	144,000	0.2228	6.3088	22.71
125	7,500	180,000	0.2785	7.8860	28.39
150	9,000	216,000	0.3342	9.4632	34.07
175	10,500	252,000	0.3899	11.040	39.74
200	12,000	288,000	0.4456	12.618	45.42
250	15,000	360,000	0.5570	15.772	56.78
300	18,000	432,000	0.6684	18.926	68.13
350	21,000	504,000	0.7798	22.081	79.49
400	24,000	576,000	0.8912	25.235	90.84
450	27,000	648,000	1.0026	28.390	102.2
500	30,000	720,000	1.1140	31.544	113.6
600	36,000	864,000	1.3368	37.853	136.3
700	42,000	1,008,000	1.5596	44.162	159.0
800	48,000	1,152,000	1.7824	50.471	181.7
900	54,000	1,296,000	2.0052	56.779	204.4
1,000	60,000	1,440,000	2.2280	63.088	227.1
1,100	66,000	1,584,000	2.4508	69.397	249.8
1,200	72,000	1,728,000	2.6736	75.706	272.5
1,300	78,000	1,872,000	2.8964	82.015	295.2
1,400	84,000	2,016,000	3.1192	88.324	317.9
1,500	90,000	2,160,000	3.3420	94.632	340.7
2,000	120,000	2,880,000	4.4560	126.18	454.2
2,500	150,000	3,600,000	5.5700	157.72	567.8
3,000	180,000	4,320,000	6.6840	189.26	681.3
4,000	240,000	5,760,000	8.9120	252.35	908.4
5,000	300,000	7,200,000	11.1400	315.44	1,135.5

Discharge vs Runoff

Discharge (cfs/sq mile)	Run-off (Depth in Inches)					
	1 day	28 days	29 days	30 days	31 days	365 days
1	0.0372	1.041	1.079	1.116	1.153	13.57
2	0.0748	2.095	2.170	2.245	2.320	27.31
3	0.1116	3.124	3.236	3.347	3.459	40.72
4	0.1488	4.165	4.314	4.463	4.612	54.30
5	0.1860	5.207	5.393	5.579	5.764	67.87
6	0.2231	6.248	6.471	6.694	6.917	81.45
7	0.2603	7.289	7.550	7.810	8.070	95.02
8	0.2975	8.331	8.628	8.926	9.223	108.59
9	0.3347	9.372	9.707	10.041	10.376	122.17

Discharge (cu ft/sec)	Run-off (Acre-feet)					
	1 day	28 days	29 days	30 days	31 days	365 days
1	1.983	55.54	57.52	59.50	61.49	724.0
2	3.967	111.1	115.0	119.0	123.0	1,448
3	5.950	166.6	172.6	178.5	184.5	2,172
4	7.934	222.1	230.1	238.0	246.0	2,896
5	9.917	277.7	287.6	297.5	307.4	3,620
6	11.90	333.2	345.1	357.0	368.9	4,344
7	13.88	388.8	402.6	416.5	430.4	5,068
8	15.87	444.3	460.2	476.0	491.9	5,792
9	17.85	499.8	517.7	535.5	553.4	6,516

Discharge (cu ft/sec)	Run-off (Million Cubic Feet)					
	1 day	28 days	29 days	30 days	31 days	365 days
1	0.0864	2.419	2.506	2.592	2.678	31.54
2	0.1728	4.838	5.011	5.184	5.357	63.07
3	0.2592	7.258	7.517	7.776	8.035	94.61
4	0.3456	9.677	10.02	10.37	10.71	126.1
5	0.4320	12.10	12.53	12.96	13.39	157.7
6	0.5184	14.52	15.03	15.55	16.07	189.2
7	0.6048	16.93	17.54	18.14	18.75	220.8
8	0.6912	19.35	20.04	20.74	21.43	252.3
9	0.7776	21.77	22.55	23.33	24.11	283.8

Discharge (cu ft/sec)	Run-off (Million Gallons)					
	1 day	28 days	29 days	30 days	31 days	365 days
1	0.6463	18.10	18.74	19.39	20.04	235.9
2	1.293	36.19	37.49	38.78	40.07	471.8
3	1.939	54.29	56.23	58.17	60.11	707.7
4	2.585	72.39	74.97	77.56	80.14	943.6
5	3.232	90.48	93.71	96.95	100.2	1,179
6	3.878	108.6	112.5	116.3	120.2	1,415
7	4.524	126.7	131.2	135.7	140.2	1,651
8	5.170	144.8	149.9	155.1	160.3	1,887
9	5.817	162.9	168.7	174.5	180.3	2,123

One Miner's Inch

The Miner's Inch is frequently used for irrigation flow measurements, generally defined as flow through a one-inch square orifice under a head of one foot of water.

The precise definition by law, for various localities, is given below:

Cubic Ft/Second	**1 / 50**	**1 / 40**	**1 / 38.4**	**1 / 35.7**
Approx. GPM	**9.0**	**11.2**	**11.7**	**12.6**
Locality	Idaho Kansas Nebraska Nevada New Mexico North Dakota South Dakota Utah Washington	Arizona California Montana Oregon	Colorado	British Columbia

Theoretical Velocity of Water (Ft/Sec) for Heads of 0 to 10 Ft

Head (Ft)	0.00	0.01	0.02	0.03	0.04	0.05	0.06	0.07	0.08	0.09
0.0		0.80	1.13	1.39	1.60	1.79	1.96	2.12	2.27	2.41
0.1	2.54	2.66	2.78	2.89	3.00	3.11	3.21	3.31	3.40	3.50
0.2	3.59	3.68	3.76	3.85	3.93	4.01	4.09	4.17	4.24	4.32
0.3	4.39	4.47	4.54	4.61	4.68	4.74	4.81	4.88	4.94	5.01
0.4	5.07	5.14	5.20	5.26	5.32	5.38	5.44	5.50	5.56	5.61
0.5	5.67	5.73	5.78	5.84	5.89	5.95	6.00	6.05	6.11	6.16
0.6	6.21	6.26	6.31	6.37	6.42	6.47	6.52	6.56	6.61	6.66
0.7	6.71	6.76	6.81	6.85	6.90	6.95	6.99	7.04	7.08	7.13
0.8	7.17	7.22	7.26	7.31	7.35	7.39	7.44	7.48	7.52	7.57
0.9	7.61	7.65	7.69	7.73	7.78	7.82	7.86	7.90	7.94	7.98
Head (Ft)	**0.0**	**0.1**	**0.2**	**0.3**	**0.4**	**0.5**	**0.6**	**0.7**	**0.8**	**0.9**
1	8.0	8.4	8.8	9.1	9.5	9.8	10.1	10.5	10.8	11.1
2	11.3	11.6	11.9	12.2	12.4	12.7	12.9	13.2	13.4	13.7
3	13.9	14.1	14.3	14.6	14.8	15.0	15.2	15.4	15.6	15.8
4	16.0	16.2	16.4	16.6	16.8	17.0	17.2	17.4	17.6	17.8
5	17.9	18.1	18.3	18.5	18.6	18.8	19.0	19.1	19.3	19.5
6	19.6	19.8	20.0	20.1	20.3	20.4	20.6	20.8	20.9	21.1
7	21.2	21.4	21.5	21.7	21.8	22.0	22.1	22.3	22.4	22.5
8	22.7	22.8	23.0	23.1	23.2	23.4	23.5	23.7	23.8	23.9
9	24.1	24.2	24.3	24.5	24.6	24.7	24.8	25.0	25.1	25.2

Theoretical Head (Ft) for Velocities of 0 to 20 Ft/Sec

Velocity (Ft/Sec)	0.0	0.1	0.2	0.3	0.4	0.5	0.6	0.7	0.8	0.9
0	0.000	0.000	0.001	0.001	0.002	0.004	0.006	0.008	0.010	0.013
1	0.016	0.019	0.022	0.026	0.030	0.035	0.040	0.045	0.050	0.056
2	0.062	0.069	0.075	0.082	0.090	0.097	0.105	0.113	0.122	0.131
3	0.140	0.149	0.159	0.169	0.180	0.190	0.201	0.213	0.225	0.236
4	0.249	0.261	0.274	0.287	0.301	0.315	0.329	0.343	0.358	0.373
5	0.389	0.404	0.420	0.437	0.453	0.470	0.488	0.505	0.523	0.541
6	0.560	0.579	0.598	0.617	0.637	0.657	0.677	0.698	0.719	0.740
7	0.762	0.784	0.806	0.829	0.851	0.875	0.898	0.922	0.946	0.970
8	0.995	1.020	1.045	1.071	1.097	1.123	1.150	1.177	1.204	1.231
9	1.259	1.287	1.316	1.345	1.374	1.403	1.433	1.463	1.493	1.524
10	1.555	1.586	1.618	1.649	1.682	1.714	1.747	1.780	1.813	1.847
11	1.881	1.916	1.950	1.985	2.021	2.056	2.092	2.128	2.165	2.202
12	2.239	2.276	2.314	2.352	2.391	2.429	2.468	2.508	2.547	2.587
13	2.627	2.668	2.709	2.750	2.792	2.833	2.876	2.918	2.961	3.004
14	3.047	3.091	3.135	3.179	3.224	3.269	3.314	3.360	3.405	3.452
15	3.498	3.545	3.592	3.639	3.687	3.735	3.784	3.832	3.881	3.931
16	3.980	4.030	4.080	4.131	4.182	4.233	4.284	4.336	4.388	4.440
17	4.493	4.546	4.600	4.653	4.707	4.761	4.816	4.871	4.926	4.981
18	5.037	5.093	5.150	5.207	5.264	5.321	5.379	5.437	5.495	5.554
19	5.613	5.672	5.731	5.791	5.851	5.912	5.973	6.034	6.095	6.157

$V = \sqrt{2gH}$ $\qquad$ $H = \frac{V^2}{2g}$

Vapor Pressure of Water

Temperature Degrees F	Vapor Pressure Ft of Water	Temperature Degrees F	Vapor Presure Ft of Water	Temperature Degrees F	Vapor Pressure Ft of Water
40	0.28	55	0.50	70	0.84
41	0.29	56	0.51	71	0.87
42	0.30	57	0.53	72	0.90
43	0.32	58	0.55	73	0.93
44	0.33	59	0.57	74	0.96
45	0.34	60	0.59	75	0.99
46	0.35	61	0.62	76	1.02
47	0.37	62	0.64	77	1.06
48	0.38	63	0.66	78	1.10
49	0.40	64	0.68	79	1.14
50	0.41	65	0.71	80	1.17
51	0.43	66	0.73		
52	0.44	67	0.75		
53	0.46	68	0.78		
54	0.48	69	0.81		

Viscosity and Density of Water

Temperature		Viscosity	Density*	Temperature		Viscosity	Density*
Deg. C.	Deg. F	(Centipoises)	(grams/ml)	Deg. C	Deg. F	(Centipoises)	(grams/ml)
0	32	1.79	0.99987	50	122	0.55	0.98807
5	41	1.52	0.99999	55	131	0.51	0.98573
10	50	1.31	0.99973	60	140	0.47	0.98324
15	59	1.14	0.99913	65	149	0.44	0.98059
20	68	1.01	0.99823	70	158	0.41	0.97781
25	77	0.89	0.99707	75	167	0.38	0.97489
30	86	0.80	0.99567	80	176	0.36	0.97183
35	95	0.72	0.99406	85	185	0.34	0.96865
40	104	0.66	0.99224	90	194	0.32	0.96534
45	113	0.60	0.99025	95	203	0.30	0.96192

*Based on a density of 1.00000 g/ml at 4 degrees C.

Atmospheric Pressure and Boiling Point of Water

Elevation		Atmospheric Pressure				Boiling Pt. of Water	
Feet	Meters	In. Hg.	Mm. Hg.	PSI	Ft. Water	Degrees F	Degrees C
−1,000	−305	31.0	788	15.2	35.2	214.0	101.1
−500	−152	30.5	775	15.0	34.6	212.9	100.5
Sea Level	0	29.9	760	14.7	33.9	212.0	100.0
500	152	29.4	747	14.4	33.4	211.1	99.5
1,000	305	28.9	734	14.2	32.8	210.2	99.0
1,500	457	28.3	719	13.9	32.1	209.3	98.5
2,000	610	27.8	706	13.7	31.5	208.2	97.9
2,500	762	27.3	694	13.4	31.0	207.5	97.5
3,000	914	26.8	681	13.2	30.4	206.6	97.0
3,500	1,067	26.3	668	12.9	29.8	205.5	96.4
4,000	1,219	25.8	655	12.7	29.2	204.6	95.9
4,500	1,372	25.4	645	12.5	28.8	203.9	95.5
5,000	1,524	24.9	633	12.2	28.3	203.0	95.0
6,000	1,829	24.0	610	11.8	27.2	201.2	94.0
7,000	2,134	23.1	587	11.4	26.2	199.2	92.9
8,000	2,438	22.2	564	10.9	25.2	197.2	91.8
9,000	2,743	21.4	544	10.5	24.3	195.6	90.9
10,000	3,048	20.6	523	10.1	23.4	193.8	89.9

Specific Gravity Conversion

The specific gravity of liquids heavier than water is often expressed in degrees Baumé. For liquids lighter than water the specific gravity is often expressed in degrees API. The following tables gives the conversions and the resulting weight in pounds per U.S. gallon.

The general conversion formulas are as follows:

To convert degrees Baumé to specific gravity:

$$\text{Sp. Gr.} = \frac{145}{(145 - \text{Degrees Baumé})}$$

To convert degrees API to specific gravity:

$$\text{Sp. Gr.} = \frac{141.5}{(131.5 + \text{Degrees API})}$$

Specific Gravity Conversion Table

Liquids Heavier Than Water (Degrees Baumé)

Baumé	Sp. Gravity	Lbs/US Gal	Baumé	Sp. Gravity	Lbs/US Gal
0	1.000	8.33	10	1.074	8.95
1	1.007	8.39	11	1.082	9.01
2	1.014	8.45	12	1.090	9.08
3	1.021	8.51	13	1.098	9.15
4	1.028	8.57	14	1.107	9.22
5	1.036	8.63	15	1.115	9.29
6	1.043	8.69	16	1.124	9.36
7	1.051	8.75	17	1.133	9.44
8	1.058	8.82	18	1.142	9.51
9	1.066	8.88	19	1.151	9.59

Liquids Lighter Than Water (Degrees API)

API	Sp. Gravity	Lbs/US Gal	API	Sp. Gravity	Lbs/US Gal
10	1.000	8.33	20	0.934	7.78
11	0.993	8.27	21	0.928	7.73
12	0.986	8.21	22	0.922	7.68
13	0.979	8.16	23	0.916	7.63
14	0.973	8.10	24	0.910	7.58
15	0.966	8.05	25	0.904	7.53
16	0.959	7.99	26	0.898	7.48
17	0.953	7.94	27	0.893	7.44
18	0.946	7.88	28	0.887	7.39
19	0.940	7.83	29	0.882	7.34

Friction Loss in Pipe Flow

Friction losses for flow of water in pipes is generally calculated by either the Hazen-Williams or Manning formulas. Each formula includes a coefficient to account for the relative "roughness" of the pipe interior. Tables of various powers of numbers are included in this Chapter, or elsewhere in this handbook, to assist in making computations from these two formulas.

Hazen-Williams

The Hazen-Williams formula for flow in pipes is:

$$V = 1.318\ C\ R^{0.63}\ S^{0.54}$$

where: V = velocity, in feet per second
C = coefficient of roughness
R = hydraulic radius (cross-section area divided by wetted perimeter; for pipes flowing full, R = one-fourth the inside diameter, in feet)
S = energy (friction) loss, as a slope (ft/ft)

Common values used for the coefficient "C" are:

150 – Extremely smooth pipe, such as PVC
140 – Very smooth pipe, asbestos-cement (A/C)
130 – Very smooth pipe, new steel or wrought iron
120 – Smooth pipe, wood or wood-stave
110 – New, riveted steel
100 – Old steel or cast-iron

A graph for calculating the friction (head) loss, in feet per 100 feet of pipe (= meters per 100 meters), is given in this Chapter for the Hazen-Williams formula with C = 130. Values of friction loss from the C = 130 graph can be converted to other values of "C" by multiplying by the following factors:

C of 150	= C of 130 ×	0.77
140		0.87
130		1.00
120		1.16
110		1.36
100		1.63
90		1.98
80		2.46

Manning

The Manning formula for flow in pipe (or open channels) is:

$$Q = \frac{1.486}{n} A R^{2/3} S^{1/2}$$

or:

$$V = \frac{1.486}{n} R^{2/3} S^{1/2}$$

(For pipes flowing full):

$$V = \frac{0.590}{n} D^{2/3} S^{1/2}$$

where: Q = discharge, in cubic feet per second
A = cross-section area of flow, in square feet
R = hydraulic radius (area divided by wetted perimeter) in feet; equal to 1/4D for pipes flowing full
D = pipe diameter, in feet
S = energy (friction) loss, expressed as a slope (ft/ft)

Common values used for the coefficient "n" are:

0.011 – concrete cylinder pipe
0.012 – new steel or wrought-iron pipe
0.012 – planed wooden flumes
0.013 – concrete pipe (short lengths)
0.015 – spiral weld steel pipe, corrugated plastic pipe
0.025 – corrugated metal pipe
0.022-0.025 – earth canals
0.030 – canals, weedy or stony

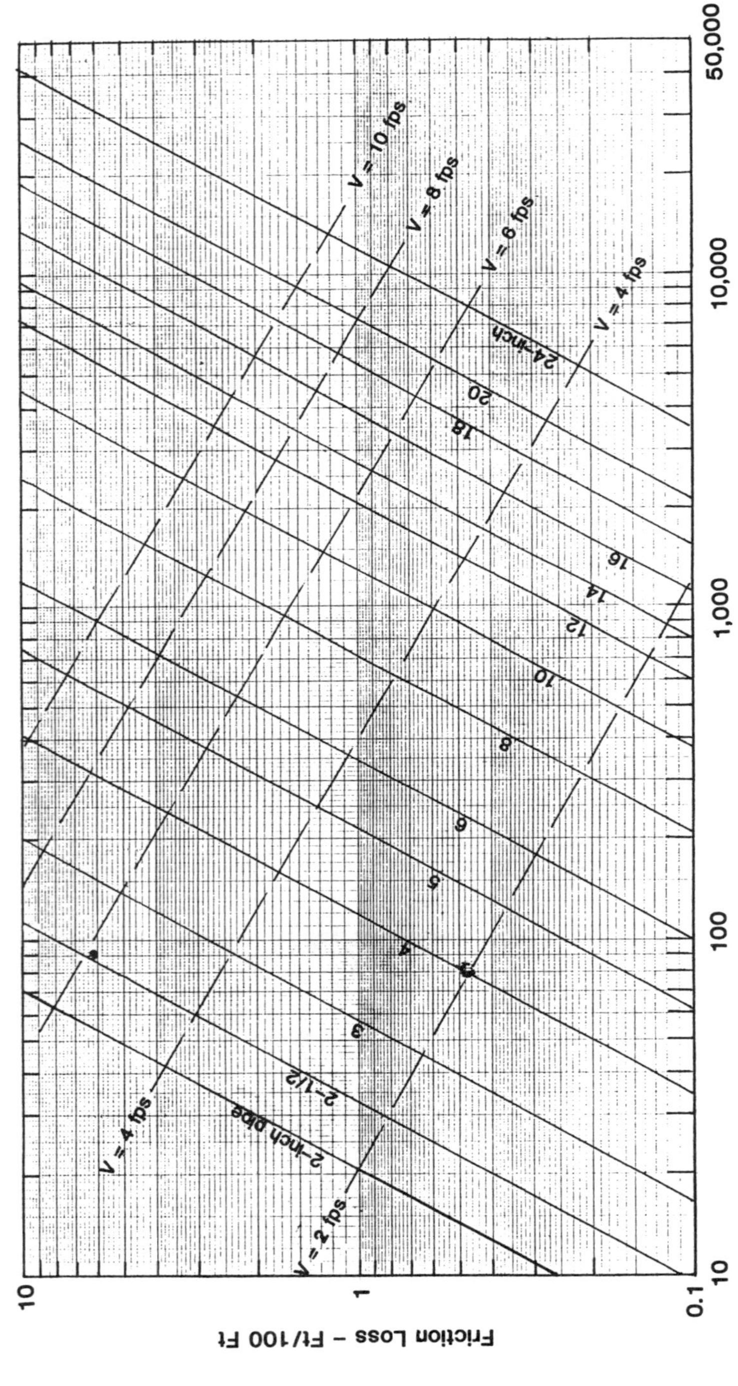

Friction Loss in Standard Steel Pipe (Hazen-Williams C = 130)

0.54 Powers of Decimal Numbers (For Use In Hazen-Williams Formula)

Number	.- -0	.- -1	.- -2	.- -3	.- -4	.- -5	.- -6	.- -7	.- -8	.- -9
0.00001	0.001995	0.002101	0.002202	0.002299	0.002393	0.002484	0.002572	0.002657	0.002741	0.002822
0.00002	0.002901	0.002979	0.003054	0.003128	0.003201	0.003273	0.003343	0.003411	0.003479	0.003546
0.00003	0.003611	0.003676	0.003739	0.003802	0.003864	0.003925	0.003985	0.004044	0.004103	0.004161
0.00004	0.004218	0.004275	0.004331	0.004386	0.004441	0.004495	0.004549	0.004602	0.004654	0.004707
0.00005	0.004758	0.004809	0.004860	0.004910	0.004960	0.005010	0.005058	0.005107	0.005155	0.005203
0.00006	0.005251	0.005298	0.005344	0.005391	0.005437	0.005482	0.005528	0.005573	0.005618	0.005662
0.00007	0.005706	0.005750	0.005794	0.005837	0.005880	0.005923	0.005965	0.006008	0.006050	0.006091
0.00008	0.006133	0.006174	0.006215	0.006256	0.006297	0.006337	0.006377	0.006417	0.006457	0.006496
0.00009	0.006536	0.006575	0.006614	0.006652	0.006691	0.006729	0.006767	0.006805	0.006843	0.006881
0.00010	0.006918	0.006956	0.006993	0.007030	0.007066	0.007103	0.007139	0.007176	0.007212	0.007248
0.0001	0.006918	0.006956	0.006993	0.007030	0.007066	0.007103	0.007139	0.007176	0.007212	0.007248
0.0002	0.010059	0.010086	0.010113	0.010140	0.010167	0.010194	0.010221	0.010248	0.010274	0.010301
0.0003	0.012521	0.012544	0.012566	0.012589	0.012611	0.012633	0.012656	0.012678	0.012700	0.012723
0.0004	0.014626	0.014645	0.014665	0.014685	0.014704	0.014724	0.014744	0.014763	0.014783	0.014802
0.0005	0.016498	0.016516	0.016534	0.016552	0.016570	0.016587	0.016605	0.016623	0.016641	0.016658
0.0006	0.018205	0.018222	0.018238	0.018255	0.018271	0.018287	0.018304	0.018320	0.018336	0.018352
0.0007	0.019786	0.019801	0.019816	0.019832	0.019847	0.019862	0.019877	0.019892	0.019908	0.019923
0.0008	0.021265	0.021280	0.021294	0.021308	0.021323	0.021337	0.021351	0.021365	0.021380	0.021394
0.0009	0.022662	0.022675	0.022689	0.022702	0.022716	0.022730	0.022743	0.022757	0.022770	0.022784
0.0010	0.023988	0.024001	0.024014	0.024027	0.024040	0.024053	0.024066	0.024079	0.024092	0.024105

0.54 Powers of Decimal Numbers (Cont'd)
(For Use In Hazen-Williams Formula)

Number	.--0	.--1	.--2	.--3	.--4	.--5	.--6	.--7	.--8	.--9
0.001	0.02399	0.02526	0.02647	0.02764	0.02877	0.02986	0.03092	0.03195	0.03295	0.03393
0.002	0.03488	0.03581	0.03672	0.03761	0.03849	0.03934	0.04019	0.04101	0.04183	0.04263
0.003	0.04342	0.04419	0.04496	0.04571	0.04645	0.04718	0.04791	0.04862	0.04933	0.05002
0.004	0.05071	0.05139	0.05207	0.05273	0.05339	0.05404	0.05469	0.05533	0.05596	0.05659
0.005	0.05721	0.05782	0.05843	0.05903	0.05963	0.06023	0.06082	0.06140	0.06198	0.06255
0.006	0.06313	0.06369	0.06425	0.06481	0.06536	0.06591	0.06646	0.06700	0.06754	0.06807
0.007	0.06860	0.06913	0.06966	0.07018	0.07069	0.07121	0.07172	0.07223	0.07273	0.07323
0.008	0.07373	0.07423	0.07472	0.07521	0.07570	0.07619	0.07667	0.07715	0.07763	0.07810
0.009	0.07858	0.07905	0.07951	0.07998	0.08044	0.08090	0.08136	0.08182	0.08227	0.08273
0.010	0.08318	0.08362	0.08407	0.08451	0.08496	0.08540	0.08584	0.08627	0.08671	0.08714
0.01	0.08318	0.08362	0.08407	0.08451	0.08496	0.08540	0.08584	0.08627	0.08671	0.08714
0.02	0.12094	0.12126	0.12159	0.12191	0.12224	0.12256	0.12288	0.12320	0.12352	0.12385
0.03	0.15054	0.15081	0.15108	0.15135	0.15162	0.15189	0.15216	0.15242	0.15269	0.15296
0.04	0.17584	0.17608	0.17631	0.17655	0.17679	0.17702	0.17726	0.17749	0.17773	0.17796
0.05	0.19836	0.19857	0.19878	0.19900	0.19921	0.19942	0.19964	0.19985	0.20006	0.20028
0.06	0.21888	0.21907	0.21927	0.21947	0.21966	0.21986	0.22006	0.22025	0.22045	0.22064
0.07	0.23788	0.23806	0.23824	0.23843	0.23861	0.23879	0.23898	0.23916	0.23934	0.23952
0.08	0.25566	0.25584	0.25601	0.25618	0.25635	0.25653	0.25670	0.25687	0.25704	0.25721
0.09	0.27245	0.27262	0.27278	0.27294	0.27311	0.27327	0.27343	0.27359	0.27376	0.27392
0.10	0.28840	0.28856	0.28871	0.28887	0.28903	0.28918	0.28934	0.28949	0.28965	0.28980

0.63 Power of Numbers (For Use In Hazen-Williams Formula)

n	0.00	0.01	0.02	0.03	0.04	0.05	0.06	0.07	0.08	0.09
0.0	0.000	0.055	0.085	0.110	0.132	0.151	0.170	0.187	0.204	0.219
0.1	0.234	0.249	0.263	0.277	0.290	0.303	0.315	0.327	0.339	0.351
0.2	0.363	0.374	0.385	0.396	0.407	0.418	0.428	0.438	0.448	0.458
0.3	0.468	0.478	0.488	0.497	0.507	0.516	0.525	0.535	0.544	0.553
0.4	0.561	0.570	0.579	0.588	0.596	0.605	0.613	0.621	0.630	0.638
0.5	0.646	0.654	0.662	0.670	0.678	0.686	0.694	0.702	0.710	0.717
0.6	0.725	0.732	0.740	0.747	0.755	0.762	0.770	0.777	0.784	0.792
0.7	0.799	0.806	0.813	0.820	0.827	0.834	0.841	0.848	0.855	0.862
0.8	0.869	0.876	0.882	0.889	0.896	0.903	0.909	0.916	0.923	0.929
0.9	0.936	0.942	0.949	0.955	0.962	0.968	0.975	0.981	0.987	0.994
1.0	1.000	1.006	1.013	1.019	1.025	1.031	1.037	1.044	1.050	1.056
1.1	1.062	1.068	1.074	1.080	1.086	1.092	1.098	1.104	1.110	1.116
1.2	1.122	1.128	1.133	1.139	1.145	1.151	1.157	1.163	1.168	1.174
1.3	1.180	1.185	1.191	1.197	1.202	1.208	1.214	1.219	1.225	1.231
1.4	1.236	1.242	1.247	1.253	1.258	1.264	1.269	1.275	1.280	1.286
1.5	1.291	1.296	1.302	1.307	1.313	1.318	1.323	1.329	1.334	1.339
1.6	1.345	1.350	1.355	1.360	1.366	1.371	1.376	1.381	1.387	1.392
1.7	1.397	1.402	1.407	1.412	1.418	1.423	1.428	1.433	1.438	1.443
1.8	1.448	1.453	1.458	1.463	1.468	1.473	1.478	1.483	1.488	1.493
1.9	1.498	1.503	1.508	1.513	1.518	1.523	1.528	1.533	1.538	1.543

Square Roots of Decimal Numbers (For Use In Manning's Formula)

Number	.- -0	.- -1	.- -2	.- -3	.- -4	.- -5	.- -6	.- -7	.- -8	.- -9
0.00001	0.003162	0.003317	0.003464	0.003606	0.003742	0.003873	0.004000	0.004123	0.004243	0.004359
0.00002	0.004472	0.004583	0.004690	0.004796	0.004899	0.005000	0.005099	0.005196	0.005292	0.005385
0.00003	0.005477	0.005568	0.005657	0.005745	0.005831	0.005916	0.006000	0.006083	0.006164	0.006245
0.00004	0.006325	0.006403	0.006481	0.006557	0.006633	0.006708	0.006782	0.006856	0.006928	0.007000
0.00005	0.007071	0.007141	0.007211	0.007280	0.007348	0.007416	0.007483	0.007550	0.007616	0.007681
0.00006	0.007746	0.007810	0.007874	0.007937	0.008000	0.008062	0.008124	0.008185	0.008246	0.008307
0.00007	0.008367	0.008426	0.008485	0.008544	0.008602	0.008660	0.008718	0.008775	0.008832	0.008888
0.00008	0.008944	0.009000	0.009055	0.009110	0.009165	0.009220	0.009274	0.009327	0.009381	0.009434
0.00009	0.009487	0.009539	0.009592	0.009644	0.009695	0.009747	0.009798	0.009849	0.009899	0.009950
0.00010	0.010000	0.010050	0.010100	0.010149	0.010198	0.010247	0.010296	0.010344	0.010392	0.010440
0.0001	0.01000	0.01049	0.01095	0.01140	0.01183	0.01225	0.01265	0.01304	0.01342	0.01378
0.0002	0.01414	0.01449	0.01483	0.01517	0.01549	0.01581	0.01612	0.01643	0.01673	0.01703
0.0003	0.01732	0.01761	0.01789	0.01817	0.01844	0.01871	0.01897	0.01924	0.01949	0.01975
0.0004	0.02000	0.02025	0.02049	0.02074	0.02098	0.02121	0.02145	0.02168	0.02191	0.02214
0.0005	0.02236	0.02258	0.02280	0.02302	0.02324	0.02345	0.02366	0.02387	0.02408	0.02429
0.0006	0.02449	0.02470	0.02490	0.02510	0.02530	0.02550	0.02569	0.02588	0.02608	0.02627
0.0007	0.02646	0.02665	0.02683	0.02702	0.02720	0.02739	0.02757	0.02775	0.02793	0.02811
0.0008	0.02828	0.02846	0.02864	0.02881	0.02898	0.02915	0.02933	0.02950	0.02966	0.02983
0.0009	0.03000	0.03017	0.03033	0.03050	0.03066	0.03082	0.03098	0.03114	0.03130	0.03146
0.0010	0.03162	0.03178	0.03194	0.03209	0.03225	0.03240	0.03256	0.03271	0.03286	0.03302

Square Roots of Decimal Numbers (Cont'd)
(For Use In Manning's Formula)

Number	.- -0	.- -1	.- -2	.- -3	.- -4	.- -5	.- -6	.- -7	.- -8	.- -9
0.001	0.03162	0.03317	0.03464	0.03606	0.03742	0.03873	0.04000	0.04123	0.04243	0.04359
0.002	0.04472	0.04583	0.04690	0.04796	0.04899	0.05000	0.05099	0.05196	0.05292	0.05385
0.003	0.05477	0.05568	0.05657	0.05745	0.05831	0.05916	0.06000	0.06083	0.06164	0.06245
0.004	0.06325	0.06403	0.06481	0.06557	0.06633	0.06708	0.06782	0.06856	0.06928	0.07000
0.005	0.07071	0.07141	0.07211	0.07280	0.07348	0.07416	0.07483	0.07550	0.07616	0.07681
0.006	0.07746	0.07810	0.07874	0.07937	0.08000	0.08062	0.08124	0.08185	0.08246	0.08307
0.007	0.08367	0.08426	0.08485	0.08544	0.08602	0.08660	0.08718	0.08775	0.08832	0.08888
0.008	0.08944	0.09000	0.09055	0.09110	0.09165	0.09220	0.09274	0.09327	0.09381	0.09434
0.009	0.09487	0.09539	0.09592	0.09644	0.09695	0.09747	0.09798	0.09849	0.09899	0.09950
0.010	0.10000	0.10050	0.10100	0.10149	0.10198	0.10247	0.10296	0.10344	0.10392	0.10440
0.01	0.1000	0.1049	0.1095	0.1140	0.1183	0.1225	0.1265	0.1304	0.1342	0.1378
0.02	0.1414	0.1449	0.1483	0.1517	0.1549	0.1581	0.1612	0.1643	0.1673	0.1703
0.03	0.1732	0.1761	0.1789	0.1817	0.1844	0.1871	0.1897	0.1924	0.1949	0.1975
0.04	0.2000	0.2025	0.2049	0.2074	0.2098	0.2121	0.2145	0.2168	0.2191	0.2214
0.05	0.2236	0.2258	0.2280	0.2302	0.2324	0.2345	0.2366	0.2387	0.2408	0.2429
0.06	0.2449	0.2470	0.2490	0.2510	0.2530	0.2550	0.2569	0.2588	0.2608	0.2627
0.07	0.2646	0.2665	0.2683	0.2702	0.2720	0.2739	0.2757	0.2775	0.2793	0.2811
0.08	0.2828	0.2846	0.2864	0.2881	0.2898	0.2915	0.2933	0.2950	0.2966	0.2983
0.09	0.3000	0.3017	0.3033	0.3050	0.3066	0.3082	0.3098	0.3114	0.3130	0.3146
0.10	0.3162	0.3178	0.3194	0.3209	0.3225	0.3240	0.3256	0.3271	0.3286	0.3302

Flow in Parallel Pipes

When additional flow in a pipeline is required, the question often arises as to how much additional flow can be obtained in a parallel pipe of specific diameter, or what size parallel pipe should be selected to obtain a specific increase in flow (assuming in each case that the parallel pipes are of equal length and have the same inlet and outlet pressures).

In both cases, the friction loss in each pipe must be the same, and the solution can be obtained from the Hazen-Williams friction loss graph in this Chapter of the handbook.

Examples are:

- If a 24-inch pipe is carrying 10,000 gpm, how much added flow can be obtained through a parallel 10-inch pipe? Find the intersection of 10,000 gpm with the 24-inch pipe size (friction loss of 0.7 ft per 100 ft), then move horizontally to intersect the 10-inch pipe size and read the flow at that intersection to be about 1,080 gpm.
- If a 12-inch pipe is carrying 2,000 gpm, what size parallel pipe should be chosen to carry an additional 500 gpm? Find the intersection of 2,000 gpm with the 12-inch pipe size (friction loss of 0.9 ft per 100 ft), then move horizontally to intersect the line for a flow of 500 gpm. This point falls between the 6-inch and 8-inch pipe sizes and the larger size (8-inch) must be chosen.

The table "Approximate Carrying Capactiy of Pipes," also in this Chapter, can be used in answering questions about parallel flow in pipes.

Minor Losses in Pipes

Minor head losses occur at entrance to pipes, and at sudden enlargement or contraction of pipe sizes. These losses are a function of flow velocity and, in the case of entrances, the entrance condition and shape. The accompanying tables give head loss (in feet of water) for various situations.

Minor Losses in Pipe Flow – (Feet of Water)

Entrance To Pipes

Condition	Velocity in Pipe (Feet per Second)										
	2	3	4	6	8	10	12	14	16	18	20
Inward Projecting	0.05	0.11	0.19	0.44	0.78	1.21	1.75	2.38	3.10	3.93	4.85
Sharp Cornered	0.03	0.07	0.12	0.28	0.50	0.78	1.12	1.52	1.99	2.52	3.11
Slightly Rounded	0.01	0.03	0.06	0.13	0.23	0.36	0.51	0.70	0.92	1.16	1.43
Bell-Mouth	0.00	0.01	0.01	0.02	0.04	0.06	0.09	0.12	0.16	0.20	0.25

Sudden Contraction

Diameter Ratio	Velocity in Smaller Pipe (Feet per Second)										
(large/small)	2	3	4	6	8	10	12	15	20	30	40
1.20	0.00	0.01	0.02	0.04	0.07	0.12	0.18	0.28	0.54	1.38	2.74
1.40	0.01	0.02	0.04	0.10	0.17	0.27	0.40	0.65	1.14	2.68	4.98
1.60	0.02	0.04	0.06	0.14	0.26	0.40	0.67	0.89	1.56	3.44	5.97
1.80	0.02	0.05	0.08	0.19	0.33	0.51	0.73	1.12	1.92	4.05	6.72
2.00	0.02	0.05	0.09	0.21	0.36	0.55	0.79	1.19	2.06	4.28	7.09
2.50	0.03	0.06	0.10	0.23	0.40	0.62	0.88	1.34	2.30	4.76	7.71
3.00	0.03	0.06	0.11	0.24	0.42	0.65	0.92	1.40	2.41	4.98	8.11
4.00	0.03	0.06	0.12	0.25	0.44	0.69	0.97	1.48	2.53	5.24	8.48
5.00	0.03	0.07	0.12	0.26	0.46	0.70	1.00	1.52	2.60	5.36	8.67

Sudden Enlargement

Diameter Ratio	Velocity in Smaller Pipe (Feet per Second)										
(large/small)	2	3	4	6	8	10	12	15	20	30	40
1.20	0.01	0.01	0.02	0.06	0.10	0.14	0.21	0.32	0.55	1.20	2.08
1.40	0.02	0.04	0.06	0.14	0.23	0.36	0.51	0.78	1.36	2.96	5.14
1.60	0.02	0.05	0.09	0.20	0.36	0.55	0.78	1.19	2.07	4.50	7.82
1.80	0.03	0.07	0.12	0.26	0.45	0.70	0.99	1.52	2.64	5.74	9.97
2.00	0.04	0.08	0.14	0.31	0.53	0.81	1.16	1.77	3.08	6.71	11.7
2.50	0.05	0.10	0.17	0.38	0.66	1.01	1.44	2.20	3.83	8.34	14.5
3.00	0.05	0.11	0.19	0.42	0.74	1.13	1.60	2.46	4.27	9.29	16.1
4.00	0.06	0.12	0.22	0.47	0.82	1.25	1.78	2.76	4.73	10.3	17.9
5.00	0.06	0.13	0.23	0.49	0.85	1.31	1.86	2.85	4.95	10.8	18.7

Hydraulic Elements of Circular Conduits

Depth Ratio* (d/D)	Area (A)	Wetted Perimeter (P)	Hydraulic Radius (R)	Depth Ratio* (d/D)	Area (A)	Wetted Perimeter (P)	Hydraulic Radius (R)
0.05	0.0187 × A	0.1436 × P	0.130 × R	0.55	0.5635 × A	0.5319 × P	1.060 × R
0.10	0.0520	0.2048	0.254	0.60	0.6265	0.5641	1.111
0.15	0.0941	0.2531	0.372	0.65	0.6881	0.5969	1.153
0.20	0.1424	0.2952	0.482	0.70	0.7477	0.6310	1.185
0.25	0.1955	0.3333	0.587	0.75	0.8045	0.6667	1.207
0.30	0.2523	0.3690	0.684	0.80	0.8576	0.7048	1.217
0.35	0.3119	0.4031	0.774	0.85	0.9059	0.7469	1.213
0.40	0.3735	0.4359	0.857	0.90	0.9480	0.7952	1.192
0.45	0.4365	0.4681	0.932	0.95	0.9813	0.8564	1.146
0.50	0.5000	0.5000	1.000	1.00	1.0000	1.0000	1.000

* - where d = depth of flow; D = diameter of conduit, and A, P, and R are values for full conduit

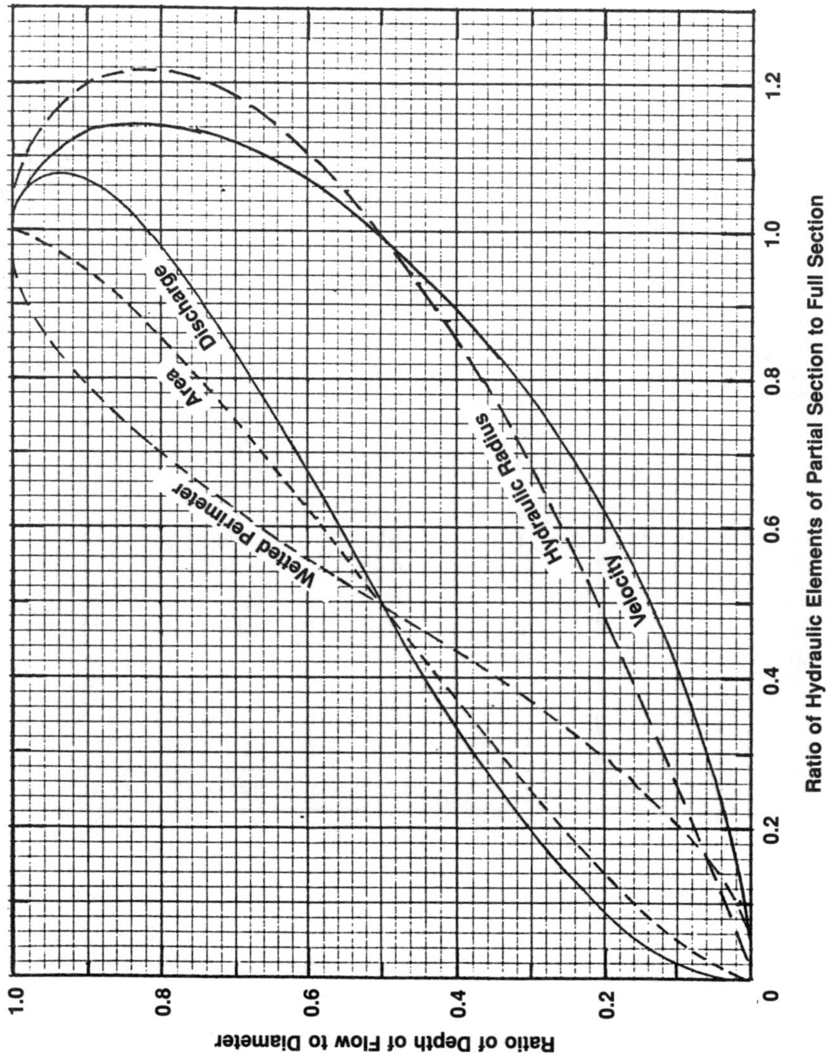

Hydraulic Elements of Circular Sections

Steel Pipe Sizes for Reasonable Friction Loss

Flow Rate		Nominal Pipe	Friction Loss*	Velocity	
(US Gal/Min)	(Liters/Sec)	Size (Inches)	(Ft/100 Ft)	(Ft/Sec)	(Meters/Sec)
100	6.3	4	0.73	2.52	0.77
150	9.5	5	0.51	2.41	0.73
200	12.6	5	0.87	3.21	0.98
250	15.8	6	0.54	2.78	0.85
300	18.9	6	0.76	3.33	1.02
350	22.1	8	0.26	2.24	0.68
400	25.2	8	0.34	2.57	0.78
500	31.5	8	0.51	3.21	0.98
600	37.9	10	0.24	2.44	0.74
700	44.2	10	0.32	2.85	0.87
800	50.5	10	0.40	3.25	0.99
900	56.8	12	0.21	2.55	0.78
1,000	63.1	12	0.25	2.84	0.86
1,100	69.4	12	0.30	3.12	0.95
1,200	75.7	12	0.36	3.40	1.04
1,300	82.0	14	0.25	3.02	0.92
1,400	88.3	14	0.29	3.26	0.99
1,500	94.6	14	0.33	3.49	1.06
1,600	100.9	14	0.37	3.72	1.13
1,700	107.3	14	0.42	3.96	1.21
1,800	113.6	14	0.46	4.19	1.28
1,900	119.9	16	0.26	3.34	1.02
2,000	126.2	16	0.28	3.51	1.07
2,500	157.7	18	0.24	3.43	1.05
3,000	189.3	18	0.33	4.12	1.26
4,000	252.4	20	0.33	4.41	1.34

*Friction loss based on standard weight steel pipe; Hazen-Williams formula with C = 130

Approximate Carrying Capacity of Pipes

(Example: One 48-inch diameter pipe carries approximately the same flow as six 24-inch pipes, with same friction loss)

Pipe Diam.(In)	8	10	12	14	16	18	20	22	24	26	28	30	36	42	48	54	60	66	72
72							27.1	21.3	17.4	14.0	11.6	9.5	6.0	4.0	2.9	2.1	1.6	1.24	1.0
66						29.0	21.8	17.1	14.0	11.2	9.3	7.6	4.8	3.2	2.3	1.7	1.28	1.0	
60					30.5	22.6	17.0	13.4	10.9	8.7	7.3	6.0	3.7	2.5	1.8	1.3	1.0		
54				32.2	23.2	17.2	13.0	10.2	8.3	6.7	5.6	5.55	2.85	1.92	1.37	1.0			
48			34.7	23.5	16.9	12.5	9.5	7.4	6.0	4.8	4.0	3.3	2.1	1.4	1.0				
42		40.0	24.7	16.8	12.2	9.0	6.8	5.3	4.3	3.5	2.9	2.4	1.5	1.0					
36	48.0	26.9	16.7	11.3	8.1	6.0	4.5	3.6	2.9	2.3	1.9	1.6	1.0						
30	30.0	16.9	10.5	7.1	5.1	3.8	2.8	2.3	1.8	1.5	1.2	1.0							
28	25.0	13.8	8.6	6.0	4.2	3.1	2.3	1.8	1.5	1.2	1.0								
26	20.5	11.5	7.1	4.8	3.5	2.6	1.95	1.55	1.25	1.0									
24	17.0	9.5	6.0	4.0	2.9	2.1	1.60	1.22	1.0										
22	13.4	7.5	4.7	3.2	2.3	1.7	1.27	1.0											
20	10.7	6.0	3.7	2.5	1.8	1.35	1.0												
18	8.0	4.5	2.75	1.9	1.4	1.0													
16	6.0	3.3	2.05	1.4	1.0														
14	4.2	2.4	1.5	1.0															
12	2.9	1.6	1.0																
10	1.8	1.0																	
8	1.0																		

Maximum Flow of Water Through Pipes

(Flow In U.S. Gallons/Minute Through 100 Feet Of New, Schedule 40, Standard Steel Pipe)

Pressure		Pipe Diameter (Inches)								
PSI	Ft. of Water	½	¾	1	1¼	1½	2	2½	3	4
10	23.1	4.6	10	19	40	60	115	185	330	640
20	46.2	6.7	14	27	57	85	165	265	470	930
30	69.3	8.4	18	34	70	106	205	325	570	1,150
40	92.4	9.7	21	39	81	123	235	375	660	1,330
50	115.5	11	23	44	91	138	265	420	740	1,500
60	138.5	12	25	48	100	151	290	465	820	1,670
70	161.6	13	27	52	109	165	315	505	890	1,800
80	184.7	14	29	56	118	177	340	540	950	1,950
90	207.8	15	31	60	125	188	360	575	1,010	2,090
100	230.9	16	33	63	130	199	380	605	1,070	2,200

Friction Loss Through Fittings and Valves

(Approximate lengths of straight pipe giving equivalent friction loss to flow through fittings and valves)

Nominal Pipe Diam. (Inches)	Standard Elbow	45-Degree Elbow	Long Radius Elbow	Tee	Gate Valve (Open)	Globe Valve (Open)	Angle Valve (Open)
1	2.7	1.3	1.7	5.8	0.6	27	14
1¼	3.7	1.6	2.4	7.8	0.8	37	18
1½	4.3	2.0	2.8	9.0	1.0	44	22
2	5.5	2.5	3.5	11	1.2	57	28
2½	6.5	3.0	4.2	14	1.4	66	33
3	8.1	3.8	5.1	17	1.7	85	42
4	11	5.0	7.0	22	2.3	110	56
5	14	6.1	8.9	27	2.9	140	70
6	16	7.7	11	33	3.5	160	80
8	21	10	14	43	4.5	220	100
10	26	13	17	56	5.7	280	140
12	32	15	20	66	6.7	340	170
14	37	17	23	78	8.1		
16	42	19	26	89	9.5		
18	47	22	30	101	11		
20	52	24	34	112	12		
24	62	29	40	133	14		

Flow of Water through Small-Diameter Hose (U.S. Gal/Min)

Hose I.D. (Inches)	Water Pressure (Pounds Per Square Inch)								
	20	30	40	50	60	70	80	90	100
½	4.2	5.2	5.9	6.6	7.3	8.0	8.5	9.0	9.4
⅝	7.4	9.1	10	12	13	14	15	16	16
¾	11	14	16	18	20	22	23	25	26
1	23	28	33	37	40	43	46	49	52
1¼	40	50	57	64	70	76	81	86	90
1½	64	78	90	101	111	120	128	135	143
2	130	159	184	206	227	242	262	275	292
2½	226	278	322	358	394	425	455	482	509
3	356	437	504	570	620	665	715	755	800
4	745	910	1,055	1,180	1,292	1,395	1,492	1,582	1,670

Flow data based on 100-foot length of straight hose with open-end discharge.

Theoretical Discharge Through Nozzles – (U.S. Gallons/Minute)

Pressure		Discharge Velocity (Ft/Sec)	Nozzle Diameter (Inches)									
PSI	Ft of Water		1/16	1/8	3/16	1/4	3/8	1/2	5/8	3/4	7/8	1
10	23.1	38.5	0.36	1.46	3.28	5.82	13.1	23.2	36.3	52.2	70.9	92.7
20	46.2	54.5	0.52	2.06	4.64	8.23	18.5	32.9	51.3	73.8	100	131
30	69.3	66.8	0.63	2.53	5.68	10.1	22.7	40.2	62.8	90.4	123	160
40	92.4	77.1	0.73	2.92	6.56	11.6	26.2	46.5	72.5	104	142	185
50	115.5	86.2	0.82	3.26	7.33	13.0	29.3	52.0	81.1	117	159	207
60	138.5	94.4	0.89	3.57	8.03	14.3	32.0	56.9	88.8	128	174	227
70	161.6	102	0.97	3.86	8.67	15.4	34.6	61.5	96.0	138	188	245
80	184.7	109	1.03	4.12	9.27	16.5	37.0	65.7	103	148	201	262
90	207.8	116	1.09	4.37	9.83	17.5	39.3	69.7	109	157	213	278
100	230.9	122	1.15	4.61	10.4	18.4	41.4	73.5	115	165	224	293
110	254.0	128	1.21	4.84	10.9	19.3	43.4	77.1	120	173	235	307
120	277.1	134	1.26	5.05	11.4	20.2	45.3	80.5	126	181	246	321
130	300.2	139	1.32	5.26	11.8	21.0	47.2	83.8	131	188	256	334
140	323.3	144	1.37	5.46	12.3	21.8	49.0	86.9	136	195	265	347
150	346.4	149	1.41	5.65	12.7	22.5	50.7	90.0	140	202	275	359
160	369.4	154	1.46	5.83	13.1	23.3	52.3	92.9	145	209	284	371
170	392.5	159	1.50	6.01	13.5	24.0	53.9	95.8	150	215	293	382
180	415.6	164	1.55	6.19	13.9	24.7	55.5	98.6	154	221	301	393
190	438.7	168	1.59	6.36	14.3	25.4	57.0	101	158	227	309	404
200	461.8	172	1.63	6.52	1[illegible].7	26.0	58.5	104	162	233	317	414

Actual discharge will be somewhat less, depending upon nozzle shape and pressure measuring point. Smooth tapered nozzles will have a discharge about 95% of values in the table.

Hydraulic Jet

The power available from a hydraulic jet is determined by the formula:

$$P = \frac{W V^3 A}{2g}$$

where: P = power in ft/lbs/sec
W = unit weight of water (62.4 lbs/cu ft)
V = velocity of jet discharge (ft/sec)
A = cross-sectional area of jet (sq ft)
g = acceleration due to gravity

Water Hammer

Water hammer is the shock caused by suddenly stopping the flow of water in a pipeline (such as from rapid valve closure, or delayed check-valve closing when water is returning toward a pump as a result of power failure). The excess pressure resulting from water hammer can be severe. Water hammer effects are typically reduced by the use of relief valves, air chambers, air valves, surge tanks, or slow-closing valves.

The maximum excess pressure, resulting from instantaneously stopping flow, can be determined from the following formula:

$$P = \frac{A V}{g}$$

where: P = excess pressure above normal (static) in psi
V = velocity of water in pipe (ft/sec) before closure
g = acceleration due to gravity

$$A = \frac{4660}{\sqrt{1 + \frac{K d}{E t}}}$$

where: K = modulus of elasticity of water (psi) = 294,000
E = modulus of elasticity of pipe wall (psi) (use 29,400,000 for steel)
d = pipe inside diameter (inches)
t = pipe wall thickness (inches)

Pumping vs Generation

Assumptions: Overall pumping plant efficiency = 75%
Overall power plant efficiency = 85%

Then: 100 acre-feet lifted 100 feet = 136.8 Kwhr
100 acre-feet falling 100 feet = 87.2 Kwhr

And: 100 Kwhr = 1 acre-foot lifted 73.2 feet
100 Kwhr = 1 acre-foot falling 114.7 feet

For Pumping

$$\text{Pump horsepower} = \frac{\text{Flow (gpm)} \times \text{Total head (ft)}}{3960} \quad (@100\%\ \text{eff})$$

$$= \frac{\text{Flow} \times \text{Total head}}{3960 \times \text{Efficiency}}$$

For Generation

1 horsepower = 1 cfs falling 8.81 feet(@ 100% eff.)
1 kilowatt = flow (cfs) x head (ft) x 0.0846(@ 100% eff.)
KW capacity = cfs × head × 0.072(@ 85% eff.)

Freezing of Water in Pipes

The time required for water standing in pipes (in air) to freeze is:

$T_1 + T_2$ where T_1 is the time (hours) to cool the water down to 0 degrees (C); and T_2 is the time (hours) for the water to freeze after reaching 0 degrees (C)

$$T_1 = 2.3 \frac{q}{k} \log \frac{t_2}{t_1}$$

where q is the volume (*not* flow rate) in kilograms of water per meter of pipe; k is the heat loss in kilogram-calories/hour for one meter of pipe,as insulated, for 1 degree (C) change; t_1 is the original difference between water and air temperature (degrees C); and t_2 is the final difference between water and air temperature (degrees C). For bare steel pipe, k approaches 1.0

$$T_2 = \frac{80q}{k\ T}$$

where T is the outside air temperature in degrees C below 0; and q and k are as in the above formula

Chapter 5

Pumps and Water Systems

Vertical Turbine Pump Selection

Vertical turbine pumps for use in wells are of either the line-shaft or submersible type.A typical line-shaft pumping unit comprises a driver (power unit), discharge head, column (discharge) pipe and shaft, and one or more stages of impellers in a "bowl assembly." The driver is commonly a vertical hollow-shaft electric motor, but can be an internal combustion engine with a gear head (or belt-driven from a power takeoff with a pulley head). The shaft is centered in the column pipe, either as an open line-shaft (bearings lubricated with water), or as an enclosed line-shaft with the shaft inside tubing (bearings lubricated with oil). A typical submersible pumping unit comprises a submersible electric motor directly coupled to the bowl assembly, a column (discharge) pipe, and electrical cable from the pump and motor (in the well) to the motor controls (at land surface).

Small capacity turbine pumps, such as for domestic wells, are typically submersible units; pumps of larger capacity, such as for irrigation, industrial, or municipal use, may be either of the line-shaft or submersible type.

Performance curves are available from manufacturers for each individual impeller design and for various rotation speeds (such as 1170, 1750 and 3500 rpm). These curves typically include (1) a head-capacity curve relating the lift to the discharge rate for an individual impeller stage, (2) a curve relating efficiency to discharge rate, and (3) a curve relating brake-horsepower requirement to discharge rate. A particular impeller design can be selected by examining these curves and choosing one that gives the desired discharge at reasonable efficiency. If the total head requirement is greater than the lift of a single stage, additional stages must be added (e.g. if the head-capacity curve for a single stage shows 500 gpm at 40 ft head, and the total head requirement is 200 feet, then five stages will be needed). The impeller (bowl) diameter selected must, of course, be such that it can be installed in the well.

In selecting a deep-well turbine pump it is usually desirable to pick an impeller type that has a "steep" rather than "flat" head-capacity curve. This is particularly true if the pumping lift will change significantly during seasons of the year, or if the pumping lift is expected to increase over future years. With a "steep" curve, a moderate increase in pumping lift will not result in a large decrease in the discharge rate; with a "flat" curve, even a slight increase in pumping lift can reduce the discharge by 50 percent or more. When the performance (pump discharge) of a well has decreased significantly, it may not be the result of a problem with the well itself, such as plugging of the screen by corrosion or incrustation, but simply that the water level has lowered somewhat and the pump is not designed to deliver the desired discharge from the new pumping level. In such a case, the remedy is to correct the pump design—not attempt to rehabilitate or re-develop the well.

In ordering or specifying turbine pump installations, the following data should be furnished:

1. The minimum diameter of the well to the depth to which the pump will be set
2. The non-pumping (static) depth to water below ground level
3. The desired discharge rate of the pump
4. The drawdown when pumping (or depth to water below ground when pumping)
5. The required pressure at ground level
6. The type of power to be used (if electricity, specify fully such as: 220 v, 3 phase, 60 hz, a.c.)

The manufacturer or supplier can make specific recommendations for a pumping unit once the requirements have been furnished. Different diameters of bowl assemblies can be chosen for a particular requirement for discharge rate and total lift, depending upon the manufacturer's particular impeller design. As a guide, and without regard to well diameter, the following nominal bowl diameters should give an efficient selection for lifts up to 300 feet at 1750 rpm (bowl diameter will decrease at higher speed or increase at lower speed):

Nominal Bowl Diam. (In.)	Discharge (GPM)
6	50-150
8	200-300
10	400-800
12	900-1200
14	1400-2500

Pump Efficiency

In discussions or measurements of vertical turbine pump efficiency, the following terms or definitions are commonly used:

- WHP = Water horsepower (or theoretical horsepower), the actual work performed
- BHP = Brake horsepower, or power applied to the pump shaft
- IHP = Input horsepower, or power input to the motor
- PE = Pump efficiency ("water-to-water")
- ME = Motor efficiency, or ratio of brake horsepower output to electrical input $= \frac{BHP}{IHP}$
- OAE = Overall efficiency ("wire-to-water")
- H = Total head, in feet, including suction lift, difference in elevation between pump and discharge point, friction losses, and pressure at discharge point (if any).
- TDH = Total dynamic head, or total head plus shaft and bearing losses, etc.
- GPM = Pump discharge, in U.S. gallons per minute

From which:

$$WHP = \frac{GPM \times H}{3960}$$

$$BHP = \frac{GPM \times TDH}{3960 \times PE}$$

$$PE = \frac{WHP}{BHP}$$

$$IHP = \frac{BHP}{ME} \quad \text{or: Input in } \textit{Kilowatts} = \frac{BHP \times 0.746}{ME}$$

$$OAE = \frac{WHP}{IHP} = PE \times ME = \frac{WHP \times 0.746}{\text{KW input to motor}}$$

$$\text{KW input to 3-phase motor} = 0.746 \times IHP = \frac{E \times I \times 1.732 \times PF}{1000}$$

where E = volts, I = amperes, PF = power factor (generally 0.85-.90)

For most installations, the overall, or "wire-to-water," efficiency will not exceed 75 percent.

Pump Performance at Different Speeds

Pump performance curves show the interrelationship between discharge (or capacity), head (or lift), horsepower input, and efficiency. Performance curves for vertical turbine pumps typically show the relationship between these factors for a single stage (one impeller and bowl) and for a specified speed of rotation. The head (or lift) produced is dependent upon the number of stages; if a pump is desired that will lift 400 gpm a total head of 250 feet, and one stage of a particular size and speed will deliver 400 gpm at a head of 50 feet, then five stages will be required for the total head of 250 feet.

As the speed of the pump varies: the capacity will vary directly as the speed; the head will vary as the square of the speed; and the horsepower required will vary as the cube of the speed. If a performance curve is available for a particular speed, a new curve can be constructed for a different speed from these relationships. As an example, consider a pump with the following per-stage characteristics:

Speed	1760 rpm
Capacity	400 gpm
Head	50 feet
Power input	6 BHP

Performance characteristics for the same pump, if operated at 1160 rpm, would then be:

1. *Capacity* varies directly with speed:

$$Q = \frac{1160}{1760} \times 400 = 0.659 \times 400 = 264 \text{ gpm}$$

2. *Head* varies as the square of speed:

$$H = \frac{(1160)^2}{(1760)^2} \times 50$$

$$= (0.659 \times 0.659) \times 50 = 0.434 \times 50 = 21.7 \text{ feet}$$

3. *Horsepower* input varies as the cube of speed:

$$\text{BHP} = \frac{(1160)^3}{(1760)^3} \times 6 = (0.659 \times 0.659 \times 0.659) \times 6$$

$$= 0.286 \times 6 = 1.7 \text{ bhp}$$

By selecting other points on the 1760 rpm performance curve, and performing the same calculations as above, a new performance curve can be drawn for the 1160 rpm speed. The efficiency at the new speed can be expected to change slightly, depending upon mechanical losses and other factors.

Pump Shaft and Tube Weights			
Shaft Diam. (Inches)	Shaft Weight (Lbs/Ft)	Tube Diam. (Inches)	Tube Weight (Lbs/Ft)
3/4	1.50	1 1/4	3.00
1	2.67	1 1/2	3.63
1 3/16	3.77	2	5.02
1 1/2	6.01	2 1/2	7.66
1 11/16	7.60	3	10.25
1 15/16	10.02	3	10.25
2 3/16	12.78	3 1/2	12.50
2 7/16	15.86	4	14.98
2 11/16	19.29	5	20.78
2 15/16	23.04	5	20.78
3 3/16	27.13	5	20.78
3 7/16	31.56	5	20.78
3 11/16	36.31	6	28.57
3 15/16	41.40	6	28.57

Shaft Horsepower Loss					
Shaft Diam. (inches)	Horsepower Loss per 100 Feet of Shaft*				
	Shaft Speed – RPM				
	870	1170	1450	1750	3450
7/8	—	0.27	0.35	0.41	0.82
1	—	0.36	0.45	0.53	1.05
1 3/16	0.36	0.50	0.61	0.73	1.45
1 1/2	0.56	0.78	0.96	1.14	2.25
1 11/16	0.70	0.96	1.17	1.41	—
1 15/16	0.91	1.26	1.55	1.84	—
2 3/16	1.14	1.56	1.92	2.31	—
2 7/16	1.40	1.93	2.41	2.85	—
2 11/16	1.68	2.32	2.84	3.40	—
2 15/16	2.00	2.75	3.40	4.04	—
3 3/16	2.30	3.20	3.90	4.65	—

*For enclosed shaft with oil lubrication, or open shaft with water lubrication.

Vertical Turbine Column Assembly – Friction Loss
(Loss of Head in Feet per 100 Feet of Column — Open or Enclosed Shaft)

Pipe × Shaft	Pump Discharge – US Gal/Min												
(Inches)	50	100	150	200	300	400	500	600	800	1000	1500	2000	3000
3 × 7/8	4.4	14.0											
4 × 1 3/16	1.7	5.2											
4 × 1	0.9	2.9	5.8	9.5									
4 × 7/8	0.7	2.2	4.5	7.4	15.0								
5 × 1 3/16		0.9	1.9	3.1	6.3	10.5							
5 × 1		0.7	1.4	2.3	4.6	7.5	11.0						
6 × 1 11/16			0.8	1.4	2.8	4.6	6.9	9.5					
6 × 1 3/16				1.0	2.0	3.4	5.0	6.8	11.3				
6 × 1				0.7	1.5	2.5	3.7	5.1	8.6	12.9			
8 × 1 15/16					0.6	1.0	1.5	2.1	3.6	5.4	11.2		
8 × 1 11/16						0.7	1.1	1.5	2.6	3.8	8.2	14.0	
8 × 1 3/16							0.9	1.2	2.1	3.2	6.8	11.6	
10 × 2 7/16									0.9	1.4	3.0	5.0	10.9
10 × 1 15/16									0.8	1.2	2.4	4.1	8.8
10 × 1 11/16									0.7	1.0	2.0	3.4	7.3
10 × 1 3/16										0.9	1.8	3.0	6.3
12 × 2 7/16											1.1	1.8	3.8
12 × 1 11/16											0.8	1.4	2.9

Net Positive Suction Head

Net Positive Suction Head (NPSH) is the absolute pressure, expressed in feet of water, that is required to induce the flow of water through the suction piping into the impeller of the pump. Two types of NPSH must be considered in any pump selection and installation. The *required* NPSH (or NPSHR) is a characteristic of a particular pump and depends upon such factors as the impeller eye diameter, the number and shape of vanes, and the shape of the suction passage. The NPSHR varies with each pump and with capacity and speed. The NPSHR is always given by the manufacturer, based upon extensive laboratory testing. The *available* NPSH (or NPSHA) is a characteristic of the system design and installation and depends upon such factors as atmospheric pressure, static suction lift or pressure, vapor pressure at the operating temperature, and total losses in the suction piping including both friction and velocity head.

In all pump installations, the NPSHA must always exceed the NPSHR to prevent cavitation, noise, vibration, loss of efficiency, and reduction of capacity.

For installations where the water is open to the atmosphere (wells, open tanks, or bodies of surface water) the NPSHA is determined by:

(Static suction head (if any) + atmospheric pressure)

minus

(Static suction lift (if any) + vapor pressure + all friction loss, including velocity head)

Tables or formulas for atmospheric pressure vs. elevation, vapor pressure vs. temperature, velocity head, and pipe friction loss are included in Chapter 4 of this handbook.

Pump Suction Lifts at Different Altitudes

Altitude		Barometric Pressure	Equiv. Head	Typical Suction Lift (Ft)*	
Feet	Meters	(Lbs/Sq.In.)	(Ft. of Water)	Reciprocating	Centrifugal
Sea Level	0	14.7	33.9	22	15
500	152	14.4	33.2	22	15
1,000	305	14.2	32.8	21	15
1,500	457	13.9	32.1	21	14
2,000	610	13.7	31.6	21	14
2,500	762	13.4	30.9	20	14
3,000	914	13.2	30.4	20	14
4,000	1,219	12.7	29.3	19	13
5,000	1,524	12.2	28.1	18	13
6,000	1,829	11.8	27.2	18	12
8,000	2,438	10.9	25.1	16	11
10,000	3,048	10.1	23.3	15	10

*Pump suction lift depends not only on barometric pressure (altitude) but also on water temperature and the individual pump design.

Centrifugal Pump Capacity

Centrifugal pumps are frequently designated by a number which represents the diameter of the discharge (in inches). The capacities given in the table should be considered as only representative, with variations possible both above and below the figures given. Consult the manufacturer for specific information for a particular application.

No. of pump (Discharge diameter – inches)	Capacity (US gpm)
2	100
2½	150
3	225
3½	300
4	400
5	700
6	900
7	1200
8	1600
10	3000
12	4500

Air-Lift Pumping

The features of an air-lift pump are shown on the accompanying drawing. These pumps operate on the principle that the mixture of air and water inside the eductor (discharge) pipe has less weight than the column of water outside, causing the air-water mixture to rise. Air-lift pumping is frequently used in well development and well testing. Air-lift pumps are sometimes used for a permanent pump installation with the advantages that they can be used in crooked wells, have no moving parts below ground, can have high rates of discharge, and are unaffected if the water contains some sand. Their principal disadvantages are the relatively low efficiency and, in some cases, the requirement for a considerable length of submergence of the eductor pipe.

There are two types of "submergence" in air-lift pumping. Static submergence is the distance from the static water level in the well to the bottom of the air pipe (usually expressed in feet). Pumping submergence is the distance from the pumping level in the well to the bottom of the air pipe (expressed either in feet, or in "percent submergence"). "Percent submergence" is the distance from the pumping level to the bottom of the air pipe divided by the length of the air pipe below the discharge point. The percent pumping submergence required depends upon the total lift, typically ranging from 70% for low lifts (up to 100 ft) to 40% for high lifts (500 to 700 ft).

There are three typical types of air-lift pump installations: (1) In small diameter wells, the well casing or well bore serves as the eductor pipe, with a single air pipe inside the well; this does not allow measurement of the depth to water while pumping, however, unless a separate tube for a measuring line can be placed to a greater depth (below the air-water mixture). (2) In larger diameter wells, the air pipe can be installed either inside the eductor pipe, or (3) outside the eductor pipe. For temporary installations, method (2) above is typical.

The required capacity of the air compressor can be approximately determined from the following formula:

$$Va = \frac{H}{C \log \frac{(S + 34)}{34}}$$

where: Va = volume of free air (cu. ft.) per gallon of water
H = total lift (ft.)
S = submergence (ft.) when pumping
C = a constant related to % submergence

Pumping Submergence (Percent)	C
40	185
50	240
60	285
70	325

Where a separate eductor (discharge) pipe is used, it should typically extend 3 to 5 feet below the bottom end of the air pipe. The size of the eductor pipe should be such that the cross-sectional area of the pipe (in square inches) is approximately equal to the water discharge (in US gpm) multiplied by a factor of 0.7 to 0.8. If the air pipe is placed inside the eductor pipe, the area of the annulus between air pipe and eductor should be calculated the same way.

Usual practice has the air velocity in the air pipe at 1800 to 2400 feet per minute. The required diameter of air pipe can be estimated from the formula:

$$D = 13.54 \sqrt{\frac{Q}{V}}$$

where: D = diameter of air pipe (inches)
Q = free air volume (cu ft/min)
V = velocity of air (ft/min) in air pipe

The required compressor pressure rating should be somewhat greater than the static submergence. Multiply the static submergence (ft) by 0.433 to obtain pressure (psi) and add about 10% to obtain minimum compressor rating needed.

To measure the flow rate from an air-lift pump, the air must first be allowed to separate from the discharging water-air mixture. This can be accomplished by discharging into a weir box, flume, or open channel.

Air-Lift Pumping

Total Lift		Pumping Submergence (Percent)	
(Feet)	(Meters)	Allowable	Optimum
20	6	55	65 - 70
40	12	50	65 - 70
60	18	50	65 - 70
80	24	50	65 - 70
100	30	45	65 - 70
150	46	40	60 - 65
200	61	40	55 - 60
250	76	40	55 - 60
300	91	37	50 - 55
350	107	37	50 - 55
400	122	37	45 - 50
500	152	35	40 - 45
600	183	35	40 - 45
700	213	35	40

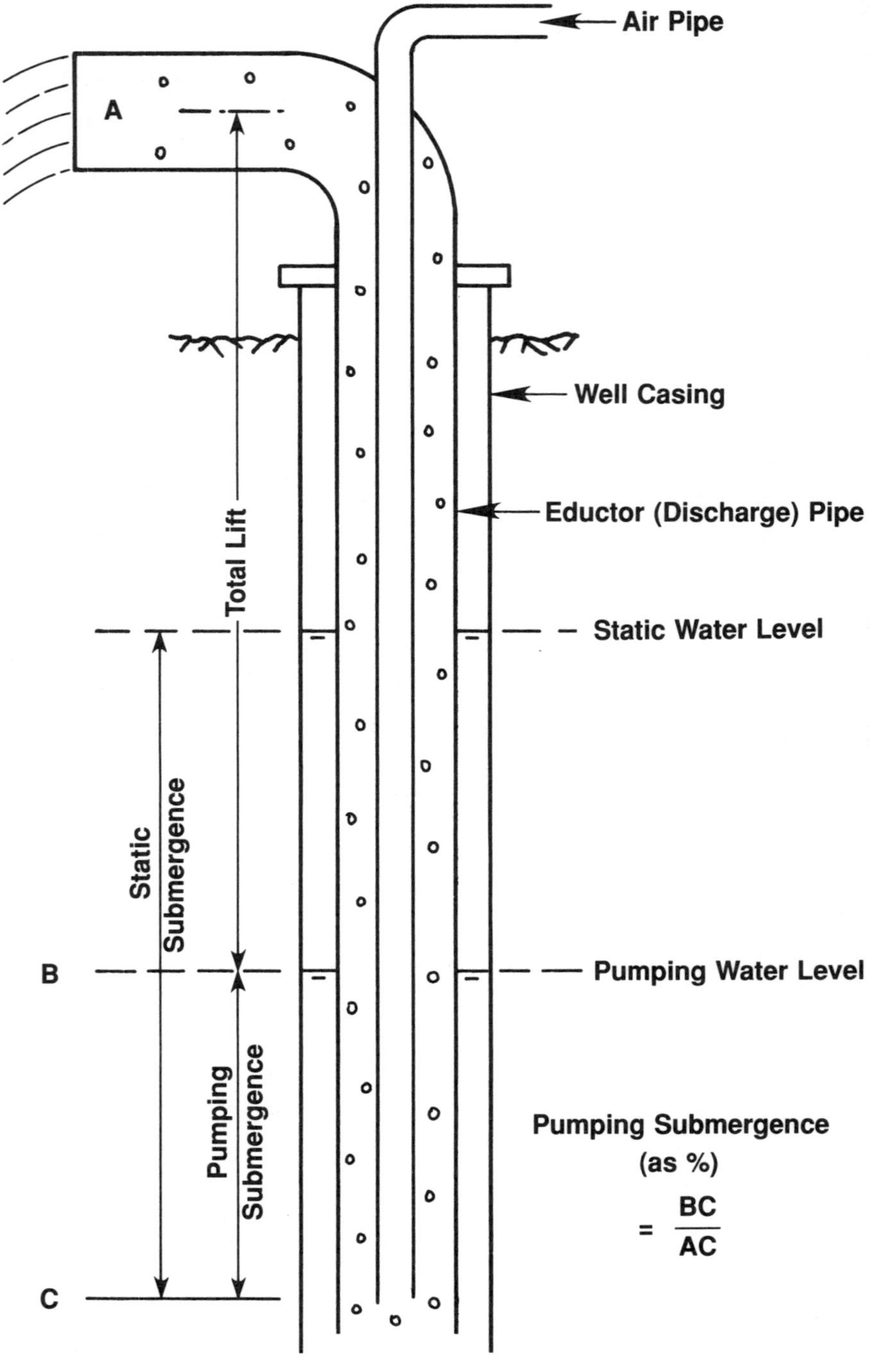

Air-Lift Pumping

Hydro-Pneumatic (Pressure) Tanks

Hydro-pneumatic (pressure) tanks are commonly used to allow intermittent operation of a pump while maintaining a continuous (but variable) pressure on the distribution system. As water is pumped into the tank, the air in the tank is compressed until the maximum desired system pressure is reached, at which point the pump shuts off. As water is withdrawn from the tank, the air expands until the minimum desired pressure is reached, at which point the pump restarts and the entire cycle is repeated. Pressure tanks are not a means of providing water-supply storage — but are tanks of limited size to furnish water, within a regulated range of pressure, to meet an intermittent demand.

Most smaller pressure tank systems, such as for domestic supply to one or more households, utilize a diaphragm or bladder within the tank that separates the air from the water. This prevents air from gradually dissolving in the water which causes "waterlogging" of the tank that requires periodic re-pressurization. Large pressure tanks, such as for commercial or industrial use, have "sight gages" on the side (or end) of the tank to show the water level at all times. These tanks also require a provision for maintaining the proper air volume in the tank. Starting and stopping of the pump supplying the tank can be governed by pressure switches or by electrodes installed in the tank at levels corresponding to the desired pressure range of operation.

Typical installations provide for an operating range of 20 to 30 psi, such as between 20-40, or 30-60 pounds per square inch. Selection of the operating range allows calculation of the percentage of tank volume that can be used in each cycle of operation. The accompanying table shows the percentage of tank volume that is filled with water for various gage pressures — for conditions of both zero and 10 psi initial air pressure (with the tank empty). As an example, assume the desired operating range is 30-50 psi and the tank has no initial air pressure prior to filling. The table indicates that at 50 psi air pressure the water fills 77.3 percent of the tank volume, and at 30 psi the water fills 67.1 percent of the tank. The operating range is then 77.3 – 67.1 or 10.2 percent of the tank volume. If the total tank volume was 550 gallons, this would allow drawing off 550 × .102 or 56 gallons of water for each cycle of the pump operation.

The same table shows the water levels within the tank at various gage pressures. For tanks installed vertically, the percentage of the tank depth filled with water is the same as the percentage of volume filled. For tanks installed horizontally, however, the percentage of tank diameter varies as shown.

Tanks should be sized so that the number of pumping cycles per hour is not excessive for the horsepower rating of the pump motor. The maximum cycling rate occurs when the average rate of withdrawal is equal to one-half of the average pump discharge rate. Suggested maximum cycles per hour are:

Motor HP	Cycles/Hour
5 or less	25
7½ to 15	20
20 to 30	15
40 and 50	10
60 or more	6

The pump size should be chosen to have a discharge rate large enough to meet the peak demand rate and the tank does not have to provide significant storage. The pump must be designed to deliver its selected discharge against the maximum tank operating pressure.

The size of tank required depends upon the operating range (as percent of tank volume) and the number of cycles per hour desired. Reducing the number of cycles requires a larger tank. Tank volume required can be calculated from:

$$V = \frac{1500 \times Q}{P \times C}$$

where: V = tank volume in gallons
Q = pump discharge in gallons/minute
P = operating tank range (as % of tank volume)
C = maximum cycles per hour

Selecting a tank with a volume (in gallons) that is ten times the pump discharge rate (in gallons per minute) is a conservative approximation.

Hydro-Pneumatic Tank Data

Tank Air Gage Pressure	No Initial Air Pressure in Tank			10 PSI Initial Air Pressure in Tank		
	Water – % of Tank Volume	Water – % of Vertical Tank Height	Water – % of Horizontal Tank Diameter	Water – % of Tank Volume	Water – % of Vertical Tank Height	Water – % of Horizontal Tank Diameter
0	0	0	0			
5	25.4	25.4	30.1			
10	40.5	40.5	42.5	0	0	0
15	50.5	50.5	50.4	16.8	16.8	22.5
20	57.6	57.6	56.0	28.8	28.8	33.0
25	63.0	63.0	60.3	37.8	37.8	40.4
30	67.1	67.1	63.6	44.7	44.7	45.8
35	70.4	70.4	66.3	50.3	50.3	50.3
40	73.1	73.1	68.6	54.8	54.8	53.8
45	75.4	75.4	70.6	58.6	58.6	56.8
50	77.3	77.3	72.2	61.8	61.8	59.4
60	80.3	80.3	74.9	66.9	66.9	63.5
70	82.6	82.6	77.0	70.8	70.8	66.7
80	84.5	84.5	78.8	73.9	73.9	69.3
90	86.0	86.0	80.3	76.4	76.4	71.4
100	87.2	87.2	81.4	78.5	78.5	73.3

Typical Hydro-Pneumatic Tank Dimensions

Tank Dimensions		Approximate	Tank Dimensions		Approximate
Diameter (Inches)	Length (Feet)	Total Capacity (U.S. Gallons)	Diameter (Inches)	Length (Feet)	Total Capacity (U.S. Gallons)
16	4	40	42	12	900
20	2½	40	42	13	980
20	5	80	42	14	1050
24	5	120	48	10	1080
30	6	220	48	13	1500
30	7	270	48	16	1640
30	8	310	48	20	1920
30	10	380	54	16	2070
36	6	310	54	20	2400
36	7	400	60	14	2180
36	8	450	60	18	2800
36	10	550	66	20	3740
42	8	620	72	23	5100
42	9	700	72	24	5270
42	10	760			

Theoretical Horsepower Required to Lift Water

U.S. Gal/Min	Lift (Feet) 10	20	30	40	50	60	70	80	90	100	125	150	175	200	250	300	400
10	0.025	0.051	0.076	0.101	0.126	0.152	0.177	0.202	0.227	0.253	0.316	0.379	0.442	0.505	0.631	0.758	1.01
20	0.051	0.101	0.152	0.202	0.253	0.303	0.354	0.404	0.455	0.505	0.631	0.758	0.884	1.01	1.26	1.52	2.02
30	0.076	0.152	0.227	0.303	0.379	0.455	0.530	0.606	0.682	0.758	0.947	1.14	1.33	1.52	1.89	2.27	3.03
40	0.101	0.202	0.303	0.404	0.505	0.606	0.707	0.808	0.909	1.01	1.26	1.52	1.77	2.02	2.53	3.03	4.04
50	0.126	0.253	0.379	0.505	0.631	0.758	0.884	1.01	1.14	1.26	1.58	1.89	2.21	2.53	3.16	3.79	5.05
60	0.152	0.303	0.455	0.606	0.758	0.909	1.06	1.21	1.36	1.52	1.89	2.27	2.65	3.03	3.79	4.55	6.06
70	0.177	0.354	0.530	0.707	0.884	1.06	1.24	1.41	1.59	1.77	2.21	2.65	3.09	3.54	4.42	5.30	7.07
80	0.202	0.404	0.606	0.808	1.01	1.21	1.41	1.62	1.82	2.02	2.53	3.03	3.54	4.04	5.05	6.06	8.08
90	0.227	0.455	0.682	0.909	1.14	1.36	1.59	1.82	2.05	2.27	2.84	3.41	3.98	4.55	5.68	6.82	9.09
100	0.253	0.505	0.758	1.01	1.26	1.52	1.77	2.02	2.27	2.53	3.16	3.79	4.42	5.05	6.31	7.58	10.1
125	0.316	0.631	0.947	1.26	1.58	1.89	2.21	2.53	2.84	3.16	3.95	4.73	5.52	6.31	7.89	9.47	12.6
150	0.379	0.758	1.14	1.52	1.89	2.27	2.65	3.03	3.41	3.79	4.73	5.68	6.63	7.58	9.47	11.4	15.2
175	0.442	0.884	1.33	1.77	2.21	2.65	3.09	3.54	3.98	4.42	5.52	6.63	7.73	8.84	11.0	13.3	17.7
200	0.505	1.01	1.52	2.02	2.53	3.03	3.54	4.04	4.55	5.05	6.31	7.58	8.84	10.1	12.6	15.2	20.2
250	0.631	1.26	1.89	2.53	3.16	3.79	4.42	5.05	5.68	6.31	7.89	9.47	11.0	12.6	15.8	18.9	25.3
300	0.758	1.52	2.27	3.03	3.79	4.55	5.30	6.06	6.82	7.58	9.47	11.4	13.3	15.2	18.9	22.7	30.3
350	0.884	1.77	2.65	3.54	4.42	5.30	6.19	7.07	7.95	8.84	11.0	13.3	15.5	17.7	22.1	26.5	35.4
400	1.01	2.02	3.03	4.04	5.05	6.06	7.07	8.08	9.09	10.1	12.6	15.2	17.7	20.2	25.3	30.3	40.4
450	1.14	2.27	3.41	4.55	5.68	6.82	7.95	9.09	10.2	11.4	14.2	17.0	19.9	22.7	28.4	34.1	45.5
500	1.26	2.53	3.79	5.05	6.31	7.58	8.84	10.1	11.4	12.6	15.8	18.9	22.1	25.3	31.6	37.9	50.5
600	1.52	3.03	4.55	6.06	7.58	9.09	10.6	12.1	13.6	15.2	18.9	22.7	26.5	30.3	37.9	45.5	60.6
700	1.77	3.54	5.30	7.07	8.84	10.6	12.4	14.1	15.9	17.7	22.1	26.5	30.9	35.4	44.2	53.0	70.7
800	2.02	4.04	6.06	8.08	10.1	12.1	14.1	16.2	18.2	20.2	25.3	30.3	35.4	40.4	50.5	60.6	80.8
900	2.27	4.55	6.82	9.09	11.4	13.6	15.9	18.2	20.5	22.7	28.4	34.1	39.8	45.5	56.8	68.2	90.9
1000	2.53	5.05	7.58	10.1	12.6	15.2	17.7	20.2	22.7	25.3	31.6	37.9	44.2	50.5	63.1	75.8	101.0

Theoretical horsepower required to lift water is equal to the flow rate (in GPM) multiplied by the total lift–including friction–(in feet) and the resulting product divided by 3,960. Actual horsepower required is the theoretical horsepower divided by the overall pump efficiency (as a decimal).

(Example: To lift 200 GPM a total of 60 feet with overall pump efficiency of 65%; Theoretical HP = 3.03 (from table); actual HP = 4.66 (3.03 divided by 0.65).

Kilowatt Hours Required to Lift Water 100 Feet					
Wire-to-Water Efficiency (%)	Kilowatt Hours		Wire-to-Water Efficiency (%)	Kilowatt Hours	
	Per 1,000 Gal	Per Acre-Foot		Per 1,000 Gal	Per Acre-Foot
40	0.784	256.3	65	0.482	157.7
41	0.765	250.0	66	0.475	155.3
42	0.747	244.0	67	0.468	153.0
43	0.729	238.4	68	0.461	150.7
44	0.713	233.0	69	0.454	148.6
45	0.697	227.8	70	0.448	146.4
46	0.682	222.8	71	0.442	144.4
47	0.667	218.1	72	0.436	142.4
48	0.653	213.5	73	0.430	140.4
49	0.640	209.2	74	0.424	138.5
50	0.627	205.0	75	0.418	136.7
51	0.615	201.0	76	0.413	134.9
52	0.603	197.1	77	0.407	133.1
53	0.592	193.4	78	0.402	131.4
54	0.581	189.8	79	0.397	129.7
55	0.570	186.4	80	0.392	128.1
56	0.560	183.0	81	0.387	126.5
57	0.550	179.8	82	0.382	125.0
58	0.541	176.7	83	0.378	123.5
59	0.532	173.7	84	0.373	122.0
60	0.523	170.8	85	0.369	120.6
61	0.514	168.0	86	0.365	119.2
62	0.506	165.3	87	0.360	117.8
63	0.498	162.7	88	0.356	116.5
64	0.490	160.2	89	0.352	115.2

(For lifts other than 100 feet, divide the lift by 100 and multiply the result by the value from the table for the appropriate efficiency)

Typical Fuel/Energy Requirements for Pumping

Fuel/Energy	Unit	Water Horsepower-Hours* (Per Unit)
Gasoline	U.S. Gallon	8.6
	Liter	2.27
Diesel Oil	U.S. Gallon	11.0
	Liter	2.91
Propane	U.S. Gallon	6.8
	Liter	1.80
Natural Gas	1,000 Cu.Ft.	65
	Cubic Meter	2.30
Electricity	Kilowatt-Hour	0.89

*To calculate water horsepower, multiply pump discharge rate (in U.S. gal/min) by lift (in feet) and divide by 3,960.

Example: 250 gpm lifted 350 feet = (250 × 350)/3,960 = 87,500/3,960 = 22.1 WHP

Typical fuel requirement is determined by dividing the WHP by values from the table above:

22.1/11 = 2.0 gal/hr diesel oil 22.1/0.89 = 24.8 Kwh/hr elec.

Assumes pump efficiency of about 75 percent.

Internal Combustion Engine Operation

Internal combustion engines are not operated at maximum horsepower ratings, as measured with a dynamometer, but are reduced as shown in the following table:

	(Percent of Maximum HP)	
Fuel	**Intermittent Service**	**Continuous Service**
Gasoline	80-85	70-75
LP gas	75-85	65-75
Natural gas	70-75	60-65
Diesel	80-85	65-75

The horsepower at which an engine is operated should be further reduced for conditions of altitude and temperature:

A reduction of about 3 percent for each 1,000 ft (300 m) above sea level

A reduction of about 1 percent per 10 degree F (5.6 deg C) temperature rise above 60F (16C) of air at the engine intake

Right-Angle Gear Drives

(Gear drive ratio is defined as the ratio of the horizontal input speed to the vertical output speed)

Vertical Shaft RPM	Gear Ratio and Input RPM											
	1:1	10:11	5:6	4:5	3:4	2:3	5:8	4:7	1:2	4:9	2:5	1:3
720	720	655	600	576	540	480	450	411	360	320	288	240
870	870	791	725	696	653	580	544	497	435	387	348	290
960	960	873	800	768	720	640	600	549	480	427	384	320
1160	1160	1055	967	928	870	773	725	663	580	516	464	387
1460	1460	1327	1217	1168	1095	973	913	834	730	649	584	487
1760	1760	1600	1467	1408	1320	1173	1100	1006	880	782	704	587
3460	3460	3145	2883	2768	2595	2307	2163	1977	1730	1538	1384	1153

Vertical Shaft RPM	Gear Ratio and Input RPM											
		11:10	6:5	5:4	4:3	3:2	8:5	7:4	2:1	9:4	5:2	3:1
720		792	864	900	960	1080	1152	1260	1440	1620	1800	2160
870		957	1044	1088	1160	1305	1392	1523	1740	1958	2175	2610
960		1056	1152	1200	1280	1440	1536	1680	1920	2160	2400	2880
1160		1276	1392	1450	1547	1740	1856	2030	2320	2610	2900	3480
1460		1606	1752	1825	1947	2190	2336	2555	2920	3285	3650	
1760		1936	2112	2200	2347	2640	2816	3080	3520			

Belt Drives

General

The *driver* is the driving sheave (or pulley) which is attached to the prime mover (engine or motor).

The *driven* is the sheave (or pulley) which is attached to the pump or machinery.

$$\text{Diameter of driver} = \frac{\text{Diameter of driven} \times \text{RPM of driven}}{\text{RPM of driver}}$$

$$\text{Diameter of driven} = \frac{\text{Diameter of driver} \times \text{RPM of driver}}{\text{RPM of driven}}$$

$$\text{RPM of driver} = \frac{\text{Diameter of driven} \times \text{RPM of driven}}{\text{Diameter of driver}}$$

$$\text{RPM of driven} = \frac{\text{Diameter of driver} \times \text{RPM of driver}}{\text{Diameter of driven}}$$

Belt speeds should generally not exceed a maximum of 5,000-6,000 ft/min.

$$\text{Belt speed (ft/min)} = \text{Sheave diameter (in)} \times \text{RPM} \times 0.2618$$

V-Belts

V-Belt Section	Top Width (inches)	Thickness (inches)	Recommended Min. Sheave Diam. (inches)
"A"	0.50	0.31	3.0
"B"	0.66	0.41	5.4
"C"	0.88	0.53	9.0
"D"	1.25	0.75	13.0
"E"	1.50	0.91	21.0

The accompanying table gives approximate horsepower ratings for individual V-belts. The manufacturer should be consulted for horsepower ratings per belt for specific applications.

Flat Belts

The accompanying table gives approximate horsepower ratings, per inch of belt width, for flat rubber belts. The manufacturer should be consulted for horsepower ratings per inch width for particular applications.

Approximate Horsepower Rating (Per Belt) for "V"-Belts*

Belt Speed (feet/min)	Belt Size Designation				
	"A"	"B"	"C"	"D"	"E"
	Diameter of Small Sheave – Inches				
	3.8	6.2	10	15	24
1,000	1.4	2.4	5.0	10	16
2,000	2.5	3.9	8.5	18	29
3,000	3.3	5.1	11	24	39
4,000	3.8	6.7	14	28	47
5,000	3.9	7.7	16	30	52

*Consult manufacturers for particular rating specifications.

Approximate Horsepower Rating (Per Inch) for Flat Rubber Belts*

No. of Plies	Recommended Min. Pulley Diam. (In.)	Belt Speed (Feet per Minute)			
		1000	2000	3000	4000
3	6	1.1	2.0	2.8	3.3
4	8	1.5	2.8	3.8	4.5
5	10	1.9	3.4	4.6	5.5
6	12	2.3	4.1	5.5	6.6

Above values for 180-degree arc of contact; for lesser arc use correction factors:

Arc (degrees):	170	160	150	140	130
Factor:	1.0	0.9	0.9	0.8	0.8

*Consult manufacturers for particular rating specifications

Hydraulic Ram

The hydraulic ram is a true impulse pump that utilizes a supply of falling water (through an intake pipe) to lift a portion of that supply to a higher elevation (through a discharge pipe). The intake pipe should be nearly straight, with a minimum length (at least five times the fall) of:

$$\text{Intake pipe length} = \frac{(L + 2) \times (L - F)}{F}$$

where: L = lift in feet
F = fall in feet

The lift should generally be between six and twelve times the fall. The efficiency of the ram, as a pump, is related to the ratio of lift to fall and is approximately:

Ratio (L:F)	4	6	8	12	16
Efficiency	.70	.60	.50	.35	.25

Operating characteristics of rams are given in the accompanying table.

Hydraulic Rams				
Ram Size	Intake Pipe Diam. (Inches)	Discharge Pipe Diam. (Inches	Minimum Supply (gpm)	Range of Discharge (gpm)
1	1	½	4	0.05 - 1.75
1½	1½	¾	8	0.12 - 2.75
2	2	1	15	0.25 - 5.5
3	3	1½	25	0.65 - 15
4	4	2	45	1 - 35
6	6	3	90	3 - 65
9	9	4	200	7 - 140
12	12	5	300	15 - 250

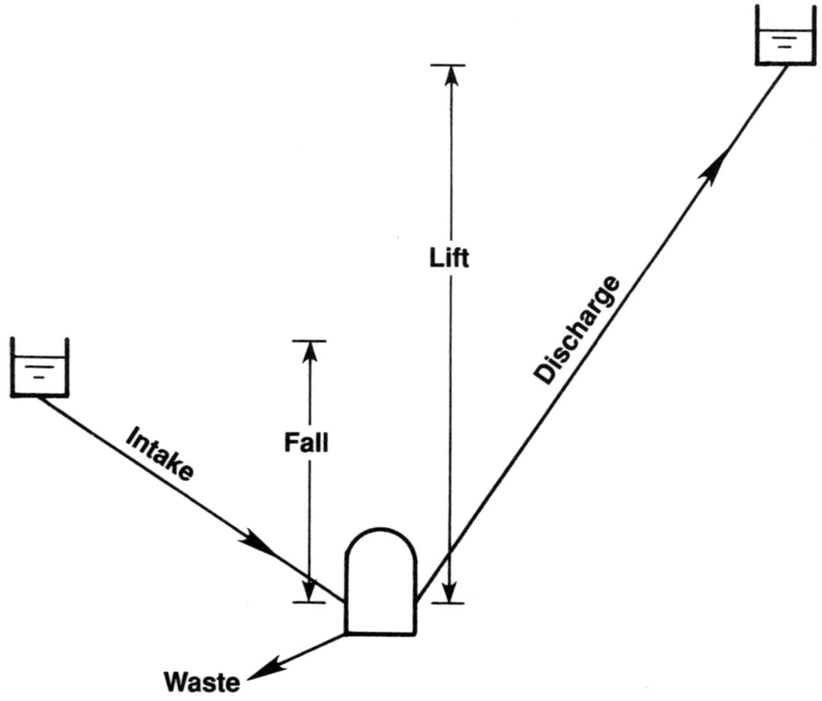

Hydraulic Ram

Cylinder Pump Discharge (U.S. Gallons/Stroke)

Pump Cylinder Diam. (Inches)	Length of Stroke (Inches) 1	2	3	4	5	6	7	8	9	10	12	14	16	18	20	22	24
1	0.003	0.007	0.010	0.014	0.017	0.020	0.024	0.027	0.031	0.034	0.041	0.048	0.054	0.061	0.068	0.075	0.082
1¼	0.005	0.011	0.016	0.021	0.027	0.032	0.037	0.042	0.048	0.053	0.064	0.074	0.085	0.096	0.106	0.117	0.127
1½	0.008	0.015	0.023	0.031	0.038	0.046	0.054	0.061	0.069	0.076	0.092	0.107	0.122	0.138	0.153	0.168	0.184
1¾	0.010	0.021	0.031	0.042	0.052	0.062	0.073	0.083	0.094	0.104	0.125	0.146	0.167	0.187	0.208	0.229	0.250
2	0.014	0.027	0.041	0.054	0.068	0.082	0.095	0.109	0.122	0.136	0.163	0.190	0.218	0.245	0.272	0.299	0.326
2¼	0.017	0.034	0.052	0.069	0.086	0.103	0.120	0.138	0.155	0.172	0.207	0.241	0.275	0.310	0.344	0.379	0.413
2½	0.021	0.042	0.064	0.085	0.106	0.127	0.149	0.170	0.191	0.212	0.255	0.297	0.340	0.382	0.425	0.467	0.510
2¾	0.026	0.051	0.077	0.103	0.129	0.154	0.180	0.206	0.231	0.257	0.309	0.360	0.411	0.463	0.514	0.566	0.617
3	0.031	0.061	0.092	0.122	0.153	0.184	0.214	0.245	0.275	0.306	0.367	0.428	0.490	0.551	0.612	0.673	0.734
3¼	0.036	0.072	0.108	0.144	0.180	0.215	0.251	0.287	0.323	0.359	0.431	0.503	0.575	0.646	0.718	0.790	0.862
3½	0.042	0.083	0.125	0.167	0.208	0.250	0.292	0.333	0.375	0.416	0.500	0.583	0.666	0.750	0.833	0.916	1.000
3¾	0.048	0.096	0.143	0.191	0.239	0.287	0.335	0.382	0.430	0.478	0.574	0.669	0.765	0.861	0.956	1.052	1.147
4	0.054	0.109	0.163	0.218	0.272	0.326	0.381	0.435	0.490	0.544	0.653	0.762	0.870	0.979	1.088	1.197	1.306
4¼	0.061	0.123	0.184	0.246	0.307	0.368	0.430	0.491	0.553	0.614	0.737	0.860	0.983	1.105	1.228	1.351	1.474
4½	0.069	0.138	0.207	0.275	0.344	0.413	0.482	0.551	0.620	0.688	0.826	0.964	1.102	1.239	1.377	1.515	1.652
4¾	0.077	0.153	0.230	0.307	0.384	0.460	0.537	0.614	0.690	0.767	0.921	1.074	1.227	1.381	1.534	1.688	1.841
5	0.085	0.170	0.255	0.340	0.425	0.510	0.595	0.680	0.765	0.850	1.020	1.190	1.360	1.530	1.700	1.870	2.040
5¼	0.094	0.187	0.281	0.375	0.469	0.562	0.656	0.750	0.843	0.937	1.125	1.312	1.499	1.687	1.874	2.062	2.249
5½	0.103	0.206	0.309	0.411	0.514	0.617	0.720	0.823	0.926	1.028	1.234	1.440	1.646	1.851	2.057	2.263	2.468
5¾	0.112	0.225	0.337	0.450	0.562	0.674	0.787	0.899	1.012	1.124	1.349	1.574	1.799	2.023	2.248	2.473	2.698
6	0.122	0.245	0.367	0.490	0.612	0.734	0.857	0.979	1.102	1.224	1.469	1.714	1.958	2.203	2.448	2.693	2.938
7	0.167	0.333	0.500	0.666	0.833	1.000	1.166	1.333	1.499	1.666	1.999	2.332	2.666	2.999	3.332	3.665	3.998
8	0.218	0.435	0.653	0.870	1.088	1.306	1.523	1.741	1.958	2.176	2.611	3.046	3.482	3.917	4.352	4.787	5.222
9	0.275	0.551	0.826	1.102	1.377	1.652	1.928	2.203	2.479	2.754	3.305	3.856	4.406	4.957	5.508	6.059	6.610
10	0.340	0.680	1.020	1.360	1.700	2.040	2.380	2.720	3.060	3.400	4.080	4.760	5.440	6.120	6.800	7.480	8.160

Estimating Water Use

It is frequently necessary to make estimates of the amount of water needed to supply the requirements of farms, homes, commercial establishments, and entire communities. Estimates are usually required for both (1) the volume of water, usually per day, that is needed, and (2) the maximum flow rate for the particular use.

Water-use rates vary greatly from place to place, depending upon many factors such as: climate, degree of development (plumbing fixtures, etc.), supply pressure, system leakage, cost of water, and whether the supply is metered. In some areas the typical water use increases considerably seasonally — such as for lawn irrigation in summer, or allowing water to run continuously to prevent freezing in winter. In all areas the average annual water use can be expected to increase during the maximum month, maximum week, maximum day, and maximum hour.

The source of supply must be sufficient to provide the anticipated daily volume of water required, unless a large storage volume is available. Distribution system piping should be sized to allow delivery of the maximum flow rate without excessive pressure drop. Maximum flow rates can exceed the rate from the original supply source if storage (such as elevated reservoir or pressure tank) is provided.

Experience with comparable systems is probably the best method of estimating water requirements. The following tables or information of average water use, however, may be useful as guides for preliminary planning purposes in the United States:

Farm-Domestic Use

Each family member (household use)	35 – 50 gal/day
Each horse, steer, or dry cow	10 – 12
Each cow producing milk	25 – 30
Each hog	2 – 5
Each sheep	2 – 3
Each 100 chickens	3 – 4

Municipal Use

For population served	150 – 250 gal/person/day

(greater use in unmetered systems and in areas with significant irrigation)

Domestic Use

Maximum instantaneous flow requirements:

Houses served	Flow rate (gpm)
1	18
5	50
10	70
20	90
50	150
100	200
200	265
300	320
500	400

Commercial Establishments

Airport (per passenger)	3 – 6 gal/day
Apartment house (per person)	80 – 130
Hospital (per person)	175 – 300
Office (per person)	10 – 15
Restaurant (per seat)	30 – 50
School (per person)	10 – 15

Standard Fixtures (Contents)

Bathtub	30 gals
Lavatory	1½
Shower (per use)	20 – 30
Water closet (toilet)	4 – 6

Standard Fixtures (Flow Rate in GPM)

	Number of Fixtures		
Type	1	4	10
Bathtubs (1 faucet)	10	32	66
Drinking fountains	2	5	8
Hose bibbs (½-in)	5	14	40
Kitchen sinks, small (per faucet)	6	16	40
Laundry trays (per faucet)	8	24	50
Lavatories (1 faucet)	3	10	24
Shower heads (rain type)	8	32	70
Shower heads (water saver)	4	16	30
Urinals (tank)	4	12	34
Urinals (flush valve)	15	45	82
Water closets (tank)	5	15	40
Water closets (flush valve)	27	70	135

Recommended Piping For Fixture Branches

Fixture	Branch Size (inches)
Bathtub	1/2 (hot and cold)
Kitchen sink	1/2 (hot and cold)
Laundry tray	1/2 (hot and cold)
Lavatory	3/8 (hot and cold)
Shower stall (per head)	1/2 (hot and cold)
Urinal (flush valve)	1/2 (cold)
Water closet (tank)	3/8 (cold)
Water closet (flush valve)	1 (cold)

Typical Variation In Water Use (Community Supplies)

	Percentage Of Annual Average
Daily average use in maximum month	120
Daily average use in maximum week	140
Maximum use in one day	180

Conservation

There is an increasing trend toward emphasizing water conservation in the United States and elsewhere. This is illustrated by the U.S. legislation, proposed in 1991, which would provide for the following maximum water-use standards:

Product	Maximum Allowable Water Use
Water closets (gallons/flush):	
Gravity, tank-type	1.6
Flushometer, valve type	1.6
tank type	1.6
Urinals (gallons/flush)	1.0
Shower heads (gallons/minute)	1.0
Faucets (gallons/minute):	
Lavatory	2.0
Kitchen	2.5

Chapter 6

Flow Measurement

Flow Measurement

A fundamental formula for steady, uniform flow of water is:

$$Q = AV$$

where Q is the discharge (flow rate), A is the cross-sectional area through which flow takes place, and V is the velocity of flow. All three values must be in consistent units (such as cubic feet per second, square feet, and feet per second). When two of these values are known, the third can be calculated from the formula.

There are many ways to measure flow of water. Some of the most common, and practical, are:

- Weirs
- Flumes
- Flow meters
- Pipe orifices
- Current meters
- Orifice Buckets
- Measuring time to fill containers
- Horizontal pipe-flow measurements
- Vertical pipe-flow measurements

Most of these methods are capable of an accuracy of ± 5 percent, depending upon the installation and the care with which measurements are taken. The pipe-flow measurement methods are usually somewhat less accurate.

Weirs

Weirs can have a variety of configuration of the opening through which the water flows. Common shapes are rectangular (with or without contractions at the sides), triangular, and Cipolletti (trapezoidal shape with side slopes at 1 h:4 v). The sharp edge over which the water flows is the "weir crest"; the height of the water surface above the crest, at a point upstream, is the "head"; the sheet of water flowing over the weir crest is the "nappe"; the width of the crest at the opening is also known as the "weir length." For triangular weirs, the opening can have any angle, but those of 90 and 60 degrees are most common; the head on a triangular weir is measured above the bottom of the V-notch.

Important considerations in a proper weir installation are:

1. The upstream side of the wall or bulkhead should be smooth, vertical, and at right angles to the direction of flow.
2. The upstream side of the crest and sides of the opening should be smooth and sharp, and be flush with the upstream face of the wall or bulkhead.
3. The entire crest width should be level, and the sides vertical (except for triangular weirs, or Cipolletti weirs with sides sloping 1 h: 4 v).
4. The height of the crest above the bottom of the approach channel should be at least twice the head on the weir, and not less than one foot.
5. The distance from the sides of the weir opening to the sides of the approach channel should be at least twice the head on the weir, and not less than one foot (except for "suppressed rectangular weirs" which have no contraction and sides are flush with the walls of the approach channel).
6. The overflow of water (nappe) should touch only the upstream edges of the crest and sides.
7. There should be free circulation of air both under and at the sides of the nappe.
8. Measurement of the head (difference in elevation between the crest and upstream water surface) should be made at a distance of 4 to 6 times the head.
9. The cross-sectional area of the approach channel should be at least 6 times that of the nappe, for a distance of at least 15 to 20 times the head.
10. The velocity of approach should be low. Velocity of approach can be approximated by dividing discharge by the cross-sectional area of the approach channel upstream of the weir; if it exceeds about 1 fps a correction should be applied to the measured head by adding the velocity head from the table in Chapter 4.

The accompanying tables give discharge for contracted and suppressed rectangular weirs and Cipolletti weirs, of various lengths, and for triangular weirs with 60 and 90 degree notches. For other sizes of weirs, the following formulas can be used to compute approximate discharge:

Rectangular (contracted): $Q = 3.33 \times (L - 0.2H) \times H^{3/2}$

Rectangular (suppressed): $Q = 3.33 \times L \times H^{3/2}$

Cipolletti: $Q = 3.367 \times L \times H^{3/2}$

Triangular: $Q = 2.5 \times \tan \beta \times H^{3/2}$

where: Q = discharge in cubic feet per second
L = length of weir crest in feet
H = head in feet
β = *one-half* the angle of the notch opening in degrees

Flumes

Open flumes can be used to measure flow using Manning's formula to calculate the flow from known measurements of the cross-sectional area (shape), depth of flow, and slope (see discussion of Manning's formula in the section "Friction Loss in Pipe Flow" in Chapter 4 of this handbook).

The Parshall flume, formerly known as "the improved Venturi flume," is a specially constructed channel with a converging section leading to a constricted "throat" and a diverging section leading downstream. The floor of the flume is sloped downward within the throat section and then upward. The size of the flume is designated by the width of the throat. The discharge through the flume is determined by measuring the depth (head) of water at a point upstream of the throat that is two-thirds of the length of the converging section (see sketch). The discharge is calculated from the formulas:

For 3-inch throat width: $Q = 0.992\ H^{1.547}$

For 6-inch throat width: $Q = 2.06\ H^{1.58}$

For 9-inch throat width: $Q = 3.07\ H^{1.53}$

For 12-inch throat width: $Q = 4.0\ H^{1.522}$

where: Q = discharge in cubic feet per second
H = upstream head, in feet

The accompany table gives Parshall flume discharges for various heads and different throat widths.

Parshall flumes can be constructed of metal, wood, or concrete but must conform to specified dimensions. Pre-fabricated flumes made of sheet metal are commercially available in a wide range of throat widths.

Small trapezoidal flumes, designed for ditches with a one-foot bottom width, can also be used to measure flows of up to 5 cfs (2250 gpm). The flumes can be permanently installed (concrete) or are commercially available in metal or fiberglass construction for portability. The dimensions for a one-foot trapezoidal flume are given on the accompanying sketch, together with a flow-rating table for various depths of water entering the flume.

Flow Meters

Flow meters are installed on pipelines flowing full and indicate the volume of water passing through the meter; by measuring the time required for a given volume, the rate of flow can be calculated. Some flow meters also give a direct reading of flow rate.

Other types of flow meters may use in-line orifices or pitot tubes which measure differential upstream and downstream pressures, with flow rate calculated from this differential and such factors as pipe diameter, etc.

Pipe Orifices

The pipe orifice method is commonly used to measure pump discharge during well tests. It consists of attaching a circular orifice on the end of the discharge pipe and measuring the pressure upstream from the orifice with a manometer. The discharge is computed from knowing the diameters of the pipe and orifice (in inches), and the pressure (in inches of water).

Important considerations with using this method are:

1. The discharge pipe must be level, and at least four feet long.
2. The orifice should be machined with accurate diameter and with a sharp upstream edge to the opening; it can be cut in a threaded pipe cap, or from a circular plate held in place with a thread protector. The orifice must be centered over the end of the pipe.
3. The manometer to measure pressure should be located two feet upstream from the end of the pipe; a small (1/8-inch) pipe should be tapped into the pipe wall but *not* projecting inside the pipe; the

manometer tap must be in the horizontal plane through the center of the pipe.

4. Measurement of the head on the orifice is made (in inches) from the center of the pipe to the water level in the observation tube.

The general formula for calculating flow through a pipe orifice is:

$$Q = C\,A\,\sqrt{2\,g\,h} = 8.02\,C\,A\,\sqrt{h}$$

where: Q = discharge in U.S. Gal/Min
A = area of orifice in square inches
h = height of water in tube
g = gravity
C = a constant depending upon ratio of diameter of orifice to pipe I.D.

The diameter of the orifice should generally be from 0.5-0.8 times the inside diameter of the pipe, giving values of C of about 0.6 to 0.8.

The accompanying tables give the discharge for various combinations of pipe and orifice diameters and for different heads.

Current Meters

Current meters are commonly used to measure flow in open channels (streams, canals, etc.). They typically consist of rotating cups on a rod or weighted line; the velocity is either read directly, or determined by timing a number of rotations and referring to a rating table for the particular meter.The channel width is divided into a number of segments and the channel depths measured for the boundaries of each segment; the average velocity is measured with the current meter at the center of each segment and the discharge computed for each segment by the formula $Q = A \times V$. The total channel discharge is the sum of the discharge of each segment.

Average velocity is measured at six-tenths of the water depth below the water surface (for shallow flows), or by averaging the measurements taken at two-tenths and eight-tenths depth below the water surface. If a current meter is not available, a crude approximation of average velocity for a given segment can be made by timing a float over a known distance and calculating the surface velocity. The average velocity for that segment will be about 0.85 times the surface velocity. Flow is equal to the average velocity multiplied by the cross-sectional area of flow.

Orifice Buckets

Orifice buckets are a convenient way to measure small pump discharges, particularly from reciprocating or air-lift pumps when the flow may be pulsating rather than steady. A typical bucket can be made from an empty 100-lb grease can (about 14 inches in diameter and 24 inches high). A number (6 to 8) of circular holes, one inch in diameter, are cut in the bottom of the can and a piezometer tube attached to the side so that the height of water above the bottom can be measured. The accompanying table gives the flow, in U.S. gallons a minute, that discharges from a single one-inch orifice for various heights of water in the bucket. Rubber corks or stoppers are used to plug or unplug holes, as necessary, so that the water level stabilizes within the bucket. Flow ranges of from 3-4 gpm (with one hole open) to as much as 100 gpm (with six holes open) can be measured. The bucket can also be made of canvas, with a metal plate bottom, for more portability.

Measuring Time To Fill Containers

A fairly accurate means of flow measurement is to measure the time required to fill a container of known volume (or to raise the water level a given distance in a container of known dimensions). An accompanying table gives flow rates for filling 55-gallon drums, as well as general formulas for use of this method.

Horizontal Or Vertical Pipe Flow

Sometimes the only practicable method of measuring approximate flow from open pipes is to measure the coordinates of the flow stream (horizontal and vertical) from horizontal (or sloping) pipes; or the height to which water rises above the open end of a vertical pipe—such as flowing artesian wells.

The procedures for these types of measurements are in this Chapter as "Estimating Flow From Open Pipes" and "Vertical Flow From Pipes."

Another method of measuring flow from partially-filled horizontal pipes, discharging freely to air, is known as the "California Pipe Method." The procedure for this type of measurement is also given in this Chapter.

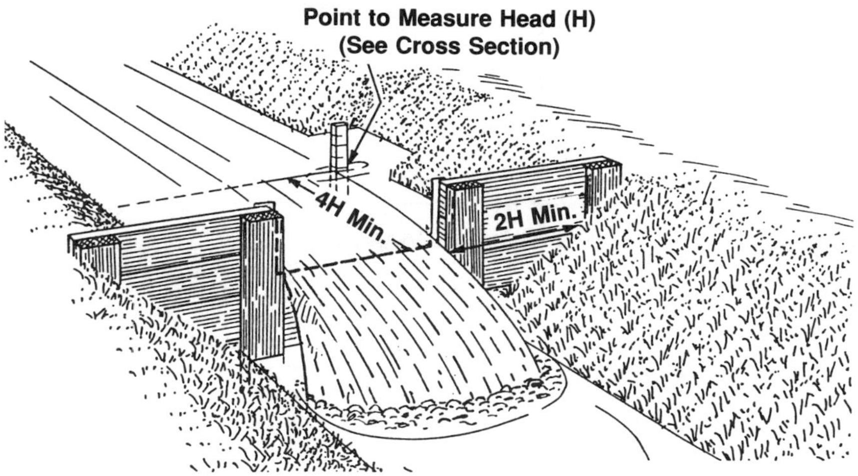

Rectangular Weir with End Contractions

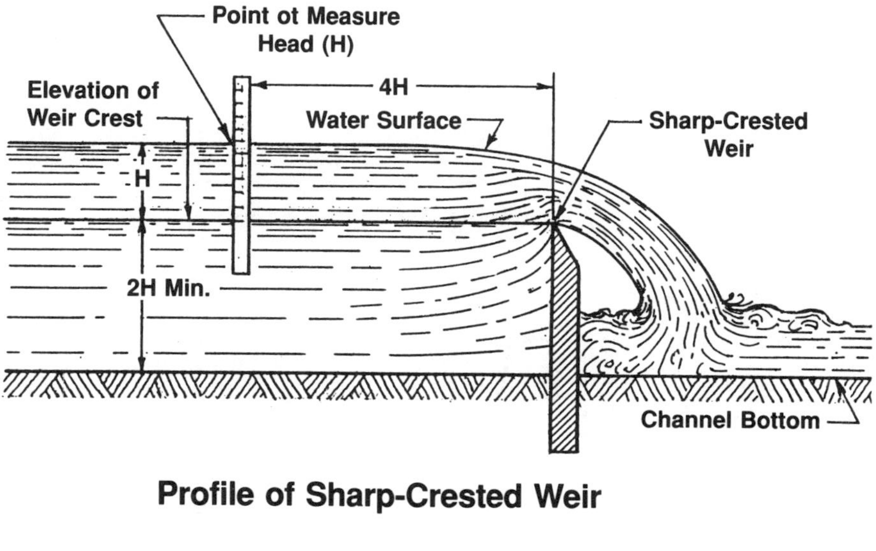

Profile of Sharp-Crested Weir

Weir Installation

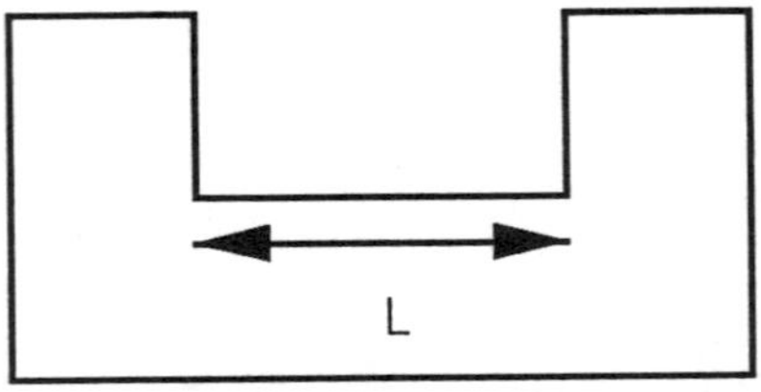

Rectangular-Contracted

Triangular

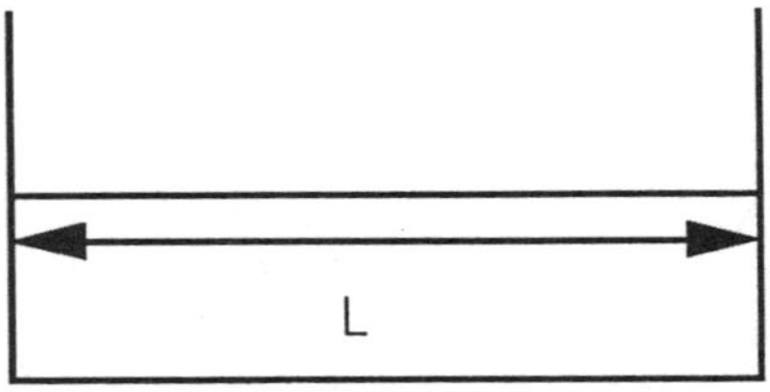

Rectangular-Suppressed

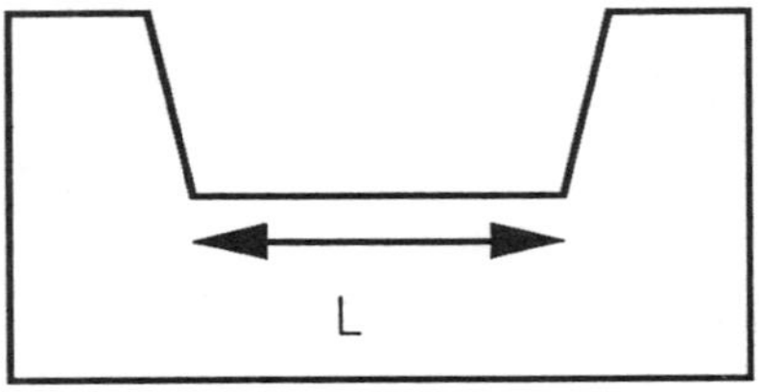

Cipolletti

Weir Plates

Rectangular Weir with End Contractions
(Discharge in Cubic Feet Per Second)

Head (Ft)	Length of Weir Crest			Head (Ft)	Length of Weir Crest	
	1 Ft.	3 Ft.	5 Ft.		3 Ft.	5 Ft.
0.02	0.009	0.028	0.047	0.62	4.68	7.93
0.04	0.026	0.080	0.133	0.64	4.90	8.31
0.06	0.048	0.146	0.244	0.66	5.12	8.69
0.08	0.074	0.225	0.376	0.68	5.35	9.08
0.10	0.103	0.314	0.52	0.70	5.58	9.48
0.12	0.135	0.412	0.69	0.72	5.81	9.88
0.14	0.170	0.52	0.87	0.74	6.05	10.29
0.16	0.206	0.63	1.06	0.76	6.28	10.70
0.18	0.245	0.75	1.26	0.78	6.52	11.11
0.20	0.286	0.88	1.48	0.80	6.77	11.53
0.22	0.329	1.02	1.70	0.82	7.01	11.96
0.24	0.373	1.16	1.94	0.84	7.26	12.39
0.26	0.419	1.30	2.18	0.86	7.51	12.82
0.28	0.466	1.45	2.44	0.88	7.76	13.26
0.30	0.51	1.61	2.70	0.90	8.02	13.70
0.32	0.56	1.77	2.98	0.95	8.66	14.83
0.34	0.62	1.94	3.26	1.00	9.32	15.98
0.36	0.67	2.11	3.54	1.05	10.00	17.16
0.38	0.72	2.28	3.84	1.10	10.68	18.36
0.40	0.78	2.46	4.14	1.15	11.38	19.59
0.42	0.83	2.64	4.46	1.20	12.08	20.84
0.44	0.89	2.83	4.77	1.25	12.80	22.11
0.46	0.94	3.02	5.10	1.30	13.52	23.40
0.48	1.00	3.22	5.43	1.35	14.26	24.71
0.50	1.06	3.41	5.77	1.40	15.00	26.04
0.52	1.12	3.62	6.11	1.45	15.76	27.39
0.54	1.18	3.82	6.46	1.50	16.52	28.75
0.56	1.24	4.03	6.82	1.55	17.29	30.14
0.58	1.30	4.24	7.18	1.60	18.06	31.54
0.60	1.36	4.46	7.55	1.65	18.84	32.96

One cubic foot per second = 448.83 U.S. gallons per minute

Rectangular Suppressed Weir
(Discharge in Cubic Feet Per Second)

Head (Feet)	Length of Weir Crest			Head (Feet)	Length of Weir Crest	
	1 Ft.	3 Ft.	5 Ft.		3 Ft.	5 Ft.
0.02	0.009	0.028	0.047	0.62	4.88	8.13
0.04	0.027	0.080	0.133	0.64	5.11	8.52
0.06	0.049	0.147	0.245	0.66	5.36	8.93
0.08	0.075	0.226	0.377	0.68	5.60	9.34
0.10	0.105	0.316	0.527	0.70	5.85	9.75
0.12	0.138	0.415	0.692	0.72	6.10	10.17
0.14	0.174	0.523	0.872	0.74	6.36	10.60
0.16	0.213	0.639	1.07	0.76	6.62	11.03
0.18	0.254	0.763	1.27	0.78	6.88	11.47
0.20	0.298	0.894	1.49	0.80	7.15	11.91
0.22	0.344	1.03	1.72	0.82	7.42	12.36
0.24	0.392	1.17	1.96	0.84	7.69	12.82
0.26	0.441	1.32	2.21	0.86	7.97	13.28
0.28	0.493	1.48	2.47	0.88	8.25	13.74
0.30	0.547	1.64	2.74	0.90	8.53	14.22
0.32	0.603	1.81	3.01	0.95	9.25	15.42
0.34	0.660	1.98	3.30	1.00	9.99	16.65
0.36	0.719	2.16	3.60	1.05	10.75	17.91
0.38	0.780	2.34	3.90	1.10	11.53	19.21
0.40	0.842	2.53	4.21	1.15	12.32	20.53
0.42	0.906	2.72	4.53	1.20	13.13	21.89
0.44	0.972	2.92	4.86	1.25	13.96	23.27
0.46	1.04	3.12	5.19	1.30	14.81	24.68
0.48	1.11	3.32	5.54	1.35	15.67	26.12
0.50	1.18	3.53	5.89	1.40	16.55	27.58
0.52	1.25	3.75	6.24	1.45	17.44	29.07
0.54	1.32	3.96	6.61	1.50	18.35	30.59
0.56	1.40	4.19	6.98	1.55	19.28	32.13
0.58	1.47	4.41	7.35	1.60	20.22	33.70
0.60	1.55	4.64	7.74	1.65	21.17	35.29

One cubic foot per second = 448.83 U.S. gallons per minute

Cipolletti Weir
(Discharge in Cubic Feet Per Second)

Head (Feet)	Length of Weir Crest (Feet)			Head (Feet)	Length of Weir Crest (Feet)		
	1	1.5	2		1	1.5	2
0.02	0.010	0.014	0.019	0.62	1.64	2.47	3.29
0.04	0.027	0.040	0.054	0.64	1.72	2.59	3.45
0.06	0.049	0.074	0.099	0.66	1.81	2.71	3.61
0.08	0.076	0.114	0.152	0.68	1.89	2.83	3.78
0.10	0.106	0.160	0.213	0.70	1.97	2.96	3.94
0.12	0.140	0.210	0.280	0.72	2.06	3.09	4.11
0.14	0.176	0.265	0.353	0.74	2.14	3.22	4.29
0.16	0.215	0.323	0.431	0.76	2.23	3.35	4.46
0.18	0.257	0.386	0.514	0.78	2.32	3.48	4.64
0.20	0.301	0.452	0.602	0.80	2.41	3.61	4.82
0.22	0.347	0.521	0.695	0.82	2.50	3.75	5.00
0.24	0.396	0.594	0.792	0.84	2.59	3.89	5.18
0.26	0.446	0.670	0.893	0.86	2.69	4.03	5.37
0.28	0.499	0.748	0.998	0.88	2.78	4.17	5.56
0.30	0.553	0.830	1.11	0.90	2.87	4.31	5.75
0.32	0.609	0.914	1.22	0.95	3.12	4.68	6.24
0.34	0.668	1.00	1.34	1.00	3.37	5.05	6.73
0.36	0.727	1.09	1.45	1.05	3.62	5.43	7.25
0.38	0.789	1.18	1.58	1.10	3.88	5.83	7.77
0.40	0.852	1.28	1.70	1.15	4.15	6.23	8.30
0.42	0.916	1.37	1.83	1.20	4.43	6.64	8.85
0.44	0.983	1.47	1.97	1.25	4.71	7.06	9.41
0.46	1.05	1.58	2.10	1.30	4.99	7.49	9.98
0.48	1.12	1.68	2.24	1.35	5.28	7.92	10.56
0.50	1.19	1.79	2.38	1.40	5.58	8.37	11.15
0.52	1.26	1.89	2.53	1.45	5.88	8.82	11.76
0.54	1.34	2.00	2.67	1.50	6.19	9.28	12.37
0.56	1.41	2.12	2.82	1.55	6.50	9.75	12.99
0.58	1.49	2.23	2.97	1.60	6.81	10.22	13.63
0.60	1.56	2.35	3.13	1.65	7.14	10.70	14.27

(One cubic foot per second = 448.83 U.S. gallons per minute)

Triangular Weir
(Discharge in Cubic Feet Per Second and Gallons Per Minute)

Head (Feet)	Weir Notch Angle (Degrees) 90		60		Head (Feet)	Weir Notch Angle (Degrees) 90		60	
	CFS	GPM	CFS	GPM		CFS	GPM	CFS	GPM
0.02		0.06		0.04	0.62	0.74	331	0.43	191
0.04		0.35		0.20	0.64	0.80	359	0.46	207
0.06	0.002	0.96	0.001	0.56	0.66	0.86	387	0.50	224
0.08	0.004	1.98	0.003	1.14	0.68	0.93	417	0.54	241
0.10	0.008	3.46	0.004	2.00	0.70	1.00	449	0.58	259
0.12	0.012	5.46	0.007	3.15	0.72	1.07	481	0.62	278
0.14	0.018	8.03	0.010	4.63	0.74	1.15	515	0.66	298
0.16	0.025	11.2	0.014	6.47	0.76	1.23	551	0.71	318
0.18	0.034	15.0	0.019	8.68	0.78	1.31	588	0.76	339
0.20	0.044	19.6	0.025	11.3	0.80	1.40	626	0.81	362
0.22	0.055	24.8	0.032	14.3	0.82	1.48	666	0.86	385
0.24	0.069	30.9	0.040	17.8	0.84	1.58	708	0.91	409
0.26	0.084	37.7	0.049	21.8	0.86	1.67	751	0.97	433
0.28	0.10	45.4	0.058	26.2	0.88	1.77	795	1.02	459
0.30	0.12	53.9	0.069	31.1	0.90	1.87	841	1.08	485
0.32	0.14	63.4	0.082	36.6	0.95	2.14	963	1.24	556
0.34	0.16	73.8	0.095	42.6	1.00	2.44	1,094	1.41	632
0.36	0.19	85.1	0.11	49.1	1.05	2.75	1,236	1.59	714
0.38	0.22	97.4	0.13	56.2	1.10	3.09	1,389	1.79	802
0.40	0.25	111	0.14	63.9	1.15	3.46	1,552	2.00	896
0.42	0.28	125	0.16	72.2	1.20	3.85	1,726	2.22	997
0.44	0.31	141	0.18	81.1	1.25	4.26	1,912	2.46	1,104
0.46	0.35	157	0.20	90.7	1.30	4.70	2,109	2.71	1,217
0.48	0.39	175	0.22	101	1.35	5.16	2,317	2.98	1,338
0.50	0.43	193	0.25	112	1.40	5.65	2,538	3.26	1,465
0.52	0.48	213	0.27	123	1.45	6.17	2,770	3.56	1,599
0.54	0.52	234	0.30	135	1.50	6.72	3,016	3.88	1,741
0.56	0.57	257	0.33	148	1.55	7.29	3,273	4.21	1,890
0.58	0.62	280	0.36	162	1.60	7.89	3,544	4.56	2,046
0.60	0.68	305	0.39	176	1.65	8.53	3,827	4.92	2,209

(One Cubic Foot Per Second = 448.83 U.S. Gallons Per Minute)

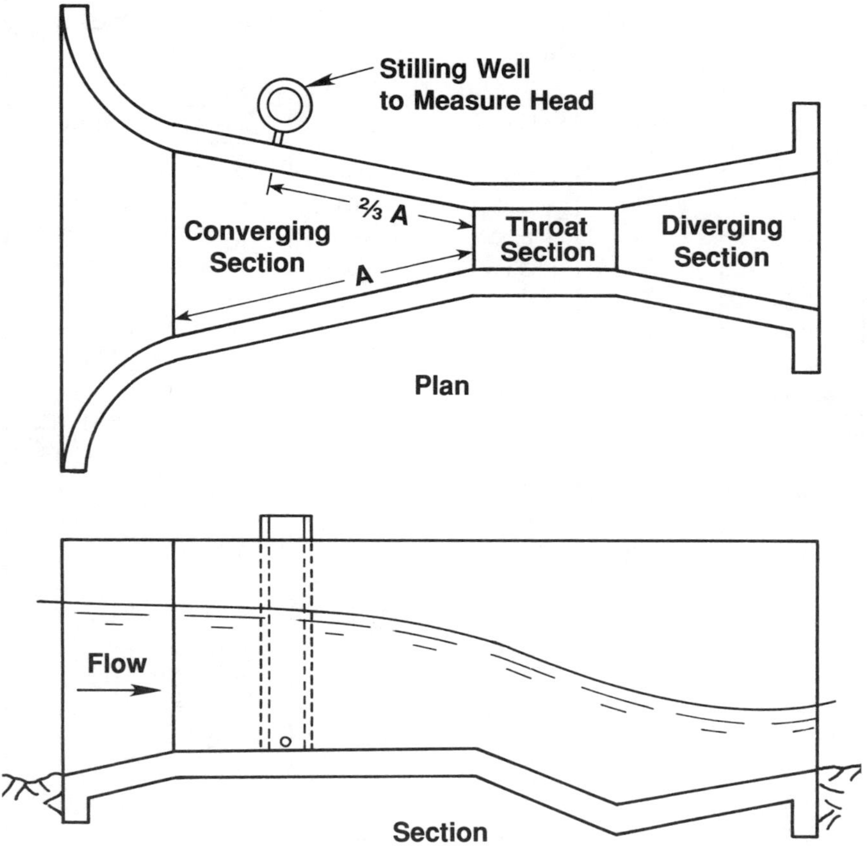

Parshall Flume

Free Flow Discharge Through Parshall Flumes
(Cubic Feet per Second)

Upper Head		Throat Width			
Feet	Inches	3-Inch	6-Inch	9-Inch	12-Inch
0.10	1 3/16	0.028	0.054	0.091	
0.15	1 13/16	0.053	0.103	0.168	
0.20	2 3/8	0.082	0.162	0.262	0.35
0.25	3	0.116	0.230	0.368	0.48
0.30	3 5/8	0.154	0.307	0.487	0.64
0.35	4 3/16	0.196	0.392	0.616	0.81
0.40	4 13/16	0.240	0.484	0.756	0.99
0.45	5 3/8	0.288	0.583	0.905	1.19
0.50	6	0.339	0.689	1.06	1.39
0.55	6 5/8	0.393	0.801	1.23	1.61
0.60	7 3/16	0.450	0.919	1.41	1.84
0.65	7 13/16	0.509	1.04	1.59	2.08
0.70	8 3/8	0.571	1.17	1.78	2.32
0.75	9	0.636	1.31	1.98	2.58
0.80	9 5/8	0.702	1.45	2.18	2.85
0.85	10 3/16	0.771	1.59	2.39	3.12
0.90	10 13/16	0.843	1.74	2.61	3.41
0.95	11 3/8	0.916	1.90	2.84	3.70
1.00	12	0.992	2.06	3.07	4.00
1.10	13 3/16	1.15	2.39	3.55	4.62
1.20	14 3/8		2.75	4.06	5.28
1.30	15 5/8			4.59	5.96
1.40	16 13/16			5.14	6.68
1.50	18			5.71	7.41

1 Cubic Foot/Second = 448.8 U.S. Gallons/Minute = 28.32 Liters/Second

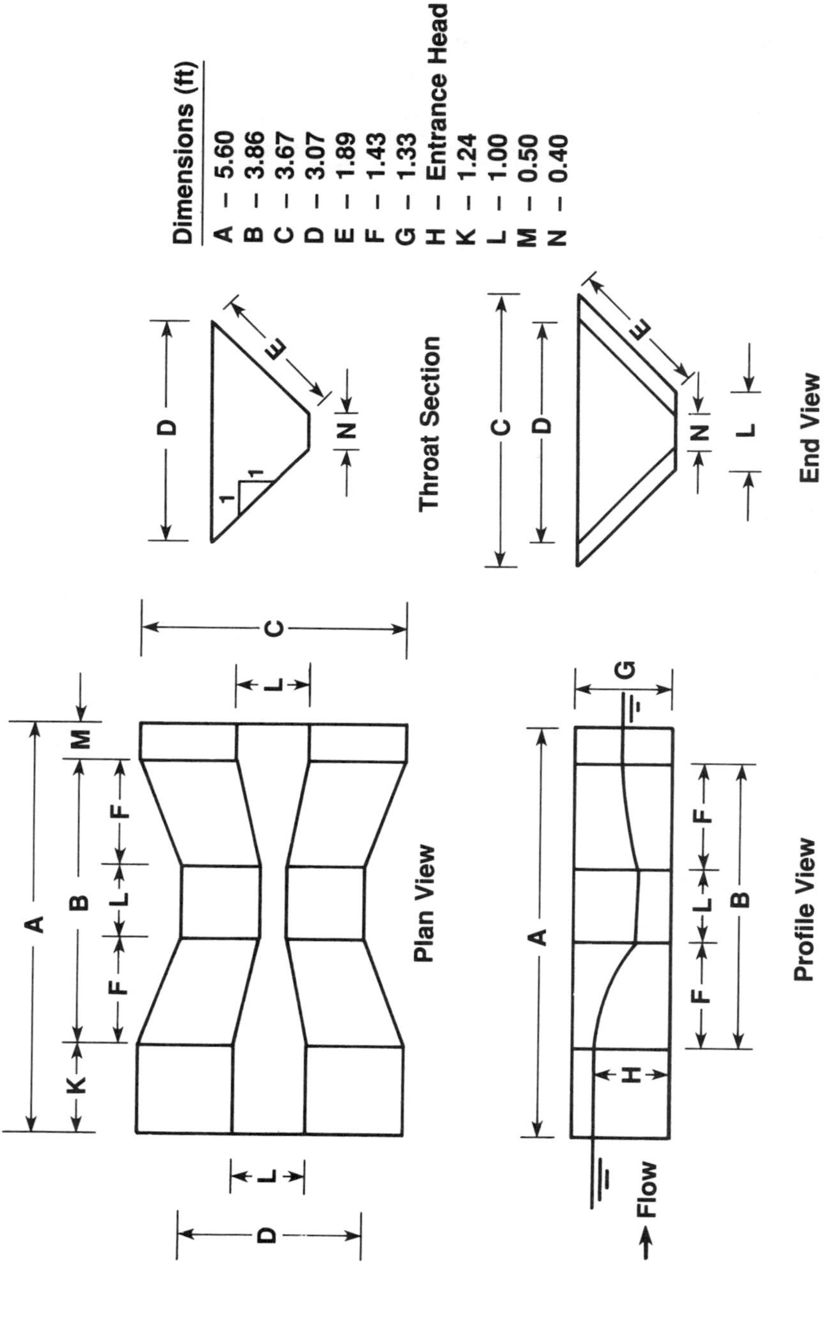

Trapezoidal Flume

Free Flow Discharge Through Trapezoidal Flume
(One Ft. Width)

Entrance Head (H)		Flume Discharge		Entrance Head (H)		Flume Discharge	
Feet	Inches	Cu/Ft/Sec	US Gal/Min	Feet	Inches	Cu/Ft/Sec	US Gal/Min
0.20	2.4	0.16	72	0.70	8.4	1.74	781
0.22	2.6	0.19	85	0.72	8.6	1.86	835
0.24	2.9	0.22	99	0.74	8.9	1.97	884
0.26	3.1	0.24	108	0.76	9.1	2.09	938
0.28	3.4	0.28	126	0.78	9.4	2.22	996
0.30	3.6	0.31	139	0.80	9.6	2.35	1055
0.32	3.8	0.35	157	0.82	9.8	2.48	1113
0.34	4.1	0.39	175	0.84	10.1	2.62	1176
0.36	4.3	0.44	197	0.86	10.3	2.77	1243
0.38	4.6	0.49	220	0.88	10.6	2.92	1311
0.40	4.8	0.54	242	0.90	10.8	3.07	1378
0.42	5.0	0.59	265	0.92	11.0	3.23	1450
0.44	5.3	0.65	292	0.94	11.3	3.39	1522
0.46	5.5	0.71	319	0.96	11.5	3.56	1598
0.48	5.8	0.78	350	0.98	11.8	3.73	1674
0.50	6.0	0.84	377	1.00	12.0	3.91	1755
0.52	6.2	0.92	413	1.02	12.2	4.09	1836
0.54	6.5	0.99	444	1.04	12.5	4.28	1921
0.56	6.7	1.07	480	1.06	12.7	4.47	2006
0.58	7.0	1.16	521	1.08	13.0	4.67	2096
0.60	7.2	1.24	557	1.10	13.2	4.88	2190
0.62	7.4	1.34	601	1.12	13.4	5.08	2280
0.64	7.7	1.43	642	1.14	13.7	5.30	2379
0.66	7.9	1.53	687	1.16	13.9	5.52	2478
0.68	8.2	1.63	732	1.18	14.2	5.74	2576

Entrance head (H) is depth of water in beginning section of flume, measured vertically.
Depth of water leaving flume should be no more than 0.75-0.8 times entrance head.

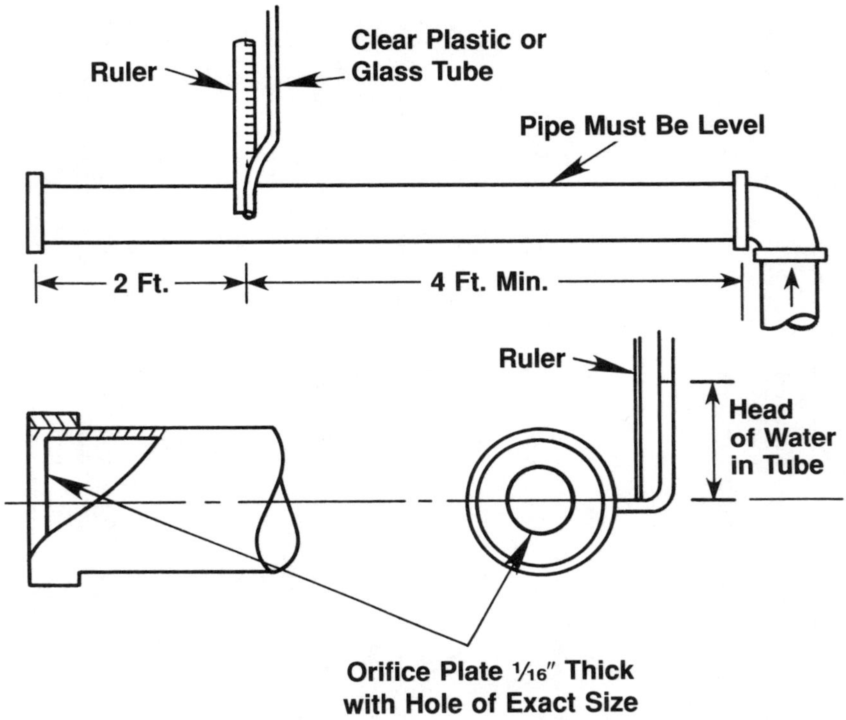

Pipe Orifice Method

Orifice Tables (Flow in U.S. Gallons per Minute)

Diameters Pipe x Orifice (Inches)	4x2	4x2.5	4x3	6x3	6x4	6x5	8x4	8x5	8x6	10x5	10x6	10x7	10x8	12x6	12x7	12x8	12x9	12x10
Head (Inches)																		
6	37	59	98	81	156	301												
7	40	64	106	87	168	325	157	262	413									
8	42	68	113	93	180	347	168	280	441	255	396	571	849					
9	45	73	120	99	191	368	178	297	468	270	420	606	900	390	555	759	1050	1470
10	47	77	126	104	201	388	188	313	493	285	443	639	949	411	585	800	1107	1550
11	50	80	133	109	211	407	197	328	517	298	464	670	995	431	614	839	1161	1625
12	52	84	139	114	220	425	205	343	540	312	485	700	1039	450	641	876	1212	1697
13	54	87	144	119	229	443	214	357	562	324	505	728	1082	469	667	912	1262	1767
14	56	91	150	123	238	459	222	370	584	337	524	756	1122	486	692	947	1310	1833
15	58	94	155	128	246	475	230	383	604	349	542	782	1162	503	717	980	1356	1898
16	60	97	160	132	254	491	237	396	624	360	560	808	1200	520	740	1012	1400	1960
17	62	100	165	136	262	506	245	408	643	371	577	833	1237	536	763	1043	1443	2020
18	64	103	170	140	269	521	252	420	662	382	594	857	1273	552	785	1073	1485	2079
19	65	105	174	144	277	535	258	432	680	392	610	880	1308	567	806	1103	1526	2136
20	67	108	179	148	284	549	265	443	698	402	626	903	1342	581	827	1131	1565	2191
21	69	111	183	151	291	563	272	454	715	412	642	926	1375	596	848	1159	1604	2245
22	70	114	188	155	298	576	278	464	732	422	657	947	1407	610	868	1187	1642	2298
23	72	116	192	158	305	589	284	475	748	432	671	969	1439	623	887	1213	1679	2350
24	73	119	196	162	311	601	291	485	764	441	686	990	1470	637	906	1239	1715	2400
25	75	121	200	165	318	614	297	495	780	450	700	1010	1500	650	925	1265	1750	2450

Orifice Tables (Flow in U.S. Gallons per Minute) [Cont'd]

Diameters Pipe x Orifice (Inches)	4x2	4x2.5	4x3	6x3	6x4	6x5	8x4	8x5	8x6	10x5	10x6	10x7	10x8	12x6	12x7	12x8	12x9	12x10
Head (Inches)																		
26	76	123	204	168	324	626	302	505	795	459	714	1030	1530	663	943	1290	1785	2499
27	78	126	208	171	330	638	308	514	811	468	727	1050	1559	675	961	1315	1819	2546
28	79	128	212	175	336	650	314	524	825	476	741	1069	1587	688	979	1339	1852	2593
29	81	130	215	178	342	661	319	533	840	485	754	1088	1616	700	996	1362	1885	2639
30	82	133	219	181	348	672	325	542	854	493	767	1106	1643	712	1013	1386	1917	2684
31	84	135	223	184	354	683	330	551	869	501	779	1125	1670	724	1030	1409	1949	2728
32	85	137	226	187	359	694	335	560	882	509	792	1143	1697	735	1047	1431	1980	2772
33	86	139	230	190	365	705	341	569	896	517	804	1160	1723	747	1063	1453	2011	2815
34	87	141	233	192	370	716	346	577	910	525	816	1178	1749	758	1079	1475	2041	2857
35	89	143	237	195	376	726	351	586	923	532	828	1195	1775	769	1094	1497	2071	2899
36	90	145	240	198	381	737	356	594	936	540	840	1212	1800	780	1110	1518	2100	2940
37	91	147	243	201	386	747	361	602	949	547	852	1229	1825	791	1125	1539	2129	2981
38	92	149	247	203	391	757	366	610	962	555	863	1245	1849	801	1140	1560	2158	3021
39	94	151	250	206	397	767	370	618	974	562	874	1261	1873	812	1155	1580	2186	3060
40	95	153	253	209	402	776	375	626	987	569	885	1278	1897	822	1170	1600	2214	3099
41	96	155	256	211	407	786	380	634	999	576	896	1293	1921	832	1185	1620	2241	3138
42	97	157	259	214	412	796	384	642	1011	583	907	1309	1944	842	1199	1640	2268	3176
43	98	159	262	216	416	805	389	649	1023	590	918	1325	1967	852	1213	1659	2295	3213
44	99	161	265	219	421	814	393	657	1035	597	929	1340	1990	862	1227	1678	2322	3250
45	101	162	268	221	426	823	398	664	1046	604	939	1355	2012	872	1241	1697	2348	3287

Orifice Tables (Flow in U.S. Gallons per Minute) [Cont'd]

Diameters Pipe x Orifice (Inches)	4x2	4x2.5	4x3	6x3	6x4	6x5	8x4	8x5	8x6	10x5	10x6	10x7	10x8	12x6	12x7	12x8	12x9	12x10
Head (Inches)																		
46	102	164	271	224	431	833	402	671	1058	610	950	1370	2035	882	1255	1716	2374	3323
47	103	166	274	226	435	842	407	679	1069	617	960	1385	2057	891	1268	1734	2399	3359
48	104	168	277	229	440	850	411	686	1081	624	970	1399	2078	901	1282	1753	2425	3395
49	105	169	280	231	445	859	415	693	1092	630	980	1414	2100	910	1295	1771	2450	3430
50	106	171	283	233	449	868	419	700	1103	636	990	1428	2121	919	1308	1789	2475	3465
51	107	173	286	236	453	877	423	707	1114	643	1000	1443	2142	928	1321	1807	2499	3499
52	108	175	288	238	458	885	428	714	1125	649	1010	1457	2163	937	1334	1824	2524	3533
53	109	176	291	240	462	894	432	721	1136	655	1019	1471	2184	946	1347	1842	2548	3567
54	110	178	294	242	467	902	436	727	1146	661	1029	1484	2205	955	1359	1859	2572	3601
55	111	179	297	245	471	910	440	734	1157	667	1038	1498	2225	964	1372	1876	2596	3634
56	112	181	299	247	475	919	444	741	1167	673	1048	1512	2245	973	1384	1893	2619	3667
57	113	183	302	249	479	927	448	747	1178	679	1057	1525	2265	981	1397	1910	2642	3699
58	114	184	305	251	484	935	452	754	1188	685	1066	1538	2285	990	1409	1927	2666	3732
59	115	186	307	253	488	943	455	760	1198	691	1075	1552	2304	999	1421	1943	2688	3764
60	116	187	310	256	492	951	459	767	1208	697	1084	1565	2324	1007	1433	1960	2711	3796
61	117	189	312	258	496	959	463	773	1218	703	1093	1578	2343	1015	1445	1976	2734	3827
62	118	191	315	260	500	967	467	780	1228	709	1102	1591	2362	1024	1457	1992	2756	3858
63	119	192	317	262	504	974	471	786	1238	714	1111	1603	2381	1032	1468	2008	2778	3889
64	120	194	320	264	508	982	474	792	1248	720	1120	1616	2400	1040	1480	2024	2800	3920
65	121	195	322	266	512	990	478	798	1258	726	1129	1629	2419	1048	1492	2040	2822	3951

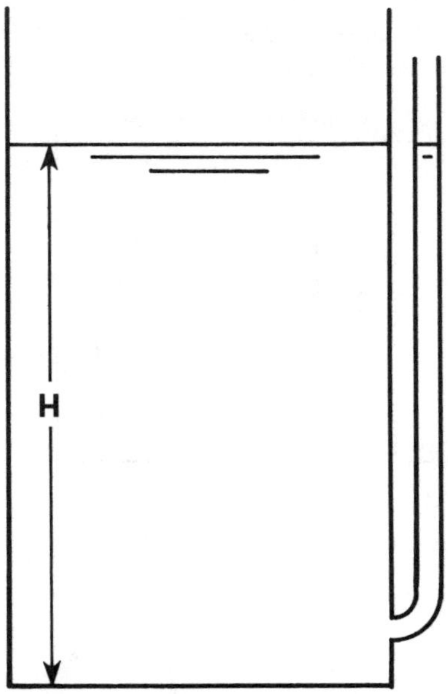

Side

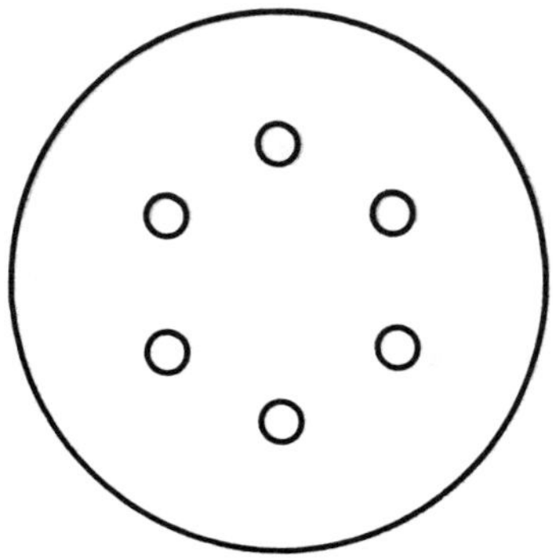

Bottom

Orifice Bucket

Orifice Bucket Flow Table

(Flow in U.S. Gallons per Minute from One 1-inch Diameter Orifice)

Height (In.)	Flow (GPM)	Height (In.)	Flow (GPM)	Height (In.)	Flow (GPM)
1	—	9	10.9	17	14.9
2	5.0	10	11.5	18	15.3
3	6.3	11	12.0	19	15.7
4	7.3	12	12.6	20	16.1
5	8.2	13	13.1	21	16.5
6	9.0	14	13.5	22	16.9
7	9.6	15	14.0	23	17.3
8	10.3	16	14.5	24	17.6

Measuring Flow Rate by Filling Containers

The flow rate from a pump or pipeline can be determined by measuring the time required to fill a container of known volume; or the time required to raise the water level a given distance. A stopwatch should be used when the time interval is short.

Small Containers

For small containers such as pails, buckets, or oil drums, the flow rate is:

$$GPM = (\text{Volume in gallons} \times 60) \div \text{Time (seconds) to fill}$$

For filling the typical 55-gallon oil drum, the following table can be used:

Time (sec)	Gals/Min	Liters/Sec	Time (sec)	Gals/Min	Liters/Sec
10	330	20.9	50	66	4.18
12	275	17.4	52	63	4.02
14	236	14.9	54	61	3.87
16	206	13.1	56	59	3.73
18	183	11.6	58	57	3.60
20	165	10.4	60	55	3.48
22	150	9.49	62	53	3.37
24	138	8.70	64	52	3.26
26	127	8.03	66	50	3.16
28	118	7.46	68	49	3.07
30	110	6.96	70	47	2.98
32	103	6.53	72	46	2.90
34	97	6.14	74	45	2.82
36	92	5.80	76	43	2.75
38	87	5.50	78	42	2.68
40	83	5.22	80	41	2.61
42	79	4.97	82	40	2.55
44	75	4.75	84	39	2.49
46	72	4.54	86	38	2.43
48	69	4.35	88	38	2.37

Large Rectangular Reservoirs

For large rectangular reservoirs or tanks, the flow rate is:

$$GPM = (\text{Width in ft x Length in ft} \times 7.48) \div \text{Minutes to raise one ft}$$

$$\text{or } GPM = (\text{Width in ft} \times \text{Length in ft} \times \text{Rise in inches} \times 0.62) \div \text{Minutes}$$

Large Circular Reservoirs

For large circular reservoirs or tanks, the flow rate is:

$$GPM = (\text{Diameter in ft} \times \text{Diameter in ft} \times 5.87) \div \text{Minutes to raise one ft}$$

$$\text{or } GPM = (\text{Diam. in ft} \times \text{Diam. in ft} \times \text{Rise in inches} \times 0.489) \div \text{Minutes}$$

"55-Gallon Drum" Calibration

Fluid Depth		Contents				Fluid Depth		Contents			
		Horizontal		Vertical				Horizontal		Vertical	
(In.)	(Cm.)	(US Gal)	(Liters)	(US Gal)	(Liters)	(In.)	(Cm.)	(US Gal)	(Liters)	(US Gal)	(Liters)
1	2.5	0.8	3.0	1.8	6.7	16	40.6	41.0	155.2	28.4	107.5
2	5.1	2.3	8.7	3.5	13.4	17	43.2	43.6	165.0	30.2	114.2
3	7.6	4.1	15.5	5.3	20.1	18	45.7	46.3	175.3	31.9	120.9
4	10.2	6.3	23.8	7.1	26.9	19	48.3	48.7	184.3	33.7	127.6
5	12.7	8.7	32.9	8.9	33.6	20	50.8	50.9	192.7	35.5	134.3
6	15.2	11.4	43.2	10.6	40.3	21	53.3	52.7	199.5	37.3	141.0
7	17.8	14.0	53.0	12.4	47.0	22	55.9	54.2	205.2	39.0	147.8
8	20.3	16.8	63.6	14.2	53.7	23	58.4	55.0	208.2	40.8	154.5
9	22.9	19.7	74.6	16.0	60.4	24	61.0			42.6	161.2
10	25.4	22.8	86.3	17.7	67.2	25	63.5			44.4	167.9
11	27.9	25.6	96.9	19.5	73.9	26	66.0			46.1	174.6
12	30.5	29.4	111.3	21.3	80.6	27	68.6			47.9	181.3
13	33.0	32.2	121.9	23.1	87.3	28	71.1			49.7	188.0
14	35.6	35.3	133.6	24.8	94.0	29	73.7			51.5	194.8
15	38.1	38.2	144.6	26.6	100.7	30	76.2			53.2	201.5

Estimating Flow from Open Pipes
(Coordinate Method)

A reasonable estimate of the flow from a horizontal or sloping pipe, *with open end and flowing full,* can be made by measuring the vertical drop of the stream of water and the distance the stream travels *parallel to the pipe* to provide that drop (see sketches).

Measure the distance, in inches, that the stream travels *parallel to the pipe* (X), and the distance, in inches, that the stream falls *vertically* (Y).

(For Full Pipes)

The discharge from a pipe flowing full is determined by the formula:

$$Q = \frac{3.61\,A\,X}{\sqrt{Y}}$$

where: Q = flow in U.S. gallons per minute
A = inside area of pipe (sq.in.)
X and Y are dimensions in inches (described above)

If the distance, in inches, that the stream travels parallel to the pipe (X) is measured for a 12-inch vertical drop, the flow can be quickly calculated by multiplying that distance (X) by the factor from the following table for standard steel pipes:

Nominal Size (In.)	Pipe I.D. (Inches)	Factor
2	2.067	3.5
4	4.026	13.3
6	6.065	30.1
8	7.981	52.1
10	10.020	82.2
12	12.000	117.9
14	13.250	143.7
16	15.250	190.3
18	17.250	243.5
20	19.250	303.3

(For Partially Filled Pipes)

If the pipe is not flowing full, measure the depth of water in the pipe and divide by the inside diameter of the pipe to obtain the depth ratio. Calculate the discharge for a full pipe (described above) and multiply by the factor for the depth ratio, from the table below, to obtain the approximate discharge from the partially filled pipe:

Depth Ratio	Factor	Depth Ratio	Factor
0.95	0.981	0.45	0.436
0.90	0.948	0.40	0.375
0.85	0.905	0.35	0.312
0.80	0.858	0.30	0.253
0.75	0.805	0.25	0.195
0.70	0.747	0.20	0.142
0.65	0.688	0.15	0.095
0.60	0.627	0.10	0.052
0.55	0.564	0.05	0.019
0.50	0.500		

Flow from partially filled pipes can also be calculated by the "California Pipe Method" described in this chapter.

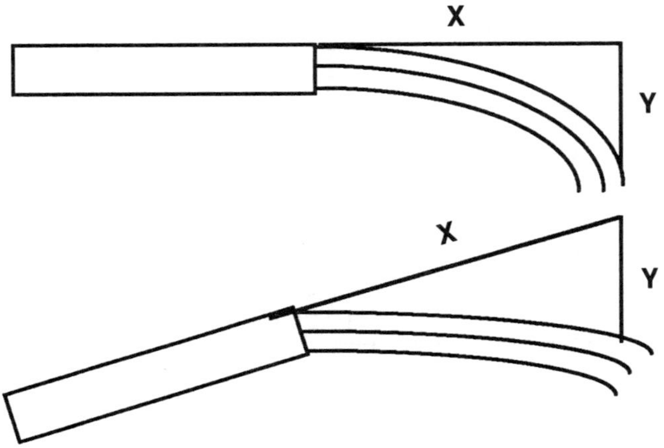

Open Pipe Flow

Vertical Flow from Pipes

The rate of flow from vertical pipes, such as artesian flow from a well casing, can be determined by measuring the height to which the jet or fountain rises above the top of the pipe. Extensive laboratory experiments at Cornell University in the early 1900's showed that, for low flow rates, the flow is similar to that over a sharp-crested circular weir and, at high flow rates, it acts as a jet. Between low and high flows there is a transitional condition.

The accompanying table gives the flow (in cubic feet per second; one cfs = 448.8 gpm) for various heads (in feet) for standard steel pipes or casings of 2 to 18 inches diameter. The head is measured as the vertical distance between the top of the pipe or casing and the top of the water column. In the table, the two solid lines indicate these three types of flow: weir flow is above the upper line, jet flow below the lower line, and transitional flow between the lines.

Many previously published tables for vertical flow from pipes have been based on a theoretical jet formula which results in appreciable error, particularly at low and intermediate heads (weir and transitional flow conditions).

Vertical Flow from Standard Steel Pipe
Cubic Feet Per Second

Head (feet)	2	4	6	8	10	12	14	16	18
Pipe I.D. (in)	2.067	4.026	6.065	7.981	10.02	12	13.25	15.25	17.25
0.10	0.042	0.10	0.17	0.24	0.31	0.39	0.44	0.53	0.62
0.15	0.059	0.17	0.29	0.41	0.54	0.68	0.77	0.92	1.07
0.20	0.071	0.22	0.41	0.60	0.80	1.00	1.13	1.35	1.58
0.25	0.081	0.27	0.53	0.80	1.08	1.35	1.53	1.83	2.13
0.30	0.089	0.31	0.63	0.98	1.38	1.73	1.96	2.34	2.73
0.35	0.096	0.35	0.72	1.14	1.65	2.13	2.41	2.88	3.36
0.40	0.10	0.38	0.81	1.28	1.90	2.55	2.89	3.45	4.02
0.45	0.11	0.42	0.89	1.42	2.12	2.89	3.37	4.04	4.71
0.50	0.12	0.44	0.97	1.55	2.33	3.22	3.82	4.66	5.43
0.55	0.12	0.46	1.03	1.68	2.53	3.52	4.22	5.29	6.18
0.60	0.13	0.48	1.08	1.80	2.72	3.80	4.58	5.80	6.78
0.65	0.13	0.50	1.13	1.90	2.90	4.07	4.92	6.24	7.35
0.70	0.14	0.52	1.19	1.99	3.07	4.32	5.25	6.66	7.90
0.75	0.14	0.54	1.23	2.08	3.23	4.56	5.57	7.06	8.44
0.80	0.15	0.56	1.27	2.17	3.38	4.77	5.84	7.44	8.97
0.85	0.15	0.58	1.31	2.26	3.52	4.98	6.08	7.80	9.49
0.90	0.16	0.60	1.35	2.34	3.65	5.17	6.29	8.14	10.0
0.95	0.16	0.62	1.39	2.41	3.76	5.34	6.47	8.45	10.5
1.00	0.17	0.63	1.43	2.47	3.87	5.50	6.63	8.74	10.9
1.10	0.18	0.67	1.51	2.60	4.08	5.84	7.06	9.34	11.8
1.20	0.19	0.70	1.58	2.72	4.29	6.13	7.45	9.75	12.6
1.30	0.19	0.73	1.65	2.84	4.47	6.42	7.81	10.2	13.2
1.40	0.20	0.76	1.71	2.96	4.65	6.66	8.16	10.6	13.8
1.50	0.21	0.79	1.78	3.07	4.82	6.91	8.41	11.0	14.4
1.60	0.22	0.81	1.84	3.17	4.99	7.15	8.73	11.4	14.9
1.70	0.22	0.84	1.90	3.28	5.15	7.38	8.99	11.8	15.4
1.80	0.23	0.87	1.96	3.38	5.31	7.61	9.26	12.2	15.8
1.90	0.24	0.89	2.01	3.48	5.47	7.83	9.53	12.6	16.2
2.00	0.24	0.92	2.07	3.57	5.62	8.04	9.80	13.0	16.6
2.20	0.26	0.96	2.18	3.76	5.91	8.46	10.3	13.6	17.4
2.40	0.27	1.01	2.28	3.93	6.19	8.86	10.8	14.3	18.2
2.60	0.28	1.05	2.38	4.11	6.46	9.24	11.3	14.9	19.0
2.80	0.29	1.09	2.47	4.27	6.71	9.61	11.7	15.5	19.8
3.00	0.30	1.13	2.56	4.43	6.96	9.97	12.1	16.1	20.5

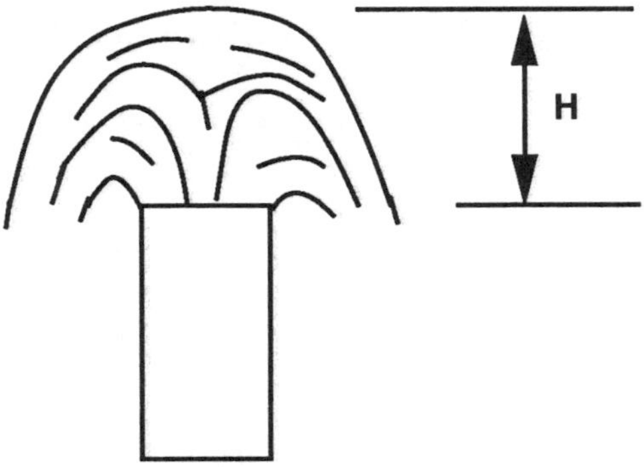

Vertical Flow

California Pipe Method for Flow Measurement

This method for calculating flow from the end of a partially-filled horizontal pipe, which discharges freely to the air, requires that the pipe be level, at least 6 or more diameters long, have free access of air to the unfilled part of the pipe, and the water should enter the pipe with a low approach velocity. Air vents, or an open standpipe, may be needed to allow free air circulation. The only measurements needed (both in feet) are: (1) the inside diameter of the pipe, and (2) the distance from the inside upper surface of the pipe down to the surface of the water flowing at the pipe outlet (see sketch).

The discharge from the pipe is computed from the formula:

$$Q = 8.69\,(1 - \frac{a}{d})^{1.88}\, d^{2.48}$$

where Q = cubic feet per second discharge
d = inside diameter of pipe, in feet
a = distance from inside top surface of pipe to water surface, in feet

Separate tables are given in this Chapter for values of:

$8.69\,(1 - \frac{a}{d})^{1.88}$ and 2.48 powers of numbers.

Values of $8.69\left(1-\frac{a}{d}\right)^{1.88}$ for California Pipe Method

a/d	0.00	0.01	0.02	0.03	0.04	0.05	0.06	0.07	0.08	0.09
0.0	8.690	8.527	8.366	8.206	8.048	7.891	7.736	7.582	7.429	7.278
0.1	7.128	6.980	6.834	6.688	6.545	6.402	6.261	6.122	5.984	5.848
0.2	5.713	5.579	5.447	5.316	5.187	5.060	4.934	4.809	4.686	4.564
0.3	4.444	4.326	4.209	4.093	3.979	3.866	3.755	3.646	3.538	3.431
0.4	3.326	3.223	3.121	3.020	2.922	2.824	2.728	2.634	2.542	2.450
0.5	2.361	2.273	2.187	2.102	2.018	1.937	1.857	1.778	1.701	1.626
0.6	1.552	1.480	1.409	1.340	1.273	1.207	1.143	1.081	1.020	0.9611
0.7	0.9037	0.8479	0.7937	0.7413	0.6905	0.6414	0.5940	0.5484	0.5044	0.4622
0.8	0.4217	0.3829	0.3459	0.3106	0.2772	0.2455	0.2156	0.1876	0.1614	0.1370
0.9	0.1146	0.0940	0.0753	0.0586	0.0438	0.0311	0.0205	0.0119	0.0056	0.0015

Numbers to the 2.48 Power for California Pipe Method

No.	0.00	0.01	0.02	0.03	0.04	0.05	0.06	0.07	0.08	0.09
0.0	—	—	—	0.0002	0.0003	0.0006	0.0009	0.0014	0.0019	0.0025
0.1	0.0033	0.0042	0.0052	0.0063	0.0076	0.0091	0.0106	0.0123	0.0142	0.0163
0.2	0.0185	0.0208	0.0234	0.0261	0.0290	0.0321	0.0354	0.0389	0.0426	0.0464
0.3	0.0505	0.0548	0.0593	0.0640	0.0689	0.0740	0.0794	0.0849	0.0908	0.0968
0.4	0.1031	0.1096	0.1163	0.1233	0.1305	0.1380	0.1458	0.1537	0.1620	0.1705
0.5	0.179	0.188	0.198	0.207	0.217	0.227	0.237	0.248	0.259	0.270
0.6	0.282	0.294	0.306	0.318	0.331	0.344	0.357	0.370	0.384	0.398
0.7	0.413	0.428	0.443	0.458	0.474	0.490	0.506	0.523	0.540	0.557
0.8	0.575	0.593	0.611	0.630	0.649	0.668	0.688	0.708	0.728	0.749
0.9	0.770	0.791	0.813	0.835	0.858	0.881	0.904	0.927	0.951	0.975
1.0	1.000	1.025	1.050	1.076	1.102	1.129	1.155	1.183	1.210	1.238
1.1	1.267	1.295	1.325	1.354	1.384	1.414	1.445	1.476	1.508	1.539
1.2	1.572	1.604	1.637	1.671	1.705	1.739	1.774	1.809	1.845	1.880
1.3	1.917	1.954	1.991	2.028	2.066	2.105	2.144	2.183	2.223	2.263
1.4	2.304	2.345	2.386	2.428	2.470	2.513	2.556	2.600	2.644	2.688
1.5	2.73	2.78	2.82	2.87	2.92	2.96	3.01	3.06	3.11	3.16
1.6	3.21	3.26	3.31	3.36	3.41	3.46	3.51	3.57	3.62	3.67
1.7	3.73	3.78	3.84	3.89	3.95	4.01	4.06	4.12	4.18	4.24
1.8	4.30	4.36	4.42	4.48	4.54	4.60	4.66	4.72	4.79	4.85
1.9	4.91	4.98	5.04	5.11	5.17	5.24	5.31	5.37	5.44	5.51

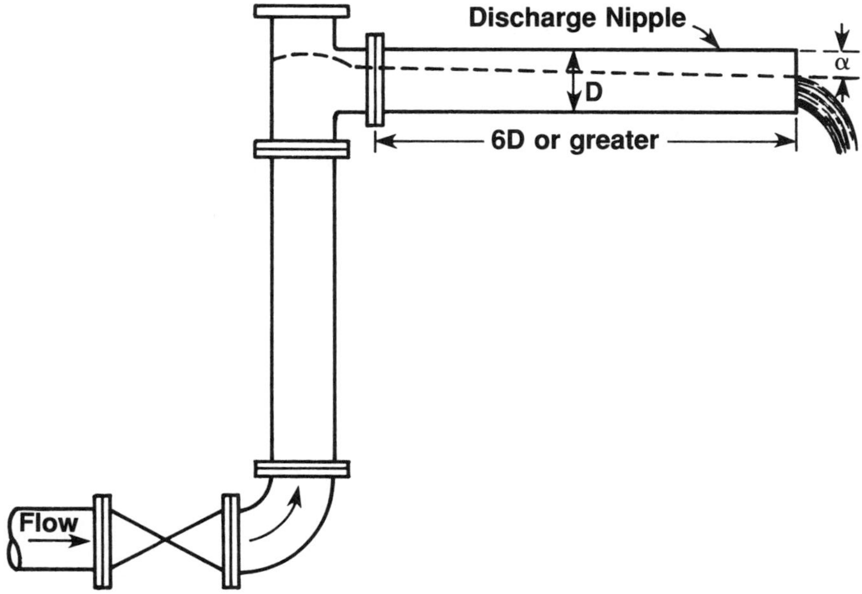

California Pipe Method

Chapter 7

Electrical Data

Electrical Formulas

To Find	Single Phase AC	Three Phase AC
Amperes when horsepower is known	$= \dfrac{hp \times 746}{E \times Eff \times pf}$	$= \dfrac{hp \times 746}{1.73 \times E \times Eff \times pf}$
Amperes when kilowatts is known	$= \dfrac{kw \times 1000}{E \times pf}$	$= \dfrac{kw \times 1000}{1.73 \times E \times pf}$
Amperes when kva is known	$= \dfrac{kva \times 1000}{E}$	$= \dfrac{kva \times 1000}{1.73 \times E}$
Kilowatts	$= \dfrac{I \times E \times pf}{1000}$	$= \dfrac{I \times E \times pf \times 1.73}{1000}$
Kva	$= \dfrac{I \times E}{1000}$	$= \dfrac{I \times E \times 1.73}{1000}$
Horsepower (output)	$= \dfrac{I \times E \times Eff \times pf}{746}$	$= \dfrac{I \times E \times 1.73 \times Eff \times pf}{746}$

Where:
- I = Amperes
- E = Volts
- Eff = Efficiency
- hp = Horsepower
- pf = Power factor
- kw = Kilowatts
- kva = Kilovolt-amperes

Useful Approximations

At 3,600 rpm, a motor develops 1.5 lb-ft per horsepower
At 1,800 rpm, a motor develops 3 lb-ft per horsepower
At 1,200 rpm, a motor develops 4.5 lb-ft per horsepower

At 550 and 575 volts, a 3-phase motor draws 1 ampere per horsepower
At 440 and 460 volts, a 3-phase motor draws 1.25 amperes per horsepower
At 220 and 230 volts, a 3-phase motor draws 2.5 amperes per horsepower

Power From Watt-Hour Meters

The electrical power input to a motor can be determined from revolving-disc watt-hour meters by timing the revolutions. The speed of the disc is directly proportional to the power being used, and the power input to the motor can be calculated from the following formulas:

$$\text{Kilowatts} = 3.6 \times K \times M \times \frac{R}{t}$$

$$\text{or, Horsepower} = 4.826 \times K \times M \times \frac{R}{t}$$

where: K (or k_h) is the disc constant (marked on the meter nameplate or the revolving disc)
M, if current or potential transformers are used, is the product of the transformer ratios.
When either is not used, the equivalent ratio is one
R is the number of revolutions of the disc
t is the time (in seconds) for that number of revolutions

The motor *output*, in Brake Horsepower, is:

$$\text{B.H.P.} = \frac{\text{Horsepower input} \times \text{Motor efficiency \%}}{100}$$

Full-Load Current of Motors (Amperes)

Single-Phase, Alternating Current

HP	115 v	200 v	208 v	230 v
1/6	4.4	2.5	2.4	2.2
1/4	5.8	3.3	3.2	2.9
1/3	7.2	4.1	4.0	3.6
1/2	9.8	5.6	5.4	4.9
3/4	13.8	7.9	7.6	6.9
1	16	9.2	8.8	8.0
1 1/2	20	12	11	10
2	24	14	13	12
3	34	20	19	17
5	56	32	31	28
7 1/2	80	46	44	40
10	100	58	55	50

Three-Phase, Alternating Current, Induction Type

HP	115 v	200 v	208 v	230 v	460 v	575 v	2300 v
1/2	4.0	2.3	2.2	2.0	1.0	8.0	
3/4	5.6	3.2	3.1	2.8	1.4	1.1	
1	7.2	4.1	4.0	3.6	1.8	1.4	
1 1/2	10.4	6.0	5.7	5.2	2.6	2.1	
2	13.6	7.8	7.5	6.8	3.4	2.7	
3		11	11	9.6	4.8	3.9	
5		18	17	15	7.6	6.1	
7 1/2		25	24	22	11	9	
10		32	31	28	14	11	
15		48	46	42	21	17	
20		62	59	54	27	22	
25		78	75	68	34	27	
30		92	88	80	40	32	
40		120	114	104	52	41	
50		150	143	130	65	52	
60		177	169	154	77	62	16
75		221	211	192	96	77	20
100		285	273	248	124	99	26
125		359	343	312	156	125	31
150		414	396	360	180	144	37
200		552	528	480	240	192	49

Typical Efficiency of Electric Motors (Full Load)

Motor	Efficiency (%)				Motor	Efficiency (%)			
HP	860 rpm	1160 rpm	1760 rpm	3600 rpm	HP	860 rpm	1160 rpm	1760 rpm	3600 rpm
7½	85	87	86		50	91	91	91	90
10	87	86.5	87.5	85.5	60	91.5	91.5	91	90.5
15	87	88	88.5	87.5	75	90	92	91.5	91
20	89	89	89.5	87.5	100	90.5	92	92	91.5
25	89	89.5	89.5	89	125	90.5	92	92.5	91.5
30	90.5	90	90	89.5	150	91.5	92	92	92.5
40	91	90	90.5	90	200	91.5	92	92.5	92.5

Efficiency of submersible motors typically about 5 percent lower than above values.

Submersible Cable Length Selection

(For Franklin Submersible Motors)

Single Phase (Two or Three Wire Cable) 60 Hz – Maximum Lengths (Feet) From Service Entrance to Motor														
Motor Rating		Copper Wire Size – AWG												
Volts	HP	14	12	10	8	6	4	3	2	1	0	00	000	0000
115	⅓	130	210	340	540	840	1300							
	½	100	160	250	390	620	960	1190	1460					
230	⅓	550	880	1390										
	½	400	650	1020	1610									
	¾	300	480	760	1200	1870								
	1	250	400	630	990	1540								
	1.5	190	310	480	770	1200	1870							
	2	150	250	390	620	970	1530							
	3		190	300	470	750	1190	1490	1850					
	5			180	280	450	710	890	1110	1390	1740			
	7.5				200	310	490	610	750	930	1140	1410	1720	
	10					250	390	490	600	750	930	1160	1430	1760
	15						270	340	430	530	660	820	1020	1260

Submersible Cable Length Selection (Cont'd)

Single Phase (Two or Three Wire Cable) 60 Hz – Maximum Lengths (Feet) From Service Entrance to Motor

Three Phase (Three Wire Cable) 60 Hz - Maximum Lengths (Feet) From Service Entrance to Motor

Motor Rating		Copper Wire Size – AWG												
Volts	HP	14	12	10	8	6	4	3	2	1	0	00	000	0000
230	1.5	420	670	1060	1670	2610								
	2	320	510	810	1280	2010								
	3	240	390	620	990	1540	2400							
	5		230	370	590	920	1430	1790	2190					
	7.5			260	420	650	1020	1270	1560	2340				
	10				310	490	760	950	1170	1440	1760	2160		
	15					330	520	650	800	980	1200	1470	1780	2150
	20						400	500	610	760	930	1140	1380	1680
	25						320	400	500	610	750	920	1120	1360
	30							330	410	510	620	760	930	1130
460	1.5	1700	2710											
	2	1300	2070											
	3	1000	1600	2520										
	5	590	950	2360										
	7.5	420	680	1070	1690	2640								
	10	310	500	790	1250	1960	3050							
	15			540	850	1340	2090	2600						
	20			410	650	1030	1610	2000	2470					
	25				530	830	1300	1620	1990	2450				
	30				430	680	1070	1330	1640	2030	2490			
	40						790	980	1210	1490	1830	2250		
	50						640	800	980	1210	1480	1810	2190	2650
	60								830	1020	1250	1540	1850	2240
	75									840	1030	1260	1520	1850
	100											940	1130	1380

Maximum lengths shown maintain 95% of service entrance voltage at motor.
Cable length from service entrance to 3-phase motor control should not exceed 25% of maximum allowable length.
Minimum size of cable is based on National Electric Code Table 310-16 for 75°C cable.

Wire Gauges

Gauge No.	U.S. Standard	Birmingham or Stubs	W. & M. and Roebling	British Std or Imperial	Brown & Sharpe or American		
	Diameter of Wire (Inches)				Diam. (inches)	Area (Sq. In.)	Area (Cir. Mils)
0000	0.406	0.454	0.393	0.400	0.46000	0.166	212,000
000	0.375	0.425	0.362	0.372	0.40964	0.132	168,000
00	0.344	0.380	0.331	0.348	0.36480	0.105	133,000
0	0.313	0.340	0.307	0.324	0.32486	0.0829	106,000
1	0.281	0.300	0.283	0.300	0.28930	0.0657	83,700
2	0.266	0.284	0.263	0.276	0.25763	0.0521	66,400
3	0.250	0.259	0.244	0.252	0.22942	0.0413	52,600
4	0.234	0.238	0.225	0.232	0.20431	0.0328	41,700
5	0.219	0.220	0.207	0.212	0.18194	0.0260	33,100
6	0.203	0.203	0.192	0.192	0.16202	0.0206	26,300
7	0.188	0.180	0.177	0.176	0.14428	0.0164	20,800
8	0.172	0.165	0.162	0.160	0.12849	0.0130	16,500
9	0.156	0.148	0.148	0.144	0.11443	0.0103	13,100
10	0.141	0.134	0.135	0.128	0.10189	0.00815	10,400
11	0.125	0.120	0.120	0.116	0.09074	0.00647	8,230
12	0.109	0.109	0.105	0.104	0.08081	0.00513	6,530
13	0.0938	0.095	0.092	0.092	0.07196	0.00407	5,180
14	0.0781	0.083	0.080	0.080	0.06408	0.00323	4,110
15	0.0703	0.072	0.072	0.072	0.05706	0.00256	3,260
16	0.0625	0.065	0.063	0.064	0.05082	0.00203	2,580
17	0.0563	0.058	0.054	0.056	0.04525	0.00161	2,050
18	0.0500	0.049	0.047	0.048	0.04030	0.00128	1,620
19	0.0438	0.042	0.041	0.040	0.03589	0.00101	1,290
20	0.0375	0.035	0.035	0.036	0.03196	0.000802	1,020
21	0.0344	0.032	0.032	0.032	0.02846	0.000636	810
22	0.0313	0.028	0.028	0.028	0.02535	0.000505	642
23	0.0281	0.025	0.025	0.024	0.02257	0.000400	509
24	0.0250	0.022	0.023	0.022	0.02010	0.000317	404
25	0.0219	0.020	0.020	0.020	0.01790	0.000252	320
26	0.0188	0.018	0.018	0.018	0.01594	0.000200	254
27	0.0172	0.016	0.017	0.0164	0.01420	0.000158	202
27	0.0156	0.014	0.016	0.0148	0.01264	0.000126	160
29	0.0141	0.013	0.015	0.0136	0.01126	0.0000995	127
30	0.0125	0.012	0.014	0.0124	0.01003	0.0000789	101
31	0.0109	0.010	0.013	0.0116	0.00893	0.0000626	79.7
32	0.0102	0.009	0.012	0.0108	0.00795	0.0000496	63.2
33	0.0094	0.008	0.011	0.0100	0.00708	0.0000394	50.1
34	0.0086	0.007	0.010	0.0092	0.00630	0.0000312	39.8
35	0.0078	0.005	0.0095	0.0084	0.00561	0.0000248	31.5
36	0.0070	0.004	0.009	0.0076	0.00500	0.0000196	25.0
37	0.0066		0.0085	0.0068	0.00445	0.0000156	19.8
38	0.0063		0.008	0.0060	0.00397	0.0000123	15.7
39			0.0075	0.0052	0.00353	0.0000098	12.5
40			0.007	0.0048	0.00314	0.0000078	9.9

Properties of Copper Wire

Wire Size (AWG or MCM)	Diameter (Inches)	Diameter (Mm)	Resistance @77F (Ohms/1000 Ft)	Weight (Lbs/1000 Ft)	Feet Length (Per Pound)
1000 MCM	1.152	29.3	0.0108	3,088	0.324
800	1.031	26.2	0.0135	2,470	0.405
700	0.964	24.5	0.0154	2,161	0.463
600	0.893	22.7	0.0180	1,853	0.540
500	0.813	20.7	0.0216	1,544	0.648
400	0.728	18.5	0.0270	1,235	0.810
250	0.575	14.6	0.0431	772	1.30
0000 AWG	0.4600	11.7	0.0509	653.3	1.53
000	0.4096	10.4	0.0642	518.1	1.93
00	0.3648	9.27	0.0811	410.9	2.43
0	0.3248	8.25	0.102	325.8	3.07
1	0.2893	7.35	0.129	258.9	3.87
2	0.2576	6.54	0.162	204.9	4.88
3	0.2294	5.83	0.205	162.5	6.15
4	0.2043	5.19	0.259	128.9	7.76
6	0.1620	4.11	0.410	81.05	12.34
8	0.1284	3.26	0.641	49.98	20.01
10	0.1018	2.59	1.018	31.43	31.82
12	0.0808	2.05	1.619	19.77	50.59
14	0.0640	1.63	2.575	12.43	80.44
16	0.0508	1.29	4.094	7.82	127.9
18	0.0403	1.02	6.510	4.92	203.4
20	0.0319	0.810	10.35	3.09	323.4
22	0.0254	0.645	16.46	1.95	514.2
24	0.0201	0.511	26.17	1.22	817.7
26	0.0159	0.404	41.62	0.77	1,300
28	0.0126	0.320	66.17	0.48	2,067
30	0.0100	0.254	105.2	0.30	3,287
32	0.0080	0.203	167.3	0.19	5,227
34	0.0063	0.160	266.0	0.12	8,310
36	0.0050	0.127	423.0	0.076	13,210
38	0.0040	0.102	672.6	0.047	21,010
40	0.0031	0.079	1,069	0.030	33,410
42	0.0025	0.064	1,701	0.019	52,800
44	0.0020	0.051	2,703	0.012	82,500
46	0.0016	0.041	4,299	0.008	128,800
48	0.0012	0.030	6,836	0.004	229,600
50	0.0010	0.025	10,870	0.003	330,000

Allowable Capacity (Amperes) of Insulated Copper Conductors

Wire Size (AWG)	Single Insulated Conductors In Free Air		Not More Than Three Conductors In Raceway Or Cable	
	Group A*	Group B*	Group A*	Group B*
14	25	30	20	20
12	30	35	25	25
10	40	50	30	35
8	60	70	40	50
6	80	95	55	65
4	105	125	70	85
3	120	145	85	100
2	140	170	95	115
1	165	195	110	130
0	195	230	125	150
00	225	265	145	175
000	260	310	165	200
0000	300	360	195	230

*Group A: Moisture-resistant thermoplastic
*Group B: Heat-resistant, and moisture- and heat-resistant rubber
Heat-resistant, and moisture- and heat-resistant thermoplastic

Chapter 8

Well Design

Introduction to Chapter 8

Material included in this Chapter is intended to be of assistance with certain aspects of well design. A comprehensive discussion of the design of wells, and the preparation of specifications for drilling, is beyond the scope of this handbook. For more specific information, the reader should consult one or more standard reference publications, some examples of which are as follows:

American Water Works Association, (1984). AWWA Standard for water wells, A 100-84. Denver, Colo.

Bennison, E.W., (1947). Ground water, its development, uses, and conservation. Edward E. Johnson Co., St. Paul, Minn.

Campbell, M.D. and Lehr, J.H., (1973). Water well technology. McGraw-Hill, New York, N.Y.

Driscoll, F.G., (1986). Groundwater and wells. Johnson Division, St. Paul, Minn.

Gibson, U.P. and Singer, R.D., (1969). Small wells manual. U.S. Agency for International Development, Washington, D.C.

Gordon, R.W., (1958). Water well drilling with cable tools.Bucyrus-Erie Co., South Milwaukee, Wisconsin.

National Water Well Association, (1971). Water well drillers beginning training manual. Columbus, Ohio.

Speed Star Division of Koehring Co., (1967). Well drilling manual. Enid, Oklahoma.

U.S. Department of the Army, (1965). Well drilling. Technical Manual 5-297, Washington, D.C.

U.S. Department of Interior, Bureau of Reclamation (rev. 1981) Ground water manual. Denver, Colo.

Water Well Journal, published monthly by Ground Water Publishing Co., Dublin, Ohio

plus

Other trade journals; manufacturers literature and catalogs.

Well Screens

Well screens are available with many types and materials of construction. Types include continuous slot (wire-wound), wire-wound on perforated pipe base, slotted (both vertical and horizontal), punched louver or shutter openings, and pre-packed. Materials include metals and metal alloys, plastics (PVC, ABS, etc.), asbestos-cement, fiberglass reinforced plastics, wood, concrete, and vitreous clay tile.

In addition to consideration of cost, the type of screen construction chosen is largely dependent upon aquifer characteristics, whether an artificial filter pack is used, and the quantity of water required. Selection of screen materials is usually governed by consideration of potential corrosion or incrustation problems.

Slot-opening widths of continuous wire-wound screens are usually expressed in thousandths of an inch: a 60-slot screen has an opening width of 0.060 inches. Openings in slotted and punched screens are usually designated by width and length of slots (or louvers or shutters), and the number of slots or openings per foot length of screen.

There is rather general agreement that inlet velocities of well screens need not be less than 0.1 ft/sec and should not exceed 0.3–0.5 ft/sec. The inlet velocity (V) is governed by the open area (A) and the inflow rate (Q) using the formula $V = Q \div A$ (in consistent units such as ft/sec, cubic ft/sec, and square ft). Low entrance velocities are desirable for such reasons as minimizing well friction losses, reducing the potential of pumping fine sand, and minimizing potential corrosion and incrustation difficulties.

For a given pumping rate and length of screen, entrance velocity is reduced as the percentage of open area is increased. Typical range in percentages of open area for various types of screen construction (with greater slot widths providing larger percentages) are:

Continuous wire-wound	4 to 60 percent
Louvered	4 to 35
Punched slots	4 to 20
Sawed slots	4 to 12

The accompanying table gives *typical* open areas (in square feet per foot of screen) for various diameters and slot widths of the widely used pipe-size continuous wire-wound screens. Individual manufacturers should be contacted for precise information.

Knowing the open area of a screen, the transmitting capacity per unit length for different entrance velocities can be calculated using the following multipliers:

Transmitting Capacity	(Multiply Open Area in Sq.Ft./Ft by:) 0.10 fps	0.20 fps	0.30 fps
US Gal/Min per foot of screen	45	90	135
Liters/Sec per meter of screen	9.2	18.5	27.7

Typical Open Area – Continuous Slot, Pipe-Size Well Screens
(Square Feet per Foot of Screen)

Pipe Size (Inches)	Surface Area (Sq.Ft./Ft.)	Approx. Wt. (Lbs./Ft.)	Slot Width Opening – Inches							
			0.010	0.020	0.030	0.040	0.060	0.080	0.100	0.125
4	1.211	7	0.174	0.306	0.403	0.500	0.625	0.708	0.778	0.799
6	1.734	14	0.174	0.313	0.431	0.535	0.694	0.819	0.910	0.938
8	2.258	19	0.229	0.410	0.521	0.688	0.903	1.042	1.111	1.174
10	2.814	26	0.285	0.507	0.625	0.833	0.938	1.125	1.292	1.424
12	3.338	35	0.257	0.479	0.535	0.681	0.938	1.125	1.292	1.486
14	3.665	38	0.285	0.521	0.590	0.743	1.014	1.243	1.431	1.618
16	4.189	43	0.306	0.486	0.667	0.854	1.167	1.424	1.653	1.861
18	4.712	54	0.319	0.528	0.750	0.951	1.299	1.583	1.826	1.875

Actual open area will differ somewhat, between different manufacturers and because of construction design;

Approximate weight is based on 0.020-inch slot size; weight will vary with slot width and wire shape

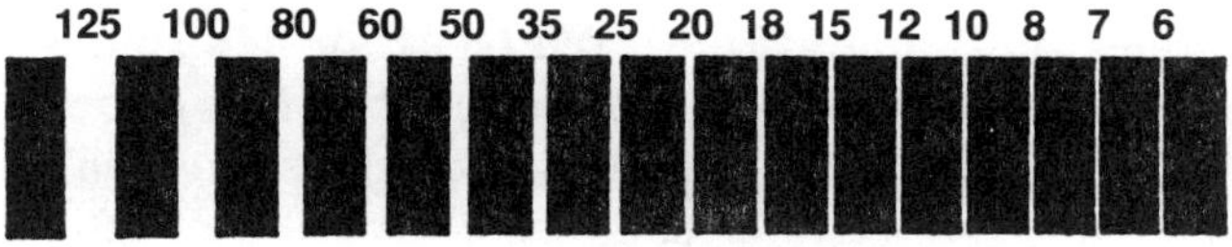

Screen Slot Widths

Filter Pack ("Gravel Pack") Design

Filter packs (often called "gravel packs" although the material is usually sand rather than gravel) are primarily used to eliminate, or minimize, sand pumping and to stabilize the aquifer; to allow use of a larger screen opening size and thus increase open area and reduce inlet velocities; and to provide an annulus of high permeability around the screen that increases the effective well radius.

The design of a filter pack and selection of screen slot width is still somewhat of an "art" rather than a "science" but there are certain principles and guidelines that can be used to advantage.

Definitions:

In this discussion, the following definitions are used:

Sieve Analysis or Mechanical Analysis (M.A.): the results of sieving a sample of the aquifer or filter material, with results expressed as percentage by weight of each size fraction.

D_{10}, D_{50}, D_{60}: the grain size for which 10, 50, or 60 percent, respectively, of the sample is smaller (percent passing).

Uniformity Coefficient (C_u): the ratio of D_{60} to D_{10}.

$$C_u = \frac{D_{60}}{D_{10}}$$

Effective Size: the D_{10} size (10 percent finer).

Uniform: material with a large percentage of the sample within a narrow range of particle size.

Graded: material with particle sizes varying uniformly over a wide range.

Skip Graded: material with both coarse- and fine-grained particles but very little of intermediate size.

Sieve Analyses:

The results of sieve analyses are the basis for determining the need for a filter pack and for the design of the filter. *It is important to keep in mind, however, that the results of sieve analyses are only as good as the samples collected.* Aquifer samples collected during drilling must be truly representative of the materials being penetrated, and the depth (or depth range) reported for each sample must be accurate.

The analysis is made by drying the sample and passing it through a stack of sieves ranging from large to small openings. The weight of sample retained on each sieve is measured, and the results plotted on a semi-logarithmic graph with a curve showing the accumulated percentage passing (or retained) on each sieve. This plot indicates the size distribution of particles, whether the material is uniform or graded, and allows calculation of the uniformity coefficient (C_u). The size of openings in standard sieves is given on an accompanying table.

Design:

The following paragraphs discuss natural development of the well, to produce a filter zone, followed by criteria for "artificial" filter packs where selected material is placed outside the screen before development begins.

"Natural" Filter Pack

If the aquifer materials are well graded, and reasonably coarse-grained, it may be possible to develop a "natural" filter pack by ordinary well development techniques such as surging or hydraulic jetting. For natural development, the following minimum criteria should exist:

$$D_{10} \geq 0.010 \text{ inches } (0.25 \text{ mm})$$

$$C_u \geq 2.0$$

Selection of screen slot size for natural development depends upon the gradation of the aquifer material, the experience and skill of the contractor, and of course the accuracy of the sampling. The percentage of the aquifer material retained by the well screen can typically vary from 40 to 70 percent (30 to 60 percent passing through the screen during development). Natural development is not advisable when there are alternating layers of fine- and coarse-grained materials.

Gradation

The results of sieve analyses of aquifer samples are the basis for selecting the proper gradation of the filter-pack materials. If aquifer samples from different depths show considerable variation in gradation, the filter design should be based to be stable against the finer-grade samples. If the aquifer material is skip-graded, the coarse-grained fraction should be ignored and the sieve analysis graph re-plotted using only the finer-grained portion of the sample.

Numerous investigators and agencies have experimented to develop formulas or criteria that will result in a stable filter-pack gradation. The following criteria have generally been found satisfactory in actual practice:

1. Aquifer material with C_u less than 2.5:
   - A. Use uniform filter material with C_u less than 2.5 and with the D_{50} of the filter 4 to 6 times the D_{50} of the aquifer
   - B. If uniform filter is not available, use a filter with C_u between 2.5 and 5 and with the D_{50} of the filter not more than 9 times the D_{50} of the aquifer
2. Aquifer material with C_u between 2.5 and 5:
   - A. Use uniform filter material with C_u less than 2.5 and with the D_{50} of the filter not more than 9 times the D_{50} of the aquifer
   - B. If uniform filter is not available, use a filter with C_u between 2.5 and 5 and with the D_{50} of the filter not more than 12 times the D_{50} of the aquifer
3. Aquifer material with C_u greater than 5:
   - A. Multiply the D_{30} of the aquifer by 6 and 9 and locate these points on the sieve analysis graph
   - B. Draw two parallel lines through these points having a C_u of 2.5 or less, and specify filter material that will fall between these lines

In all cases, the screen slot size should typically be selected that will not pass more than 10 percent of the filter material.

Filter Material

Filter materials should ideally be clean, rounded, siliceous sands or gravels. Carbonate material, shale particles, or soluble material such as gypsum should not exceed 5 percent of the total. Micaceous sands are not acceptable. Uniform sands that meet these conditions are commercially available and typical gradation specifications are given in the accompanying table. Filter material can be prepared locally from suitable pit-run deposits by screening and blending to obtain proper gradation.

Thickness

Theoretically a very thin filter pack would be successful in preventing fine aquifer materials from migrating into the well. In practice, however, a thicker filter is needed to ensure complete placement around the screen. A thickness of from 3 to 5 inches is normally adequate, depending upon conditions and method of placement. A maximum thickness should normally not exceed 8 or 9 inches; the thicker filter packs make proper well development more difficult.

Placement

There are many methods for placing the filter pack in the annulus between the screen and well bore. In shallow wells with a large annulus and a uniform filter material, it may be possible to simply shovel the filter in from the surface. Other methods may involve the use of tremie pipes or pumps. In any event, the method of placement should be such as to avoid segregation (with coarser material settling out) and to avoid "bridging" of the annulus or tremie pipe. Methods of placing filter packs are discussed more fully in the references on well-drilling techniques such as those listed in the Introduction to Chapter 2 of this handbook.

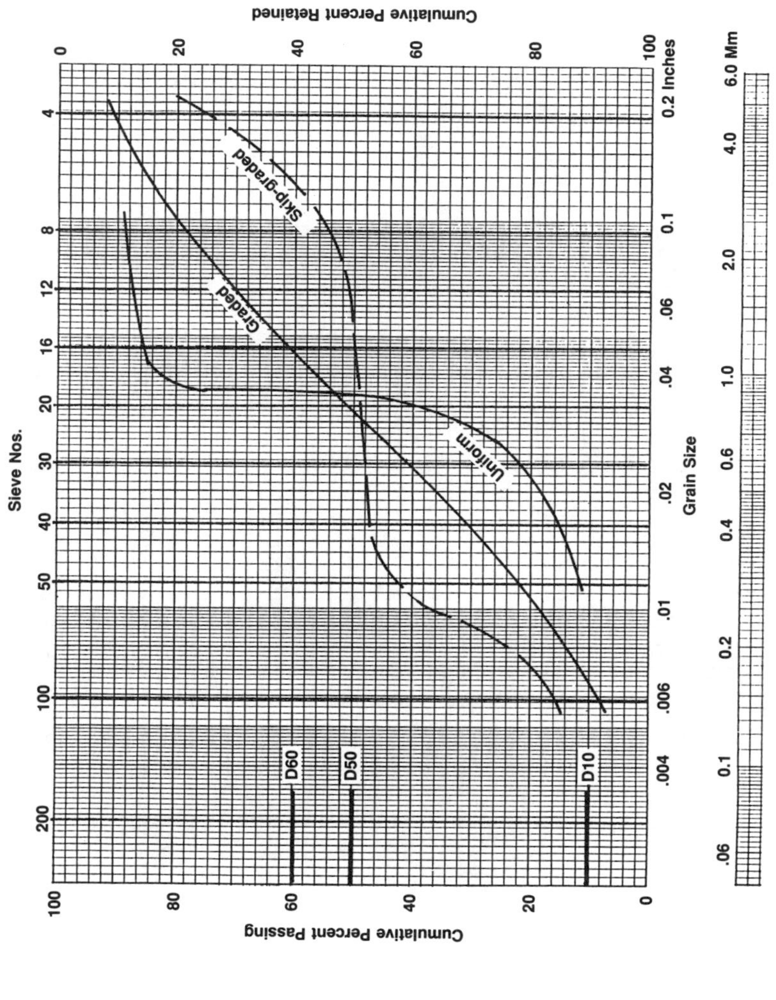

Graph for Plotting Sieve Analyses

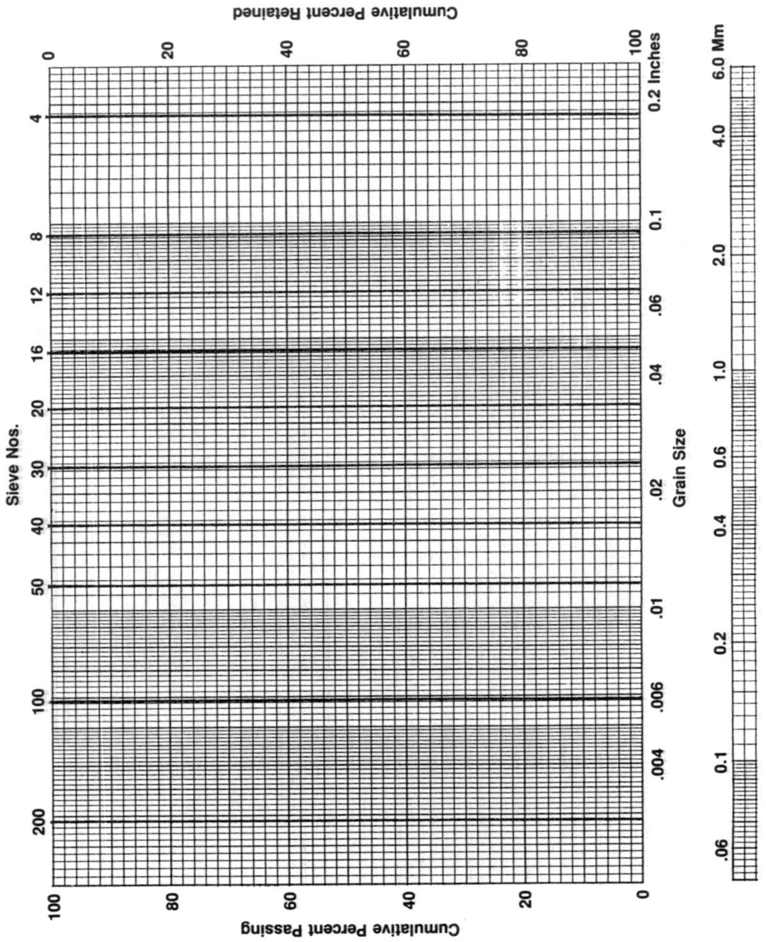

Graph for Plotting Sieve Analyses

U.S. Standard Test Sieves – (ASTM)			
Sieve Designation		**Nominal Sieve Opening**	
Standard	**Alternative**	**Inches**	**Millimeters**
25.0 mm	1″	1	25.7
22.4 mm	7/8″	0.875	22.6
19.0 mm	3/4″	0.750	19.0
16.0 mm	5/8″	0.625	16.0
12.5 mm	1/2″	0.500	12.7
11.2 mm	7/16″	0.438	11.2
9.5 mm	3/8″	0.375	9.51
8.0 mm	5/16″	0.312	8.00
6.3 mm	1/4″	0.250	6.35
5.6 mm	No. 3½	0.223	5.66
4.75 mm	No. 4	0.187	4.76
4.00 mm	No. 5	0.157	4.00
3.35 mm	No. 6	0.132	3.36
2.36 mm	No. 8	0.0937	2.38
2.00 mm	No. 10	0.0787	2.00
1.70 mm	No. 12	0.0661	1.68
1.40 mm	No. 14	0.0555	1.41
1.18 mm	No. 16	0.0469	1.19
1.00 mm	No. 18	0.0394	1.00
0.850 mm	No. 20	0.0331	0.841
0.710 mm	No. 25	0.0278	0.707
0.600 mm	No. 30	0.0234	0.595
0.500 mm	No. 35	0.0197	0.500
0.425 mm	No. 40	0.0165	0.420
0.355 mm	No. 45	0.0139	0.350
0.300 mm	No. 50	0.0117	0.297
0.250 mm	No. 60	0.0098	0.250
0.212 mm	No. 70	0.0083	0.210
0.180 mm	No. 80	0.0070	0.177
0.150 mm	No. 100	0.0059	0.149
0.125 mm	No. 120	0.0049	0.125
0.106 mm	No. 140	0.0041	0.105
0.090 mm	No. 170	0.0035	0.088
0.075 mm	No. 200	0.0029	0.074
0.063 mm	No. 230	0.0025	0.063

Typical Commercial Filter-Pack Gradation Specifications

U.S. Std. Sieve No.	Opening (Inches)	Filter Pack Designation – (In Percent Passing)				
		6 x 12	8 x 16	10 x 20	12 x 20	16 x 30
4	0.187	100				
6	0.132	95 - 100	100			
8	0.094	35 - 60	90 - 100	100		
10	0.079	5 - 20	60 - 85	95 - 100	100	
12	0.066	1 - 5	30 - 60	65 - 85	95 - 100	
14	0.055	0 - 1	5 - 20	40 - 60	60 - 85	100
16	0.047	0	2 - 8	25 - 50	30 - 50	95 - 100
18	0.039		1 - 3	8 - 30	5 - 15	60 - 85
20	0.033		0	2 - 10	1 - 10	25 - 50
25	0.0278			0 - 2	0 - 1	5 - 20
30	0.0234			0	0	1 - 10
35	0.0197					0 - 2

Gradation specifications of individual suppliers may differ slightly from those above.

Chapter 9

Irrigation

Irrigation Table

Pumping Rate		Number Of Acres Covered (To Depth Shown) In 24 Hours Pumping							
U.S. Gal/Min	Cu Ft/Sec	1 Inch	2 Inches	3 Inches	4 Inches	6 Inches	8 Inches	10 Inches	12 Inches
25	0.056	1.33	0.66	0.44	0.33	0.22	0.17	0.13	0.11
50	0.111	2.65	1.33	0.88	0.66	0.44	0.33	0.27	0.22
100	0.223	5.31	2.65	1.77	1.33	0.88	0.66	0.53	0.44
150	0.334	7.96	3.98	2.65	1.99	1.33	0.99	0.80	0.66
200	0.446	10.61	5.31	3.53	2.65	1.77	1.33	1.06	0.88
300	0.668	15.92	7.96	5.30	3.98	2.65	1.99	1.59	1.33
400	0.891	21.22	10.61	7.07	5.31	3.54	2.65	2.12	1.77
500	1.11	26.53	13.26	8.83	6.63	4.42	3.32	2.65	2.21
700	1.56	37.14	18.57	12.37	9.28	6.19	4.64	3.71	3.09
1,000	2.23	53.05	26.53	17.67	13.26	8.84	6.63	5.31	4.42
1,500	3.34	79.58	39.79	26.50	19.89	13.26	9.95	7.96	6.63
2,000	4.46	106.1	53.05	35.33	26.53	17.68	13.26	10.61	8.84
2,500	5.57	132.6	66.31	44.17	33.16	22.10	16.58	13.26	11.05
3,000	6.68	159.2	79.58	53.00	39.79	26.53	19.89	15.92	13.26
4,000	8.91	212.2	106.1	70.66	53.05	35.37	26.53	21.22	17.68
5,000	11.14	265.3	132.6	88.33	66.31	44.21	33.16	26.53	22.10
7,500	16.71	397.9	198.9	132.5	99.47	66.31	49.74	39.79	33.16
10,000	22.28	530.5	265.3	176.7	132.6	88.42	66.31	53.05	44.21
15,000	33.42	795.8	397.9	265.0	198.9	132.6	99.47	79.58	66.31
20,000	44.56	1061.0	530.5	353.3	265.3	176.8	132.6	106.1	88.42

Converting Daily Consumptive Use To Continuous Flow

Consumptive Use Of Water		Equivalent Continuous Flow Rates			
		Per Acre		Per Hectare	
Inches Per Day	Millimeters Per Day	U.S. Gallons Per Minute	Cubic Feet Per Second	Liters Per Second	Cubic Meters Per Day
0.08	2	1.5	0.0033	0.23	20
0.12	3	2.2	0.0050	0.35	30
0.16	4	3.0	0.0067	0.46	40
0.20	5	3.7	0.0083	0.58	50
0.24	6	4.5	0.0100	0.69	60
0.28	7	5.2	0.0117	0.81	70
0.32	8	6.0	0.0133	0.92	80
0.36	9	6.7	0.0150	1.04	90
0.40	10	7.5	0.0167	1.15	100

To determine required delivery rate, divide by efficiency. Example: with consumptive use of 0.20 inches per day and 75 percent irrigation efficiency, 3.7 gpm per acre is required to satisfy consumptive use and delivery rate required is 3.7/0.75 = 4.93 gpm per acre.

Friction Losses In Portable Aluminum Pipe With Couplings

(In Feet per 100 Feet of Pipe)

Flow Rate	Outside Diameter of Pipe					
(US Gpm)	3-Inch	4-Inch	5-Inch	6-Inch	8-Inch	10-Inch
40	0.66	0.06				
50	1.01	0.24				
60	1.42	0.34				
70	1.91	0.45	0.15			
80	2.46	0.58	0.19			
90	3.07	0.73	0.24			
100	3.75	0.89	0.30	0.12		
120	5.31	1.26	0.42	0.17		
140	7.11	1.69	0.56	0.23		
160	9.17	2.18	0.72	0.29		
180	11.47	2.73	0.97	0.37		
200	14.01	3.33	1.10	0.45		
220	16.79	4.00	1.32	0.54		
240	19.81	4.71	1.56	0.63		
260	23.06	5.49	1.81	0.74		
280	26.55	6.32	2.09	0.85		
300	30.27	7.20	2.38	0.97	0.24	
320	34.22	8.14	2.69	1.09	0.27	
340	38.39	9.14	3.02	1.23	0.30	
360	42.80	10.18	3.37	1.37	0.33	
380	47.43	11.29	3.73	1.52	0.37	
400		12.44	4.11	1.67	0.40	0.14
420		13.65	4.51	1.83	0.45	0.15
440		14.57	4.93	1.99	0.49	0.16
460		16.23	5.36	2.18	0.53	0.18
480		17.59	5.82	2.36	0.57	0.19
500		19.01	6.28	2.55	0.62	0.21
550		22.79	7.53	3.06	0.74	0.25
600		26.88	8.89	3.61	0.88	0.29
650		31.30	10.35	4.20	1.02	0.34
700		36.03	11.91	4.84	1.17	0.39
750			13.58	5.52	1.34	0.45
800			15.35	6.24	1.51	0.51
850			17.22	7.00	1.70	0.57
900			19.20	7.80	1.89	0.64
950			21.28	8.65	2.10	0.70
1000			23.45	9.53	2.31	0.78
1500				20.59	5.00	1.68
2000					8.63	2.89

Based on 30-ft lengths; for 20-ft, increase by 7%; for 40-ft, decrease by 3%

Friction Loss In Sprinkler Lateral Lines

The friction loss in a sprinkler lateral pipeline, with numerous equally spaced outlets, is less than the loss in an equal length pipeline with no outlets, because the flow rate is reduced each time an outlet is passed. The method developed by J. E. Christiansen to evaluate pressure loss in multiple outlet pipelines is widely used. The method involves computing the friction loss without outlets, and then reducing that loss by multiplying by a factor that is based on the number of outlets (sprinklers) on the lateral.

For example, calculate the friction loss in a 4-inch PVC pipe lateral 660 feet long, with 22 sprinkler outlets spaced 30 feet apart and discharging 10 gpm each. The total flow entering the lateral is 220 gpm and the friction loss in the line, without any outlets, would be 15.8 feet (from Hazen-Williams formula with C = 150, see Chapter 4 "Friction loss in pipe flow"). From the following table, the factor "F" for 22 outlets is 0.368 and the friction loss in the lateral — with the 22 outlets operating — becomes 15.8 x 0.368 or 5.8 feet (2.5 psi).

Number Of Outlets	Factor "F"	Number Of Outlets	Factor "F"
1	1.000	16	0.377
2	.634	17	.375
3	.528	18	.373
4	.480	19	.372
5	.451	20	.370
6	.433	21	.369
7	.419	22	.368
8	.410	23	.367
9	.402	24	.366
10	.396	25	.365
11	.392	26	.364
12	.388	27	.364
13	.384	28	.363
14	.381	29	.363
15	.379	30	.362

Agricultural Sprinkler Nozzles

Typical Discharge Per Nozzle (U.S. Gal/Min)

Nozzle Diam. (Inches)	Nozzle Pressure – Pounds/Square Inch											
	25	30	35	40	45	50	55	60	65	70	75	80
1/8	2.25	2.47	2.68	2.87	3.05	3.22	3.38	3.53	3.68	3.82	3.96	4.09
9/64	2.88	3.15	3.40	3.64	3.86	4.07	4.27	4.46	4.65	4.83	5.00	5.17
5/32	3.52	3.85	4.16	4.45	4.72	4.98	5.22	5.45	5.68	5.90	6.11	6.30
11/64	4.24	4.64	5.02	5.37	5.70	6.01	6.30	6.57	6.83	7.09	7.34	7.58
3/16	5.00	5.50	5.96	6.38	6.78	7.16	7.52	7.85	8.18	8.50	8.80	9.09
13/64	5.90	6.50	7.05	7.55	8.00	8.45	8.85	9.25	9.60	9.95	10.3	10.5
7/32	6.85	7.55	8.20	8.80	9.35	9.90	10.4	10.8	11.1	11.4	11.7	12.0
15/64	7.80	8.60	9.30	10.0	10.6	11.2	11.7	12.2	12.7	13.2	13.6	14.0
1/4	8.90	9.80	10.6	11.4	12.1	12.8	13.4	13.9	14.5	15.0	15.5	16.0
17/64	10.0	11.0	11.9	12.8	13.6	14.3	15.0	15.6	16.2	16.8	17.3	17.8
9/32	11.2	12.3	13.3	14.3	15.2	16.0	16.8	17.5	18.1	18.7	19.2	19.7

Diameter of coverage typically ranges from 80 to 130 feet, depending upon nozzle size, pressure, and trajectory angle. Consult manufacturer's specifications for particular applications.

Approximate Discharge of Aluminum Irrigation Siphon Tubes

Head		Outside Diameter of Tube – Inches									
(Inches)	(Cm)	1.0		1.25		1.5		2.0		3.0	
		(Gal/Min)	(Lit/Sec)	(Gal/Min)	(Lit/Sec)	(Gal/Min)	(Lit/Sec)	(Gal/Min)	(Lit/Sec)	(Gal/Min)	(Lit/Sec)
1	2.5	2.8	0.18	4.8	0.30	7.5	0.47	14	0.88	36	2.27
2	5.1	4.0	0.25	6.8	0.43	10.6	0.67	20	1.26	50	3.15
4	10.2	5.8	0.37	9.7	0.61	15.0	0.95	29	1.83	72	4.54
6	15.2	7.0	0.44	12.0	0.76	18.5	1.17	35	2.21	88	5.55
8	20.3	8.1	0.51	13.7	0.86	21	1.32	40	2.52	100	6.31
10	25.4	9.1	0.57	15.4	0.97	24	1.51	45	2.84	112	7.07
15	38.1	11.0	0.69	18.9	1.19	29	1.83	55	3.47	137	8.64
20	50.8	13.0	0.82	22.0	1.39	34	2.15	64	4.04	160	10.09

Head is vertical distance from water surface in supply ditch to water surface in furrow (or tube outlet if not submerged)
Based on siphon tube lengths of 5 feet (1.5 meters)

Canal Seepage Losses

Seepage losses from unlined irrigation canals or laterals are dependent primarily upon such factors as the depth of water and cross-sectional area, the depth to water table, and the character of the materials in the bed and banks.

Seepage is commonly measured as the loss in water depth over the wetted perimeter of the channel in a 24-hour period. In ordinary unlined earth channels, under normal conditions, the seepage and evaporation losses frequently amount to one-third of the total water diverted (evaporation loss is normally very small).

The Moritz formula suggests computation of seepage loss as follows:

$$S = 0.2\,C\sqrt{\frac{Q}{V}}$$

where S = loss in cfs per mile of canal
Q = discharge of canal in cfs
V = mean flow velocity in ft/sec
C = water depth (ft) lost per day through wetted area

Field observations indicate average values of “C” in earth canals as:

Material	Value of C
Cemented gravel and hardpan, with sandy loam	0.34
Clay and clayey loam	0.41
Sandy loam	0.66
Volcanic ash	0.68
Volcanic ash with sand	0.98
Sand and volcanic ash or clay	1.20
Sandy soil with rock	1.68
Sandy and gravelly soil	2.20

Irrigation Water Requirements

The *net irrigation requirement* for a particular crop is equal to the *consumptive use* of the crop minus the *effective precipitation* during the growing season. The *gross irrigation requirement* is the net irrigation requirement divided by the *irrigation efficiency.*

Consumptive use, often called evapo-transpiration, is the amount of water used by the vegetation for plant growth, transpiration, and evaporation from adjacent soil. It is usually expressed in units of depth (inches) or volume for a given area (acre-inches per acre, or acre-feet per acre).

Effective precipitation is the rainfall occurring during the growing period of the crop that is available to meet the consumptive use. It does not include precipitation that is lost to deep percolation below the root zone, nor that which is surface runoff.

Irrigation efficiency is the percentage of the applied irrigation water that is stored in the soil and available to meet the consumptive use of the crop. If the applied water is measured at the farm delivery point, it is often called "farm irrigation efficiency" and takes into account water lost on the farm by deep percolation or runoff; if the applied water is measured at a distant point of diversion, it is often called "project irrigation efficiency" and takes into account losses such as seepage from canals, laterals, or pipelines.

Many factors influence the amount of water consumed by plants, and these factors may differ not only with locality but with time. The more important factors are climate (precipitation, temperature, sunlight, humidity, wind, solar radiation, etc.), plant characteristics (type of plant, length of growing season, and stage of growth), and available water supply.

Estimating Consumptive Use:

Many methods have been developed for estimating consumptive use. Some of those commonly used include such methods as Blaney-Criddle, Penman, Jensen-Haise, and pan evaporation. Some of the methods require knowledge (or estimation) of a number of climatological factors such as wind speed, humidity, solar radiation, etc.

A discussion of the many methods for estimating consumptive use is beyond the scope of this handbook. The reader is referred to such references as the American Society of Civil Engineers Manual No. 70 "Evapotranspiration and Irrigation Water Requirements" (1990) or U.S.D.A. Soil Conservation Service Technical Release No. 21 "Irrigation Water Requirements" (rev. 1970) for more detail.

Blaney-Criddle Method

The original Blaney-Criddle formula, however, is useful in estimating *seasonal* consumptive use for a particular crop and only requires a knowledge of temperature and percentage of daytime hours (latitude). Modifications to the Blaney-Criddle formula, to better estimate consumptive use for shorter periods of the growing season, are discussed in the references listed above.

Blaney and Criddle determined that the amount of water consumptively used by crops during a normal growing season was closely related to mean monthly air temperature and daytime hours. This relationship is expressed by:

$$U = K \times F$$

where: U = Crop consumptive use *(inches)* for the growing season

K = An empirical coefficient, for the growing season, for a particular crop

F = Sum of monthly consumptive use factors (f) for the growing season

f = monthly consumptive use factor or: $\frac{(t \times p)}{100}$

where: t = mean monthly air temperature in *degrees F*

p = monthly percentage of annual daytime hours

(values of f, t, and p can be applied by periods of less than a month)

Accompanying tables give values for K (seasonal consumptive use coefficients) and for p (monthly percentage of daytime hours by latitude). World-wide data on monthly maximum and minimum temperatures and precipitation are included in the climatic tables in Chapter 12.

As an example, the consumptive use for corn, for a four-month growing season (May-August) in an arid area at 40 deg N latitude, with average daily temperatures for each month of 60, 71, 76, and 74 deg F respectively, would be estimated as follows:

Month	**t**	**p**	**f**
May	60	9.99	5.99
June	71	10.08	7.16
July	76	10.34	7.86
August	74	9.56	7.07
			F = 28.1

with seasonal factor K of 0.85, U = KF = 0.85 x 28.1 = 23.9 inches.

The net irrigation requirement would be the consumptive use minus the effective precipitation for the four-month growing season.

Pan Evaporation Method

Many observations have shown that pan evaporation is a good index of consumptive use, particularly for periods of a month or a full growing season. A standard for measurement of evaporation is the U.S. Weather Bureau Class A pan. This is a circular pan made of unpainted galvanized iron, 4 ft in diameter and 10 inches deep. The bottom is supported 6 inches above ground on a wooden frame. The water surface is maintained between 2 and 3 inches below the rim of the pan, and the water level carefully measured daily with a hook gage in a stilling well. Any precipitation must be taken into account in arriving at the total evaporation. In some arid areas the pan must be fenced or lightly screened to prevent animals or birds from using it as a source of drinking water. The evaporation from a large, open body of water (lake or reservoir) is typically about 0.7 times the evaporation from a Class A pan (ranging from 0.65 to 0.75 times).

The correlation between pan evaporation and consumptive use, as in other methods, depends upon many factors which include type of crop, stage of crop growth, and percentage of ground cover. The accompanying table gives suggested coefficients for various groups of crops and for varying stages of plant growth. The approximate consumptive use of the crop is equal to the pan evaporation multiplied by the coefficient.

Effective Precipitation:

The effective rainfall during the growing season must be known (or estimated) in order to estimate the amount of irrigation water needed to meet the crop consumptive use requirement. Effective precipitation is dependent not only on the total amount of precipitation and crop consumptive use, but on the duration and intensity of rainfall and such other factors as antecedent soil moisture, soil characteristics, slope, etc.

In arid areas, where growing season rainfall is light, the effectiveness of rainfall can be high (90 percent or more of the total). In humid areas, where growing season storms may be large and intense, much of the rainfall may run off or be lost to deep percolation and the effectiveness will be much lower (possibly 60 percent or less of the total). The U.S.D.A. Soil Conservation Service has related typical average monthly effective rainfall to mean monthly rainfall and average monthly consumptive use, as given on the accompanying table.

Blaney-Criddle Seasonal Consumptive Use Coefficients – "K"

(For Irrigated Crops)

Crop	Normal Growing Season (or Period)*	Coefficient "K"**
Alfalfa	Between frosts	0.80-0.90
Bananas	Entire year	0.80-1.00
Beans	3 months	0.60-0.70
Cocoa	Entire year	0.70-0.80
Coffee	Entire year	0.70-0.80
Corn (maize)	4 months	0.75-0.85
Cotton	7 months	0.60-0.70
Dates	Entire year	0.65-0.80
Flax	7 to 8 months	0.70-0.80
Grains, small	3 months	0.75-0.85
Oil seeds	3 to 5 months	0.65-0.75
Orchard crops:		
Avocado	Entire year	0.50-0.55
Grapefruit	Entire year	0.55-0.65
Orange/lemon	Entire year	0.45-0.55
Walnuts	Between frosts	0.60-0.70
Deciduous	Between frosts	0.60-0.70
Pasture crops:		
Grass	Between frosts	0.75-0.85
Ladino clover	Between frosts	0.80-0.85
Potatoes	3 to 5 months	0.65-0.75
Rice	3 to 5 months	1.00-1.10
Soybeans	140 days	0.65-0.70
Sugar beets	6 months	0.65-0.75
Sugar cane	Entire year	0.80-0.90
Tobacco	4 months	0.70-0.80
Tomatoes	4 months	0.65-0.70
Truck crops, small	2 to 4 months	0.60-0.70
Vineyards	5 to 7 months	0.50-0.60

* Season depends on variety and time of year when grown; annual crops grown in winter may take longer than if grown in summer.

** Lower values of "K" are for more humid climates, higher values for more arid areas.

Mean Monthly Percentage of Annual Daytime Hours

Latitude (Degrees	Jan	Feb	Mar	Apr	May	Jun	Jul	Aug	Sep	Oct	Nov	Dec
56 North	5.31	5.98	8.17	9.48	11.26	11.68	11.67	10.36	8.52	7.18	5.52	4.87
54	5.53	6.12	8.15	9.36	11.00	11.40	11.43	10.27	8.50	7.33	5.74	5.17
52	5.75	6.23	8.17	9.28	10.81	11.13	11.22	10.15	8.49	7.40	5.94	5.43
50	5.98	6.32	8.25	9.25	10.69	10.93	10.99	10.00	8.44	7.43	6.07	5.65
48	6.13	6.42	8.22	9.15	10.50	10.72	10.83	9.92	8.45	7.56	6.24	5.86
46	6.30	6.50	8.24	9.09	10.37	10.54	10.66	9.82	8.44	7.61	6.38	6.05
44	6.45	6.59	8.25	9.04	10.22	10.38	10.50	9.73	8.43	7.67	6.51	6.23
42	6.60	6.66	8.28	8.97	10.10	10.21	10.37	9.64	8.42	7.73	6.63	6.39
40	6.73	6.73	8.30	8.92	9.99	10.08	10.34	9.56	8.41	7.78	6.73	6.53
35	7.04	6.88	8.35	8.82	9.76	9.76	9.93	9.37	8.36	7.88	6.98	6.87
30	7.30	7.03	8.38	8.72	9.53	9.49	9.67	9.22	8.34	7.99	7.19	7.14
25	7.54	7.14	8.39	8.62	9.33	9.24	9.45	9.08	8.31	8.08	7.40	7.42
20	7.73	7.26	8.20	8.52	9.14	9.02	9.25	8.95	8.30	8.19	7.58	7.88
15	7.94	7.37	8.43	8.45	8.98	8.81	9.04	8.83	8.27	8.25	7.75	7.88
10	8.14	7.47	8.45	8.37	8.81	8.61	8.85	8.71	8.25	8.34	7.91	8.09
5	8.32	7.57	8.47	8.29	8.65	8.41	8.67	8.60	8.24	8.41	8.07	8.30
Equator	8.50	7.67	8.49	8.22	8.49	8.22	8.50	8.49	8.21	8.49	8.22	8.50
5 South	8.67	7.77	8.51	8.14	8.34	8.03	8.32	8.38	8.20	8.57	8.37	8.69
10	8.85	7.87	8.53	8.07	8.17	7.83	8.14	8.27	8.18	8.65	8.53	8.90
15	9.04	7.98	8.55	7.98	8.01	7.63	7.94	8.16	8.16	8.73	8.69	9.10
20	9.25	8.09	8.58	7.92	7.83	7.41	7.73	8.05	8.13	8.83	8.85	9.32
25	9.45	8.20	8.59	7.82	7.65	7.18	7.54	7.91	8.12	8.91	9.03	9.55
30	9.69	8.33	8.63	7.75	7.43	6.94	7.30	7.80	8.09	9.00	9.24	9.80
35	9.93	8.46	8.64	7.63	7.21	6.65	7.04	7.62	8.08	9.11	9.45	10.09
40	10.24	8.65	8.70	7.54	6.96	6.33	6.73	7.46	8.04	9.23	9.69	10.42

Divide by number of days in month to obtain mean daily percentage of annual daytime hours.

Average Monthly Effective Rainfall

(Related to Mean Monthly Total Rainfall and Average Monthly Consumptive Use)

Monthly Mean Total Rainfall		Average Monthly Consumptive Use																			
		In	Mm	In	Mm	In	Mm	In	Mm	In	Mm	In	Mm	In	Mm	In	Mm	In	Mm	In	Mm
In	Mm	1.0	25	2.0	51	3.0	76	4.0	102	5.0	127	6.0	152	7.0	178	8.0	203	9.0	229	10.0	254
		(Average Effective Monthly Rainfall)																			
0.5	13	0.30	8	0.32	8	0.34	9	0.36	9	0.38	10	0.40	10	0.42	11	0.45	11	0.47	12	0.50	13
1.0	25	0.63	16	0.66	17	0.70	18	0.74	19	0.78	20	0.83	21	0.88	22	0.93	24	0.98	25	1.00	25
1.5	38	0.93	24	0.98	25	1.03	26	1.09	28	1.16	29	1.22	31	1.29	33	1.37	35	1.45	37	1.50	38
2.0	51			1.27	32	1.35	34	1.43	36	1.51	38	1.59	40	1.69	43	1.78	45	1.88	48	1.99	51
2.5	64			1.56	40	1.65	42	1.74	44	1.84	47	1.95	50	2.06	52	2.18	55	2.30	58	2.44	62
3.0	76			1.83	46	1.94	49	2.05	52	2.17	55	2.29	58	2.42	61	2.56	65	2.71	69	2.86	73
3.5	89					2.22	56	2.35	60	2.48	63	2.62	67	2.77	70	2.93	74	3.10	79	3.28	83
4.0	102					2.49	63	2.63	67	2.79	71	2.95	75	3.12	79	3.29	84	3.48	88	3.68	93
4.5	114					2.76	70	2.92	74	3.09	78	3.26	83	3.45	88	3.65	93	3.86	98	4.08	104
5.0	127							3.20	81	3.38	86	3.57	91	3.78	96	4.00	102	4.23	107	4.47	114
5.5	140							3.47	88	3.67	93	3.88	99	4.10	104	4.34	110	4.59	117	4.85	123
6.0	152							3.74	95	3.95	100	4.18	106	4.42	112	4.67	119	4.94	125	5.23	133
6.5	165							4.00	102	4.23	107	4.48	114	4.73	120	5.00	127	5.29	134	5.60	142
7.0	178									4.51	115	4.77	121	5.04	128	5.33	135	5.64	143	5.96	151
7.5	191									4.78	121	5.06	129	5.35	136	5.65	144	5.98	152	6.32	161
8.0	203											5.34	136	5.65	144	5.97	152	6.32	161	6.68	170

The table of average monthly effective rainfall is based on a 3-inch net depth of application.
The net depth of application is the water required to be added to the root zone at 50% moisture to bring it up to field capacity.
For different net depth of application, multiply values in the table by the following factors:

Net depth (In)	0.75	1.0	1.5	2.0	2.5	3.0	4.0	5.0	6.0	7.0
Factor	0.72	0.77	0.86	0.93	0.97	1.00	1.02	1.04	1.06	1.07

Example: If the average monthly rainfall is 4.0 inches and average monthly consumptive use is 6.0 inches, the average effective monthly rainfall is 2.95 inches.

Pan Evaporation – Crop Consumptive Use Coefficients

Percent Of Crop Growing Season	Coefficients To Be Multiplied By Class A Pan Evaporation							
	Crop Group (see below)							
	A	B	C	D	E	F	G	Rice
0	0.20	0.15	0.12	0.08	0.90	0.60	0.50	0.80
10	0.36	0.27	0.22	0.15	0.90	0.60	0.60	0.95
20	0.64	0.48	0.38	0.27	0.90	0.60	0.70	1.05
30	0.84	0.63	0.50	0.40	0.90	0.60	0.80	1.14
40	0.97	0.73	0.58	0.52	0.90	0.60	0.90	1.21
50	1.00	0.75	0.60	0.65	0.90	0.60	1.00	1.30
60	0.99	0.74	0.60	0.77	0.90	0.60	1.00	1.30
70	0.91	0.68	0.55	0.88	0.90	0.60	0.90	1.20
80	0.75	0.56	0.45	0.90	0.90	0.60	0.80	1.10
90	0.46	0.35	0.28	0.70	0.90	0.60	0.70	0.90
100	0.20	0.20	0.17	0.20	0.90	0.60	0.50	0.20

Crop Group A: Beans, corn (maize), cotton, potatoes, sugar beets, peas, jowar
(typical average value of 1.00, with bell curve pattern)
B: Dates, olives, plums, walnuts, tomatoes, mulberry
C: Melons, onions, carrots, hops, grapes
(maximum of about 0.60, with bell curve pattern)
D: Barley, wheat, other small grains, flax, celery
(maximum of about 0.90 near end of growing season)
E: Pastures, orchards with cover crops, bananas
(fairly uniform at about 0.90)
F: Oranges, grapefruit, lemons
(fairly uniform at about 0.60)
G: Alfalfa, sugar cane
(increases with crop growth from about 0.60 to 1.00)
Rice: Rice
(typical range from 0.80 minimum to 1.30 maximum)

Guide For Relating Mean Air Temperature to Planting & Maturity Dates and Length of Growing Season

Crop	Earliest Moisture Use Or Planting Date (Related to Mean Air Temp)	Latest Moisture Use Or Maturing Date (Related to Mean Air Temp)	Growing Season (Days)
Perennial Crops			
Alfalfa	50 F mean temp	28 F frost	Variable
Grasses, cool	45 F mean temp	45 F mean temp	Variable
Orchards, deciduous	50 F mean temp	45 F mean temp	Variable
Grapes	55 F mean temp	50 F mean temp	Variable
Annual Crops			
Beans, dry	60 F mean temp	32 F frost	90 – 100
Corn	55 F mean temp	32 F frost	140 – Max
Cotton	62 F mean temp	32 F frost	240 – Max
Grain, spring	45 F mean temp	32 F frost	130 – Max
Potatoes, late	60 F mean temp	32 F frost	130 – Max
Sorghum, grain	60 F mean temp	32 F frost	130 – Max
Sugar beets	28 F frost	28 F frost	180 – Max
Wheat, winter			
(fall season)		45 F mean temp	
(spring season)	45 F mean temp		

Critical Moisture Periods for Irrigation

In addition to the requirement for sufficient soil moisture for germination and development of an adequate stand, most crops need high moisture levels during critical periods of growth to obtain high yields. These critical periods usually occur in the latter part of the growing season when the crop is near harvest. Examples for commonly irrigated crops are:

Alfalfa	Start of flowering and after cutting
Corn (field or sweet)	Tasseling through silking
Cotton	First bloom through boll-maturing stage
Melons	Blossom to harvest
Orchard	Fruit development
Pasture	After grazing
Potatoes	Blossom to harvest
Small grain	Boot to heading stage
Strawberries	Fruit development to ripening
Sugar beets	Three to four weeks after emergence
Tobacco	Knee-high to blossom

Typical Moisture-Holding Capacity of Soils

Soil Texture	Inches Water Per Ft. Soil
Very Coarse:	
Very coarse sand	0.40 – 0.75
Coarse:	
Coarse sand, fine sand, loamy sand	0.75 – 1.25
Moderately Coarse:	
Sandy loam, fine sandy loam	1.25 – 1.75
Medium:	
Very fine sandy loam, loam, silt loam	1.50 – 2.30
Moderately Fine:	
Clay loam, silty clay loam, sandy clay loam	1.75 – 2.50
Fine:	
Sandy clay, silty clay, clay	1.60 – 2.50
Peat and muck:	2.00 – 3.00

Typical Water Intake Rates (For Sprinkler Irrigation On Soils Of Different Texture)				
	Typical Maximum Water Intake Rate			
	Good Surface Structure*		Poor Surface Structure**	
Soil Texture	mm/hr	in/hr	mm/hr	in/hr
Coarse sand	20-25	0.75-1.00	12	0.50
Fine sand	12-20	0.5-0.75	10	0.35
Fine sandy loam	12	0.5	8	0.30
Silt loam	10	0.4	7	0.27
Clay loam	8	0.3	6	0.25
Clay	5	0.2	2	0.10

*Protected from drop impact by growing crop or crop residue
**Where bare soil tends to form a surface crust

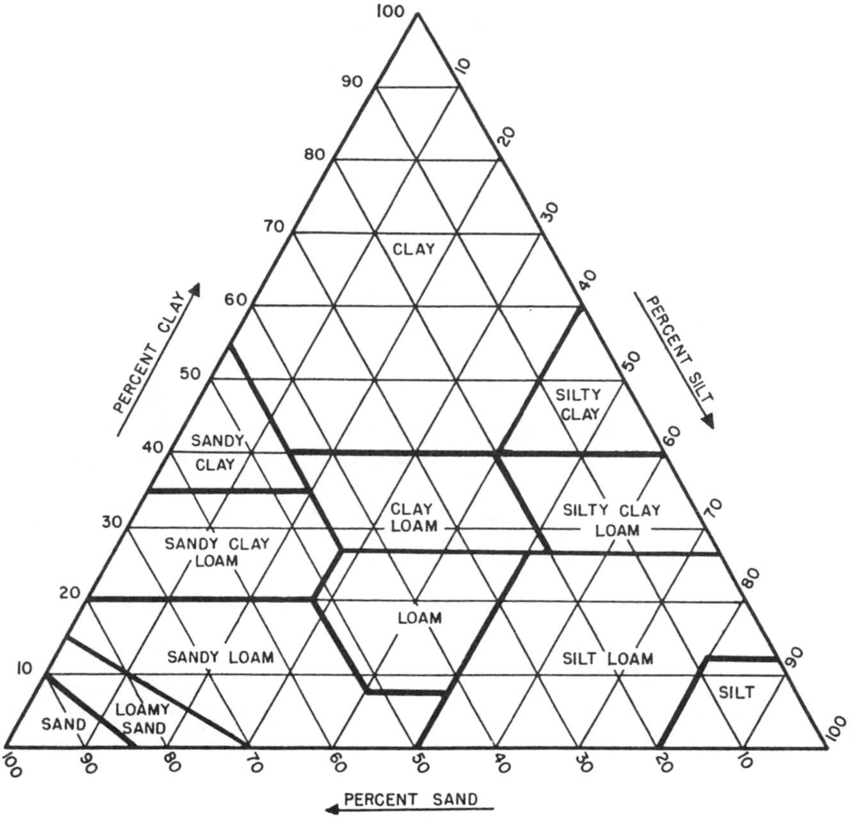

Proportions of Sand, Silt, and Clay in Basic Soil Textural Classes

Irrigation Water Quality

The suitability of water for irrigation is largely dependent upon the amount and type of dissolved mineral constituents, and the effect these constituents have upon the particular type of soil and plant species. Whether a particular quality of water is suitable for irrigation, however, also depends to a considerable extent upon such factors as soil type, drainage, climate, and irrigation practices.

The usual water-quality factors to be considered in evaluating or classifying water for irrigation use are: specific conductance (which reflects the total dissolved solids); sodium percentage; boron; residual sodium carbonate; and the sodium adsorption ratio (SAR). Specific conductance is usually expressed in micromhos (μmhos); when multiplied by 0.65 is approximately equal to total dissolved solids in milligrams per liter (mg/l). Sodium percentage is determined by dividing the sodium (plus potassium if reported) in milliequivalents per liter (meq/l) by the sum of sodium (plus potassium), calcium, and magnesium (also in meq/l):

$$\text{Sodium percentage} = \frac{\text{meq/l (Na + Mg)}}{\text{meq/l (Na + K + Ca + Mg)}}$$

Boron is expressed in milligrams per liter (mg/l). Residual sodium carbonate is expressed in milliequivalents per liter (meq/l) and is determined by subtracting the calcium plus magnesium (in meq/l) from the bicarbonate (and carbonate, if present) concentration (also in meq/l):

$$\text{Residual sodium carbonate} = \text{meq/l}\,(HCO_3 + CO_3) - \text{meq/l}\,(Ca + Mg)$$

In many irrigation water supplies the sum of the calcium and magnesium exceeds the bicarbonate and the residual sodium carbonate is therefore zero. The sodium adsorption ratio is the sodium concentration divided by the square root of one-half of the combined calcium and magnesium concentrations (all expressed in meq/l):

$$\text{SAR} = \frac{Na}{\sqrt{(Ca + Mg)/2}}$$

Classification

An early (1935) generalized classification of irrigation water quality considered sodium percentage, specific conductance, and boron. This classification was:

Class	Percent Sodium	Specific Conductance (μmhos)	Boron (mg/l)		
			Sensitive Crops	Semi-Tolerant Crops	Tolerant Crops
Excellent	<20	<250	<0.33	<0.67	<1.00
Good	20-40	250-750	0.33-0.67	0.67-1.33	1.00-2.00
Permissible	40-60	750-2000	0.67-1.00	1.33-2.00	2.00-3.00
Doubtful	60-80	2000-3000	1.00-1.25	2.00-2.50	3.00-3.75
Unsuitable	>80	>3000	>1.25	>2.50	>3.75

Sodium Adsorption Ratio (SAR)

In the mid-1950's the Salinity Laboratory of the U.S. Department of Agriculture proposed a classification system for irrigation water quality that relates specific conductance (electrical conductivity) and the sodium adsorption ratio (defined above). A particular water is then classified as to a Conductivity Class (C1 through C4) and a Salinity Class (S1 through S4).

The accompanying diagram can be used to determine the particular classes from the water analysis.

Definitions for the various classes are as follows:

Conductivity:

Low-Salinity Water (C1): can be used for irrigation with most crops on most soils with little likelihood that soil salinity will develop. Some leaching is required, but this occurs under normal irrigation practices except in soils of extremely low permeability.

Medium-Salinity Water (C2): can be used if a moderate amount of leaching occurs. Plants with moderate salt tolerance can be grown in most cases without special practices for salinity control.

High-Salinity Water (C3): cannot be used on soils with restricted drainage. Even with adequate drainage, special management for salinity control may be required and plants with good salt tolerance should be selected.

Very High-Salinity Water (C4): is not suitable for irrigation under ordinary conditions, but may be used occasionally under very special circumstances. The soils must be permeable, drainage must be adequate, irrigation water must be applied in excess to provide considerable leaching, and very salt-tolerant crops should be selected.

Sodium:

Low-Sodium Water (S1): can be used for irrigation on almost all soils with little danger of the development of harmful levels of exchangeable sodium. However, sodium-sensitive crops such as stone-fruit trees and avocados may accumulate injurious concentrations of sodium.

Medium-Sodium Water (S2): will present an appreciable sodium hazard in fine-textured soils having high cation-exchange capacity, especially under low-leaching conditions, unless gypsum is present in the soil. This water may be used on coarse-textured or organic soils with good permeability.

High-Sodium Water (S3): may produce harmful levels of exchangeable sodium in most soils and will require special management — good drainage, high leaching, and organic matter additions. Gypsiferous soils may not develop harmful levels of exchangeable sodium from such waters. Chemical amendments may be required for replacement of exchangeable sodium, except that amendments may not be feasible with waters of very high salinity.

Very High-Sodium Water (S4): is generally unsatisfactory for irrigation purposes except at low and perhaps medium salinity, where the solution of calcium from the soil or use of gypsum or other amendments may make the use of these waters feasible.

Boron

Very small quantities of boron are essential to normal growth of all plants, and a deficiency of boron can produce striking symptoms in some species. Boron can very toxic to some species, however, and the concentrations that harms sensitive plants may be about the same as that needed for normal growth of very tolerant plants. The accompanying table gives a listing of relative tolerance of certain crops to boron.

Residual Sodium Carbonate

Using the calculation of residual sodium carbonate (defined above) the following criteria have been proposed:

- Water with more than 2.5 meq/l residual sodium carbonate is not suitable for irrigation.
- Water with 1.25-2.5 meq/l residual sodium carbonate is marginal for irrigation use.
- Water with less than 1.25 meq/l residual sodium carbonate is probably safe for irrigation.

Relative Salt-Tolerance of Various Crops

The relative salt-tolerance of various crops is given in the accompanying table, with the specific conductance values of the soil saturation extract shown that are associated with a 50 percent decrease in yield as compared to the yield on a non-saline soil under similar growing conditions.

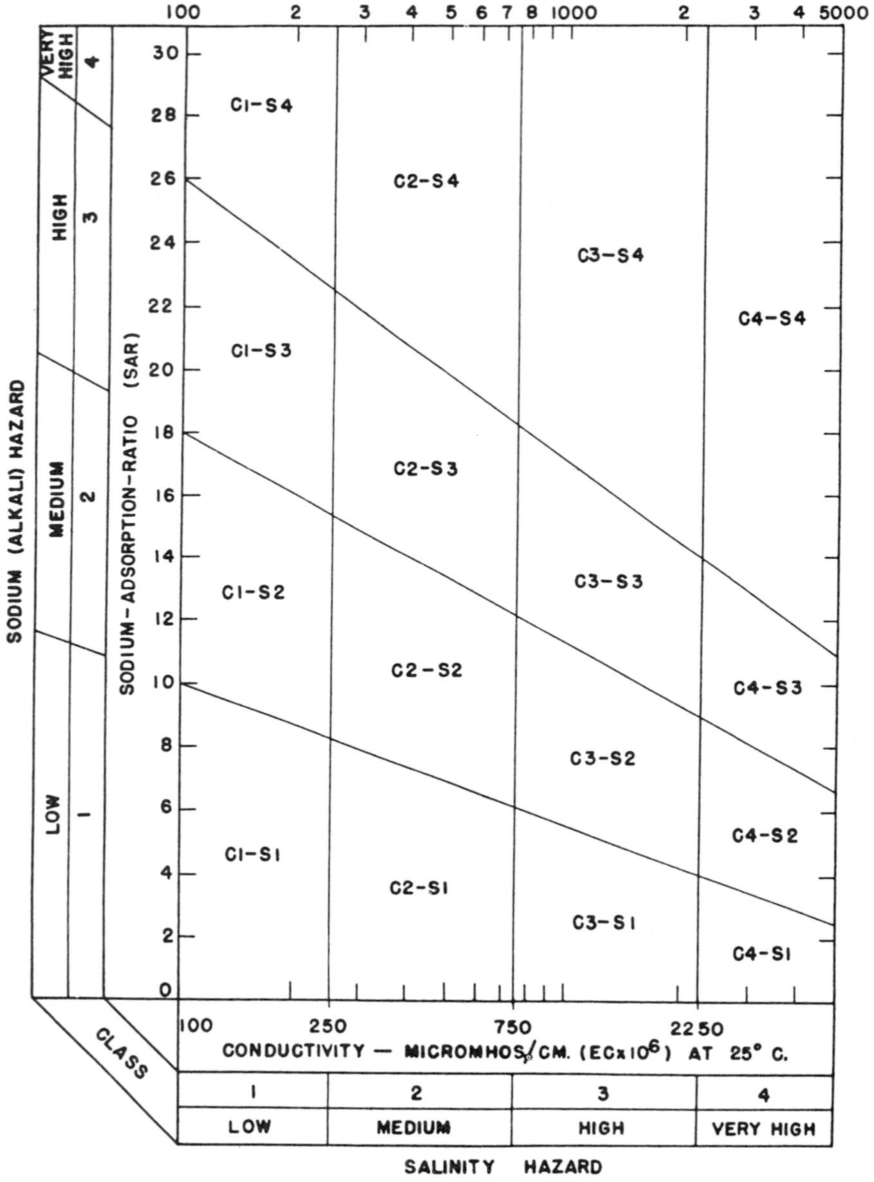

Diagram for the Classification of Irrigation Waters

Relative Salt-Tolerance of Crops

High Salt Tolerance	Medium Salt Tolerance	Low Salt Tolerance	High Salt Tolerance	Medium Salt Tolerance	Low Salt Tolerance
Fruit Crops			**Field Crops**		
Date palm	Pomegranate Fig Olive Grape Cantaloupe	Pear Apple Orange Grapefruit Prune Plum Almond Apricot Peach Strawberry Lemon Avocado	(16,000 μmhos) Barley (grain) Sugar beet Rape Cotton (10,000 μmhos)	(10,000 μmhos) Rye (grain) Wheat (grain) Oats (grain) Rice Sorghum (grain) Corn (field) Flax Sunflower Castor beans (6,000 μmhos)	(4,000 μmhos) Field beans
Forage Crops			**Vegetable Crops**		
(18,000 μmhos) Alkali sacaton Salt grass Nuttall alkali grass Bermuda grass Rhodes grass Rescue grass Canada wild rye Western wheat grass Barley (hay) Birdsfoot trefoil (12,000 μmhos)	(12,000 μmhos) White sweetclover Yellow sweetclover Perennial ryegrass Mountain brome Strawberry clover Dallis grass Sudan grass Hubam clover Alfalfa Tall fescue Rye (hay) Wheat (hay) Oats (hay) Orchard grass Blue grama Meadow fescue Reed canary Big trefoil Smooth brome Tall meadow oat grass Cicer milkvetch Sour clover Sickle milkvetch (4,000 μmhos)	(4,000 μmhos) White Dutch clover Meadow foxtail Alsike clover Red clover Ladino clover Burnet (2,000 μmhos)	(12,000 μmhos) Garden beets Kale Asparagus Spinach (10,000 μmhos)	(10,000 μmhos) Tomato Broccoli Cabbage Bell pepper Cauliflower Lettuce Sweet corn Potato Carrot Onion Peas Squash Cucumber (4,000 μmhos)	(4,000 μmhos) Radish Celery Green bean (3,000 μmhos)

Relative Tolerance of Plants to Boron		
Tolerant	**Semi-Tolerant**	**Sensitive**
Asparagus	Sunflower	Pecan
Date palm	Potato	Black walnut
Palm	Cotton	English walnut
Sugar beet	Tomato	Navy bean
Garden beet	Sweetpea	American elm
Alfalfa	Radish	Plum
Gladiolus	Field pea	Pear
Broadbean	Olive	Apple
Onion	Barley	Grape
Turnip	Wheat	Kadota fig
Cabbage	Corn	Persimmon
Lettuce	Milo	Cherry
Carrot	Oat	Peach
	Zinnia	Apricot
	Pumpkin	Blackberry
	Bell pepper	Orange
	Sweet potato	Avocado
	Lima bean	Grapefruit
		Lemon

In each column, plants listed at the top are considered more tolerant; plants listed at the bottom are more sensitive.

Chapter 10

Water Quality

Water Quality Notes

Common Constituents:

The most common constituents present in ground water are:

Cations (+)	Anions (–)
Calcium (Ca)	Bicarbonate (HCO_3)
Magnesium (Mg)	Sulfate (SO_4)
Sodium (Na)	Chloride (Cl)
Potassium (K)	Nitrate (NO_3)
Iron (Fe)	Fluoride (F)
Manganese (Mn)	Silica (SiO_2)

Less common, but which can be important, are: Boron (B), Arsenic (As), Lead (Pb), Selenium (Se), Copper (Cu), Zinc (Zn), Barium (Ba), Nitrite (NO_2), and the gases Oxygen (O_2), Carbon Dioxide (CO_2), Methane (CH_4), and Hydrogen Sulfide (H_2S).

Analytical Units

Water analyses usually report the concentration of each dissolved constituent , as weight-per-volume, in milligrams per liter (mg/l) — which are the same as parts per million (ppm). In some instances, the report may be in grains per gallon (gpg); one grain per gallon is equal to 17.118 milligrams per liter. The hydrogen-ion concentration (pH) is given in units between 0 and 14; a neutral water has a pH of 7.0. Hardness and alkalinity are usually expressed in mg/l as calcium carbonate ($CaCO_3$). The specific conductance or electrical conductivity (EC) is usually given in micromhos ($ECx10^{-6}$). Turbidity, if present, is expressed in turbidity units (NTU).

The concentration of dissolved constituents, in addition to reporting by weight, can be expressed in terms of chemical equivalence; the units are milliequivalents per liter (meq/l) — which are the same as equivalents per million (epm). One equivalent weight of a cation (such as sodium) will combine exactly with one equivalent weight of an anion (such as chloride) to give one equivalent weight of the compound sodium chloride. A table of factors to convert between mg/l and meq/l is given in this chapter. The units of equivalent weights are useful (and necessary) in some interpretation of typical water analyses — such as for the suitability of water for irrigation (calculating SAR and residual sodium carbonate). They can also be used to check the accuracy of an analysis as follows:

1. Convert the concentration of each constituent from mg/l to meq/l.
2. Add the results to determine the total cations (in meq/l) and total anions (in meq/l).
3. Total cations should be approximately equal to anions; if not there are some constituents not reported in the analysis, or the analysis is in error.
4. The percentage error is calculated by dividing the difference between total cations and anions by the sum of total cations and anions, and multiplying by 100.
5. Typical permissible analytical error depends upon total dissolved solids (TDS):

TDS (mg/l)	50	100	200	500	1000	2000
% Error	15	7	5	4	3	2

(Note: In some analyses the concentration of sodium (and potassium) has not been determined in the laboratory, but simply calculated from other constituents assuming *zero error;* in such cases the above checking is useless.)

Equivalent weights can also be used to convert some older water analyses (particularly those for industrial use) which may be reported in terms of chemical compounds. An example follows:

An analysis reports: Calcium sulfate 6.2 grains per gallon
Calcium chloride 1.9 grains per gallon

To determine the concentration of calcium, sulfate, and chloride ions in mg/l:

1. Convert grains per gallon of the compounds to mg/l:
   6.2 x 17.1 = 106.0 mg/l calcium sulfate
   1.9 x 17.1 = 32.5 mg/l calcium chloride
2. Convert mg/l of the compounds to meq/l (from conversion table):
   106.0 x 0.01469 = 1.56 meq/l calcium sulfate
   32.5 x 0.01802 = 0.59 meq/l calcium chloride
3. Convert the meq/l for each ion to mg/l (from conversion table):
   Calcium = (1.56 + 0.59) x 20.04 = 43.1 mg/l Ca
   Sulfate = 1.56 x 48.04 = 74.9 mg/l SO_4
   Chloride = 0.59 x 35.46 = 20.9 mg/l Cl

Hardness

Hardness results primarily from the calcium and magnesium in the water (and to a lesser extent iron and manganese, if present in significant amounts). Hardness is undesirable, if excessive, because of the precipitates with use of soap and because of scale formation when heated.

Hardness is usually expressed in mg/l as Total Hardness (as $CaCO_3$) and can be calculated from the calcium and magnesium (in mg/l) from a water analysis report as follows:

Total Hardness (as $CaCO_3$) = (Calcium x 2.497) + (Magnesium × 4.115) (all in mg/l)

also: Total Hardness *(in mg/l)* = 50 x (*meq/l* Ca + *meq/l* Mg)

The terms "carbonate hardness" (formerly "temporary hardness") and "non-carbonate hardness" (formerly "permanent hardness") are sometimes used. Carbonate hardness refers to that part of the total hardness attributed to the bicarbonate (and carbonate, if present) ions. Non-carbonate hardness is any part of the total hardness remaining that is attributed to sulfates or chlorides.

Some attempts have been made to classify water according to hardness, such as:

0 – 75 mg/l	Soft
75 – 150	Moderately hard
150 – 300	Hard
Over 300	Very hard

The perception of "hardness" is very subjective, however. What one person considers "hard", another may feel is "soft". In general, water with total hardness of 100 mg/l or less is entirely suitable for domestic use and most other uses.

Specific Conductance (or Electrical Conductivity) – Total Dissolved Solids

The specific conductance (or electrical conductivity – EC), expressed in micromhos, is a rapid and convenient way to determine approximately the Total Dissolved Solids (TDS) in a water sample. For most ground waters, the following relationship is valid:

Specific Conductance (micromhos) × "K" = TDS (mg/l)

where "K" is 0.65 (may range from 0.55-0.75)

and: Specific Conductance ÷ 100 = *meq/l* of cations (or anions)

For irrigation and drainage studies, where "tons of salt per acre foot" may be used:

TDS (mg/l) × 0.00136 = Tons/AF

and Tons/AF × 735 = TDS (mg/l)

Alkalinity

The alkalinity of water is determined by titrating the sample using methyl orange (MeO) or phenolphthalein (Phn) as indicators. For most ground waters, there is no carbonate present and the alkalinity is all because of bicarbonate.

When the alkalinity (MeO) is greater than zero and alkalinity (Phn) is zero:

Bicarbonate = 1.22 x Alkalinity (MeO)
Carbonate = 0

When the alkalinity (MeO) is greater than zero and alkalinity (Phn) is not zero but less than 1/2 the alkalinity (MeO):

Bicarbonate = 1.22 x (MeO - (2Phn))
Carbonate = 1.20 x Phn

In the above calculations, alkalinities are in mg/l as calcium carbonate, and concentrations of the bicarbonate and carbonate ions are also in mg/l.

Miscellaneous

Dissolved iron in excess of about 0.3 mg/l is likely to result in "red water" and cause staining of laundry and plumbing fixtures, and discoloration of tea and alcoholic beverages. Dissolved manganese in excess of about 0.05-0.1 mg/l will react similarly to iron but with a brown to black staining.

Most persons can detect chlorides in water at 250 mg/l or greater concentrations, and cannot tolerate drinking water with more than 300 mg/l carbonate, 1500 mg/l chloride, or 2,000 mg/l sulfate.

Sulfates in drinking water in excess of 500 mg/l will have a laxative effect on most persons; sulfate in excess of 250 mg/l may have a laxative effect until the user becomes accustomed to the water.

Livestock can tolerate increased mineralization in drinking water. Studies in Australia indicated approximate maximum limits of TDS to be: poultry – 2,800; pigs – 4,300; horses – 6,500; dairy cattle – 7,200; beef cattle – 10,000; adult sheep – 12,900.

Drinking Water Standards

Following passage of the Safe Drinking Water Act in 1974, the U.S. Environmental Protection Agency (EPA) began to establish standards for public water supplies. These standards include Maximum Contaminant Levels (MCL's) for certain inorganic constituents which are mandatory as levels never to be exceeded.Also established were recommended secondary (not mandatory) upper limits for other inorganic constituents. Listed below are the standards in effect in mid-1992.

NOTE: These standards are subject to revision; contact U.S. EPA or local health officials for current status.

Constituent	Maximum Limit (MCL)*	Constituent	Recommended Upper Limit*
Antimony	0.006	Aluminum	0.05-0.2
Arsenic	0.05	Chloride	250
Asbestos**	7	Color	15 color units
Barium	2	Corrosivity	Non-corrosive
Beryllium	0.004	Foaming agents	
Cadmium	0.005	(ABS)	0.5
Chromium	0.1	Iron	0.3
Copper****	1.3	Manganese	0.05
Cyanide	0.2	pH (units)	6.5-8.5
Fluoride	4	Silver	0.1
Lead****	0.015	Sulfate	250
Nickel	0.1	Total dissolved	
Nitrate (as N)	10	solids	500
Nitrite (as N)	1	Zinc	5
Selenium	0.05		
Thallium	0.002		
Adj gr alpha***	15 pCi/L		
Radium 226***	20 pCi/L		
Radium 228***	20 pCi/L		
Radon 222***	300 pCi/L		
Uranium***	30 µg/L		
Strontium 90	8 pCi/L		
Tritium	20,000 pCi/L		
Turbidity	1 NTU		

*All concentrations in mg/l except as noted.
**Million fibers/liter over 10 micron length
***Proposed standards
*****"Action levels" measured at point of use

Drinking Water Standards (Cont'd)

In mid-1992 the following Maximum Contaminant Levels (MCL's) existed, or are proposed, for Organics, Volatile Organics, and Pesticides/PCBs (with more to be added in the future):

Constituent	Mg/L	Constituent	Mg/L
Organics		**Volatile Organics**	
Benzene	0.005	o-Dichlorobenzene	0.6
Benzo(a) pyrene	0.0002	cis-1,2-Dichloroethylene	0.07
Carbon Tetrachloride	0.005	trans-1,2-Dichloroethylene	0.1
Di(ethylhexyl) adipate	0.5	1,2-Dichloropropane	0.005
Di(ethylhexyl) phthalate	0.006	Ethylbenze	0.7
p-Dichlorobenzene	0.075	Monochlorobenzene	0.1
1,2-Dichloroethane	0.005	Pentachlorophenol	0.001
1,1-Dichloroethylene	0.007	Styrene	0.1
Dichloromethane	0.005	Tetrachloroethylene (PCE)	0.005
Hexachlorobenzene	0.001	Toluene	1
Hexachlororcyclopentadiene	0.05	Xylenes (total)	10
PAHs	0.0002		
1,2-4-Trichlorobenzene	0.07	**Pesticides/PCBs**	
1,1,1-Trichloroethane	0.2	Alachlor	0.002
1,1,2-Trichloroethane	0.005	Aldicarb	0.003
Trichloroethylene (TCE)	0.005	Aldicarb sulfoxide	0.003
Trihalomethanes (total)	0.1	Aldicarb sulfone	0.003
2,3,7,8-TCDD (Dioxin)	3 × 10-8	Atrazine	0.003
Vinyl Chloride	0.002	Carbofuran	0.04
		Chlorodane	0.002
		1,2-Dibromo-3-chloropropane 9DBCP)	0.0002
		2,4-D	0.07
		Dalapon	0.2
		Dinoseb	0.007
		Diquar	0.02
		Endothall	0.1
		Endrin	0.002
		Etylene Dibromide (EBD)	0.00005
		Glyphosate	0.7
		Heptachlor	0.004
		Heptachlor epoxide	0.0002
		Lindane	0.0002
		Methoxychlor	0.04
		Oxamyl (Vydate)	0.2
		Pichloram	0.5
		Polychlorinated biphenyls (PCBs)	0.005
		Simazine	0.004
		Toxaphene	0.003
		2,4,5-TP (Silvex)	0.05

References on drinking water standards outside the United States include:

European Community (EC), "Relating to the Quality of Water Intended for Human Consumption, Council Directive 80/7778 EEC, July 15, 1980.

World Health Organization (WHO), "Guidelines for Drinking Water Quality, vols. I-III, 1984.

Conversion Factors

Between Milligrams per Liter (or Parts per Million) = Mg/l and Milliequivalents per Liter (or Equivalents per Million) = Meq/l

Constituent	A Mg/l to Meq/l	B Meq/l to Mg/l	C Mg/l to Mg/l (as $CaCO_3$)
Calcium	0.04991	20.04	2.4970
Magnesium	.08224	12.16	4.1151
Potassium	.02558	39.10	1.2798
Sodium	.04348	23.00	2.1756
Iron	.03582	27.92	1.7923
Bicarbonate	0.01639	61.01	0.8202
Carbonate	.03333	30.00	1.6680
Chloride	.02820	35.46	1.4112
Sulfate	.02082	48.04	1.0416
Hydroxide	.05879	17.01	2.9263
Nitrate	.01613	62.01	0.8070
Phosphate	.03158	31.67	1.5800
Calcium bicarbonate	0.01234	81.05	0.6174
Calcium carbonate	.01998	50.04	1.0000
Calcium sulfate	.01469	68.07	0.7351
Calcium chloride	.01802	55.50	0.9016
Magnesium bicarbonate	0.01367	73.17	0.6839
Magnesium carbonate	.02372	42.16	1.1869
Magnesium sulfate	.01661	60.20	0.6312
Magnesium chloride	.02100	47.62	1.0508
Sodium bicarbonate	0.01190	84.01	0.5956
Sodium carbonate	.01887	53.00	0.9442
Sodium sulfate	.01408	71.04	0.7044
Sodium chloride	.01711	58.46	0.8560

For any constituent:

Multiply mg/l by factors in column A to obtain meq/l
Multiply meq/l by factors in column B to obtain mg/l
Multiply mg/l by factors in column C to obtain mg/l as calcium carbonate

In water analyses, hardness (total, carbonate, or non-carbonate) and alkalinity are usually reported in terms of calcium carbonate.

Chemical Elements and Atomic Weights

Element	Symbol	Atomic Wt.	Element	Symbol	Atomic Wt.
Aluminum	Al	26.98	Manganese	Mn	54.94
Antimony	Sb	121.75	Mercury	Hg	200.59
Argon	A	39.95	Molybdenum	Mo	95.94
Arsenic	As	74.92	Neon	Ne	20.18
Barium	Ba	137.34	Nickel	Ni	58.71
Beryllium	Be	9.01	Nitrogen	N	14.01
Bismuth	Bi	208.98	Oxygen	O	16.00
Boron	B	10.81	Palladium	Pd	106.40
Bromine	Br	79.90	Phosphorus	P	30.97
Cadmium	Cd	112.40	Platinum	Pt	195.09
Calcium	Ca	40.08	Potassium	K	39.10
Carbon	C	12.01	Radium	Ra	226.00
Chlorine	Cl	35.45	Radon	Rn	222.00
Chromium	Cr	52.00	Selenium	Se	78.96
Cobalt	Co	58.93	Silicon	Si	28.09
Copper	Cu	63.55	Silver	Ag	107.87
Fluorine	F	19.00	Sodium	Na	22.99
Gold	Au	196.97	Strontium	Sr	87.62
Helium	He	4.00	Sulfur	S	32.06
Hydrogen	H	1.01	Tin	Sn	118.69
Iodine	I	126.90	Titanium	Ti	47.90
Iron	Fe	55.85	Tungsten	W	183.85
Lead	Pb	207.19	Uranium	U	238.03
Lithium	Li	6.94	Vanadium	V	50.94
Magnesium	Mg	24.31	Zinc	Zn	65.37

Water Analysis Units – Conversion Table

Units	Milligrams/ Liter* (Mg/L)	Grains/US Gallons (GPG)	Grains/ Imperial Gallion	Parts/ 100,000	Lbs/1000 US Gal
One Milligram/Liter	ONE	0.058	0.049	0.10	0.0083
One Grain/US Gallon	17.118	ONE	0.8331	1.712	0.143
One Grain/Imp Gallon	20.55	1.2003	ONE	2.06	0.172
One Part/100,000	10.00	0.583	0.486	ONE	0.0835
One Lb/1000 US Gal	119.8	6.98	5.80	11.98	ONE

*Milligrams per Liter (Mg/L) = Parts per Million (PPM).

Water Hardness Units – Conversion Table

Unit	Milligrams Per Liter*	Grains Per US Gallon	Clark Degrees	French Degrees**	German Degrees
One Milligram/Liter	One	0.058	0.07	0.10	0.056
One Grain/US Gallon	17.1	One	1.20	1.71	0.958
One Clark Degree	14.3	0.829	One	1.43	0.806
One French Degree	10.0	0.583	0.70	One	0.562
One German Degree	17.9	1.044	1.24	1.78	One

*Milligrams per liter = Parts per million
**French degrees = Hydrotimetric degrees

Milligrams per liter and grains per US gallon expressed as calcium carbonate
Clark degrees are grains per Imperial gallon as calcium carbonate
French degrees are parts per 100,000 as calcium carbonate
German degrees are parts per 100,000 as calcium oxide

Temperature Conversion Factors for Specific Conductance

Measurement Temperature			Measurement Temperature		
Degrees C	Degrees F	Factor	Degrees C	Degrees F	Factor
10.0	50.0	1.411	25.0	77.0	1.000
11.0	51.8	1.375	26.0	78.8	0.979
12.0	53.6	1.341	27.0	80.6	0.960
13.0	55.4	1.309	28.0	82.4	0.943
14.0	57.2	1.277	29.0	84.2	0.925
15.0	59.0	1.247	30.0	86.0	0.907
16.0	60.8	1.218	31.0	87.8	0.890
17.0	62.6	1.189	32.0	89.6	0.873
18.0	64.4	1.163	33.0	91.4	0.858
19.0	66.2	1.136	34.0	93.2	0.843
20.0	68.0	1.112	35.0	95.0	0.829
21.0	69.8	1.087	36.0	96.8	0.815
22.0	71.6	1.064	37.0	98.6	0.801
23.0	73.4	1.043	38.0	100.4	0.788
24.0	75.2	1.020	39.0	102.2	0.775

Multiply specific conductance observed at measurement temperature by factor in table to convert to specific conductance at standard temperature of 25C.

Dissolved Oxygen Saturation – Mg/L*

Temperature		Fresh Water	Sea Water	Temperature		Fresh Water	Sea Water
Deg. F	Deg. C			Deg. F	Deg. C		
32	0.0	14.6	11.5	60	15.6	10.0	8.1
35	1.7	14.0	11.0	65	18.3	9.5	7.7
40	4.4	13.0	10.3	70	21.1	9.0	7.4
45	7.2	12.1	9.7	75	23.9	8.5	7.0
50	10.0	11.3	9.1	80	26.7	8.1	6.6
55	12.8	10.6	8.6	85	29.4	7.7	6.3

*For fresh water and sea water exposed to air at standard atmospheric pressure at sea level.

For freshwater aquatic life, a minimum concentration of 5.0 mg/l is desirable to maintain good fish populations.

Galvanic Series of Metals

(In Order Of Increasing Corrosion Resistance)

Anode (Corroded)	Magnesium
	Magnesium alloys
	Zinc
	Aluminum 25
	Cadmium
	Aluminum 17ST
	Steel, iron, cast iron
	Chromium-iron (active)
	Ni-Resist
	18-8 Stainless steel (active)
	Lead, tin, lead-tin solder
	Nickel (active)
	Inconel (active)
	Brass, copper
	Bronze
	Monel
	Silver solder
	Nickel (passive)
	Inconel (passive)
	Chromium-iron (passive)
	18-8 Stainless steel (passive)
Cathode (Protected)	Silver
	Gold, platinum

Ryznar Stability Index

This Stability Index, proposed by J. W. Ryznar in 1944, is a means of relating the pH, calcium, total alkalinity, and total dissolved solids to appraise whether water will be corrosive or incrusting. If the Index is greater than 9.0 the water will tend to be corrosive and can be expected to cause early failure of well screens made of steel, Armco iron, silicon red brass, and Everdur metals. If the Index is less than 7.0 the water will tend to be incrusting and can be expected to deposit calcium carbonate in the screen openings and adjacent to the well. If the Index is between 7.0 and 9.0 the water is probably neither too incrusting or corrosive.

The Stability Index does *not* provide for corrosion caused by such factors as hydrogen sulfide, sulfate-reducing bacteria, or dissolved oxygen. The Index also does *not* provide for the incrusting effects resulting from such factors as iron, manganese, or iron bacteria.

To calculate the Index, it is necessary to know the following from a water analysis: Total dissolved solids (TDS); total alkalinity (methyl orange); calcium; and pH. The Index can be calculated from the following formula:

$$\text{Index} = 2\,(K - \log_{10}B - \log_{10}C) - pH$$

where: K is from the table below for total dissolved solids(TDS)
B is the calcium content in mg/l (or ppm)
C is the total alkalinity (MeO) in mg/l (or ppm)

(Note that total dissolved solids equals approximately 0.65 times the electrical conductivity (specific conductance) in micromhos; calcium is equal to 0.4 times the calcium *(not total)* hardness; total alkalinity equals the bicarbonate divided by 1.22).

TDS	K	TDS	K
100	11.60	800	11.98
150	11.65	900	12.01
200	11.71	1,000	12.03
250	11.76	1,500	12.11
300	11.78	2,000	12.17
400	11.85	2,500	12.20
500	11.90	3,000	12.24
600	11.93	4,000	12.30
700	11.96	5,000	12.33

Example: Calculate the Index for water with pH of 7.0, total dissolved solids of 400 mg/l, calcium of 50 mg/l, and total alkalinity of 200 mg/l:

For 400 mg/l TDS, K = 11.85; B = 50 and $\log_{10}50 = 1.6990$; C = 200 and $\log_{10}200 = 2.3010$

$$\begin{aligned}\text{Index} &= 2\,(11.85 - 1.699 - 2.301) - 7.0\\ &= 2\,(7.85) - 7.0\\ &= 15.7 - 7.0\\ &= 8.7\end{aligned}$$

Disinfection of Wells and Piping

A final step in well construction or pipeline installation should be disinfection with a chlorine solution. Disinfection of drilling tools, casing, and sand or gravel installed as a filter around well screens is also recommended. Chlorine solutions for disinfection are commonly made by dissolving calcium hypochlorite (70% available chlorine) such as HTH, Perchloron, etc.; chloride of lime (25% available chlorine); or using solutions of sodium hypochlorite (household bleach, 5.25% available chlorine). If dry chemicals are used, they should be thoroughly dissolved in water to form the chlorine solution. Sufficient chlorine solution should be introduced to bring the entire column of water in the well (or pipeline) up to at least a concentration of 100 mg/l chlorine. A stronger solution may have to be introduced so that after mixing in the well the final concentration will be at least 100 mg/l. The solution should remain in the well for 2 to 4 hours, preferably with some mixing or agitation (such as surging with the pump) in the last hour. The well should then be pumped until the odor of chlorine is gone.

The quantity of compounds required to prepare 100 gallons of solution of various strength is given in the accompanying table.

Material Required For 100 Gallons of Chlorine Solution

Chlorine Strength (mg/l)	Calcium Hypochlorite (70%) HTH, Perchloron, Etc) (Pounds)	Chloride of Lime (25%) (Pounds)	Sodium Hypochlorite Solutions (5.25% Bleach) (Quarts)	Sodium Hypochlorite Solutions (10%) (Quarts)
50	0.06	0.17	0.4	0.2
100	0.12	0.33	0.8	0.4
150	0.18	0.50	1.1	0.6
200	0.24	0.67	1.5	0.8
300	0.36	1.00	2.3	1.2
400	0.48	1.33	3.1	1.6
500	0.60	1.67	3.8	2.0

Tables in Chapter 3 give the volume (gal/ft) for sizes of holes and casings

Chapter 11

Hydrogeology and Ground Water Flow

Introduction to Chapter 11

Some of the material included in this chapter assumes that the user is already familiar with some of the limiting, underlying assumptions involved. This is particularly true for formulas used in analyzing pumping test data to determine hydraulic properties of an aquifer—such as homogeneity, horizontal flow, infinite extent, complete penetration, etc.

Readers wishing to obtain more detailed information on the flow of ground water and analytical methods should refer to one or more standard reference works, which include such publications as:

Bentall, R., (1963). Shortcuts and special problems in aquifer tests. U.S. Geological Survey Water-Supply Paper 1545-C, Washington, D.C.

Ferris, J.G., Knowles, D.B., Brown, R.H., and Stallman, R.W., (1962). Theory of aquifer tests. U.S. Geological Survey Water-Supply Paper 1536-E, Washington, D.C.

Kruseman, G.P. and deRidder, N.A., (1991). Analysis and evaluation of pumping test data (second edition). Publication 47, International Institute for Land Reclamation and Improvement (I.L.R.I.), Wageningen, The Netherlands.

Lohman, S.W., (1972). Ground-water hydraulics. U.S. Geological Survey Professional Paper 708, Washington, D.C.

Stallman, R.W., (1971). Aquifer test design, observation, and data analysis. U.S.Geological Survey Techniques of Water Resources Investigations, Book 3, Chapt. B1, Washington, D.C.

Todd, D.K., (1980), Groundwater hydrology (second edition). John Wiley & Sons, New York, N.Y.

U.S. Bureau of Reclamation, (1977). Ground water manual. U.S. Department of the Interior, Washington, D.C.

Walton, W.C., (1962). Selected analytical methods for well and aquifer evaluation. Illinois State Water Survey Bulletin 49, Urbana, Illinois.

Walton, W.C., (1987). Groundwater pumping tests, design and analysis. Lewis Publishers, Chelsea, Michigan.

Useful Formulas

Consistent units of both length and time must be used in these formulas. For example: drawdown and distance in *feet,* time in *days,* discharge in *cubic feet per day,* and transmissivity in *feet² per day.*

Unsteady state:

Theis (drawdown): $$s = \frac{Q}{4 \pi T} W(u)$$

where:

s = drawdown in a piezometer
r = distance from pumping well to piezometer
T = transmissivity of the aquifer
t = time since pumping started
S = storativity of the aquifer
Q = well discharge rate
$W(u)$ = "well function of u"; see table for conversion of u to W(u), or the W(u) vs u plot. Pump test data can be plotted (s on vert. axis and r^2/t on horiz. axis, then fitted to type curve

Theis (recovery): $$\Delta s = \frac{2.30\ Q}{4 \pi T}$$

where:

Δs = residual drawdown difference per log cycle of $\frac{t}{t'}$

t = time since pumping started
t' = time since pumping stopped
Q = well discharge rate
T = transmissivity of the aquifer

(plot residual drawdown against $\frac{t}{t'}$ on semi-log paper)

Cooper-Jacob ("straight-line" modification of Theis):

Time-Drawdown

$$T = \frac{2.30\ Q}{4\ \pi\ \Delta\ s}$$

$$S = \frac{2.25\ T\ t_o}{r^2}$$

Distance-Drawdown

$$T = \frac{2.30\ Q}{2\ \pi\ \Delta\ s} \text{ and } S = \frac{2.25\ T\ t}{r_0^2}$$

where:
t_0 = time intercept of semi-log plot with zero drawdown
r_0 = distance intercept of semi-log plot with zero drawdown
Δ s = drawdown difference over one log cycle
Q, T, S and t as above
(To be valid, $u \leq 0.01$, requiring than "r" be small, or "t" be large, or both)

Steady state:

Thiem (confined aquifer):

$$Q = \frac{2\ \pi\ T\ (s_1 - s_2)}{2.30 \log \frac{r_2}{r_1}}$$

where:
r_1 and r_2 are distances of piezometers from the pumping well
s_1 and s_2 are steady-state drawdowns in the piezometers
Q = rate of discharge of well
T = transmissivity of the aquifer

(If only one piezometer is available, an approximation of Q can be obtained by substituting the well radius for r_1 and using the well drawdown for s_1)

The above will also be approximately true for unconfined aquifers, if the drawdowns (s_1 and s_2) are small compared to the saturated thickness.

Well Interference:

Well interference, or the lowering of water level (drawdown) at one well caused by the pumping of a nearby well, can be calculated from the above formulas for unsteady- or steady-state flow conditions if the aquifer characteristics (transmissivity and storativity), distance from the pumping well, pump discharge rate, and duration of pumping are known.

The two accompanying semi-logarithmic graphs illustrate drawdowns surrounding a pumping well, under unsteady-state conditions (Theis formula) in a confined aquifer ($S = 1 \times 10^{-4}$), for two transmissivities ($T = 50$ and 500 ft^2/day), at a pumping rate of 19,250 ft^3/day (100 gpm), at distances from 1 to 1,000 feet, and for durations of 1, 10, and 100 days. Since drawdown is directly proportional to discharge, these graphs can be used for other discharge rates by multiplying by the appropriate factor (i.e., the drawdowns when pumping at a rate of 38,500 ft^3/day (200 gpm) would be twice those shown on the graphs—for a particular distance and duration of pumping).

Drawdowns surrounding a pumping well are not directly proportional to aquifer transmissivity, however. Another accompanying semi-logarithmic graph illustrates the relation between drawdown and transmissivity for the assumptions shown as to storativity, discharge rate, radial distance, and duration of pumping.

Values of W(u) for Values of u from 1.0 to 9.0 X 10^{-15}

u	1.0	2.0	3.0	4.0	5.0	6.0	7.0	8.0	9.0
x 1	0.219	0.049	0.013	0.0038	0.0011	0.00036	0.00012	0.000038	0.000012
x 10^{-1}	1.82	1.22	0.91	0.70	0.56	0.45	0.37	0.31	0.26
x 10^{-2}	4.04	3.35	2.96	2.68	2.47	2.30	2.15	2.03	1.92
x 10^{-3}	6.33	5.64	5.23	4.95	4.73	4.54	4.39	4.26	4.14
x 10^{-4}	8.63	7.94	7.53	7.25	7.02	6.84	6.69	6.55	6.44
x 10^{-5}	10.94	10.24	9.84	9.55	9.33	9.14	8.99	8.86	8.74
x 10^{-6}	13.24	12.55	12.14	11.85	11.63	11.45	11.29	11.16	11.04
x 10^{-7}	15.54	14.85	14.44	14.15	13.93	13.75	13.60	13.46	13.34
x 10^{-8}	17.84	17.15	16.74	16.46	16.23	16.05	15.90	15.76	15.65
x 10^{-9}	20.15	19.45	19.05	18.76	18.54	18.35	18.20	18.07	17.95
x 10^{-10}	22.45	21.76	21.35	21.06	20.84	20.66	20.50	20.37	20.25
x 10^{-11}	24.75	24.06	23.65	23.36	23.14	22.96	22.81	22.67	22.55
x 10^{-12}	27.05	26.36	25.96	25.67	25.44	25.26	25.11	24.97	24.86
x 10^{-13}	29.36	28.66	28.26	27.97	27.75	27.56	27.41	27.28	27.16
x 10^{-14}	31.66	30.97	30.56	30.27	30.05	29.87	29.71	29.58	29.46
x 10^{-15}	33.96	33.27	32.86	32.58	32.35	32.17	32.02	31.88	31.76

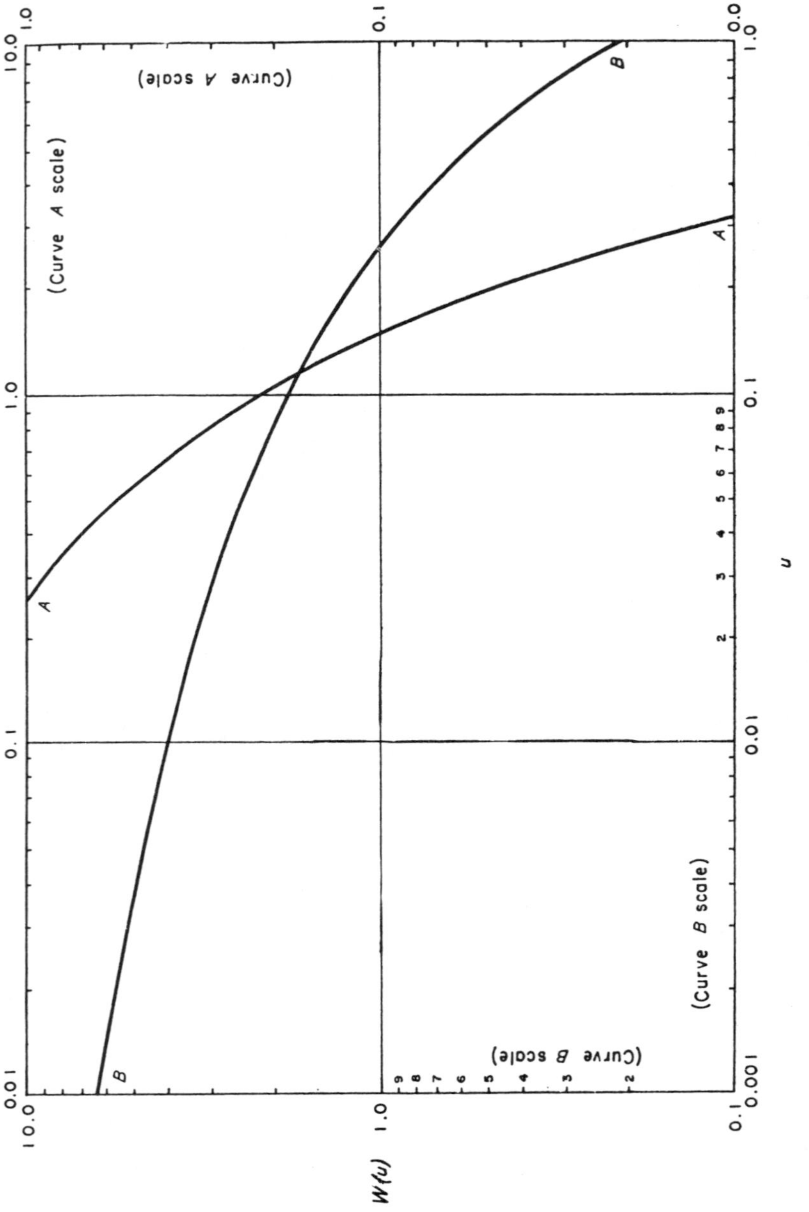

Logarithmic Graph of the Well Function W(u) — Constant Discharge

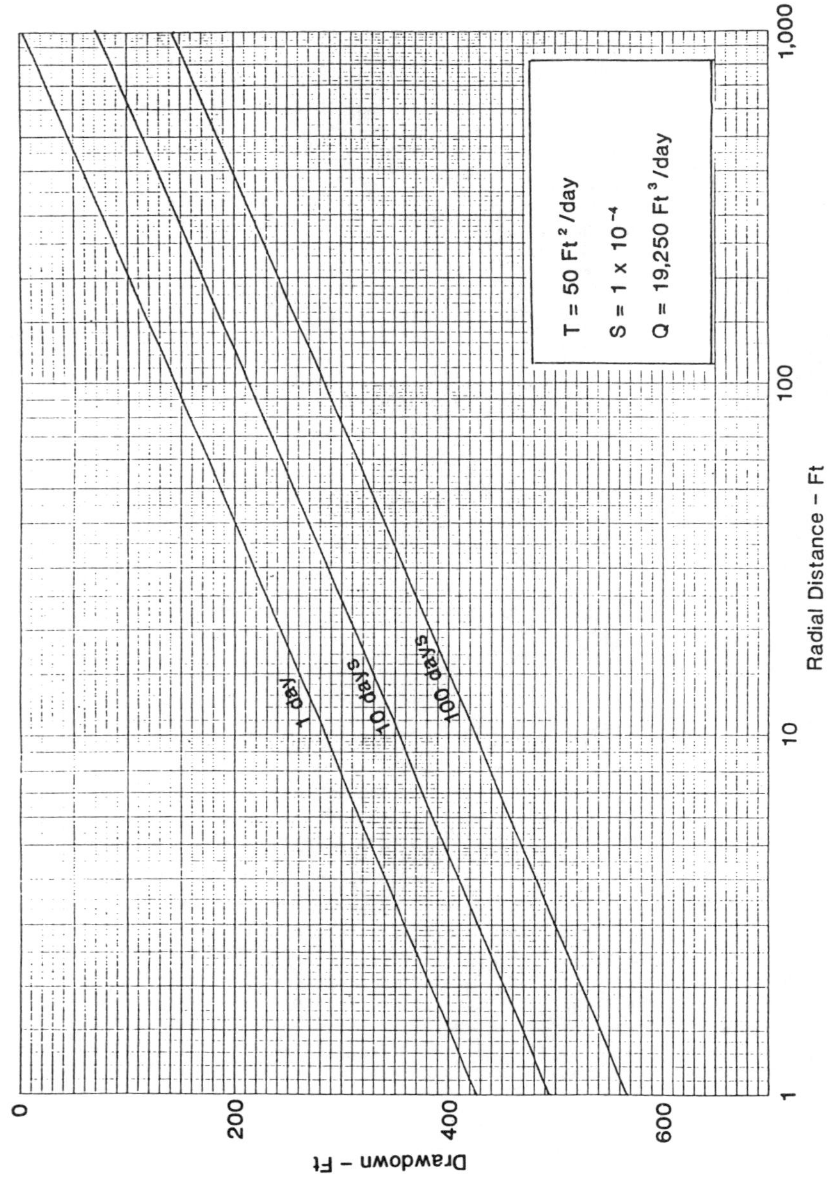

Drawdown vs. Radius, T = 50

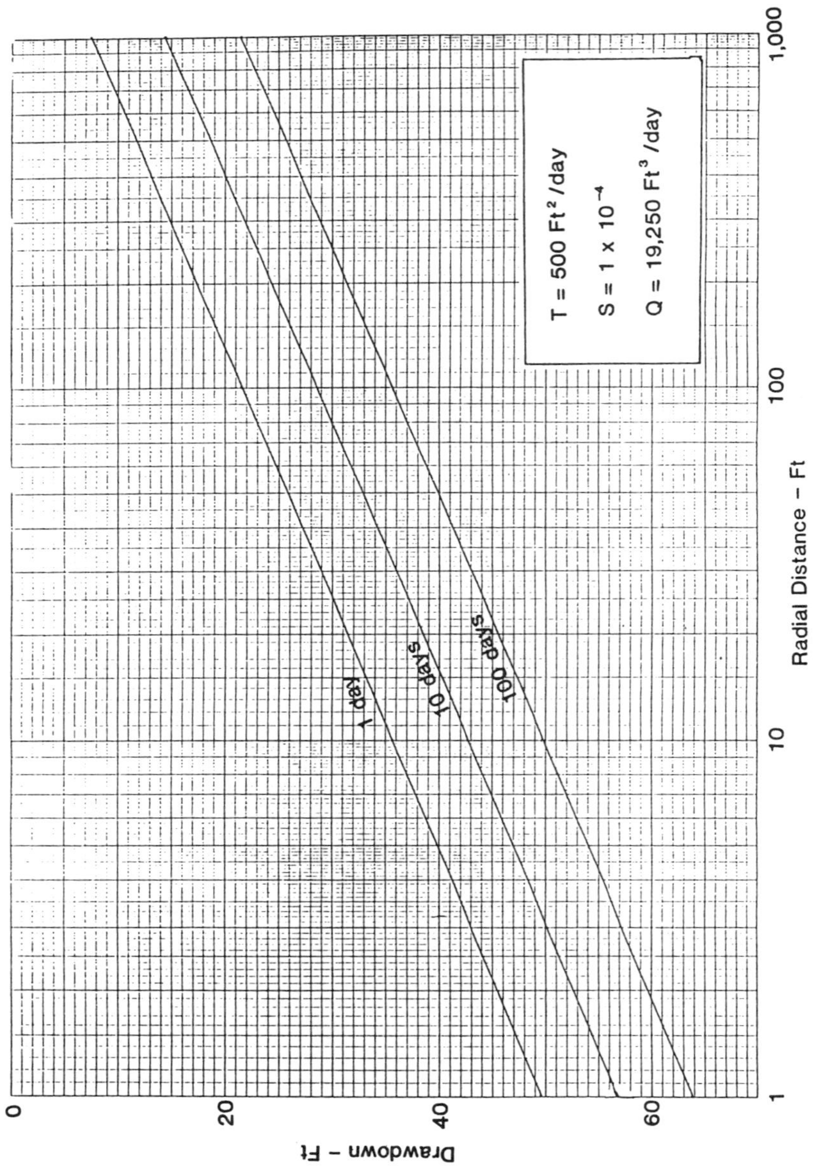

Drawdown vs. Radius, T = 500

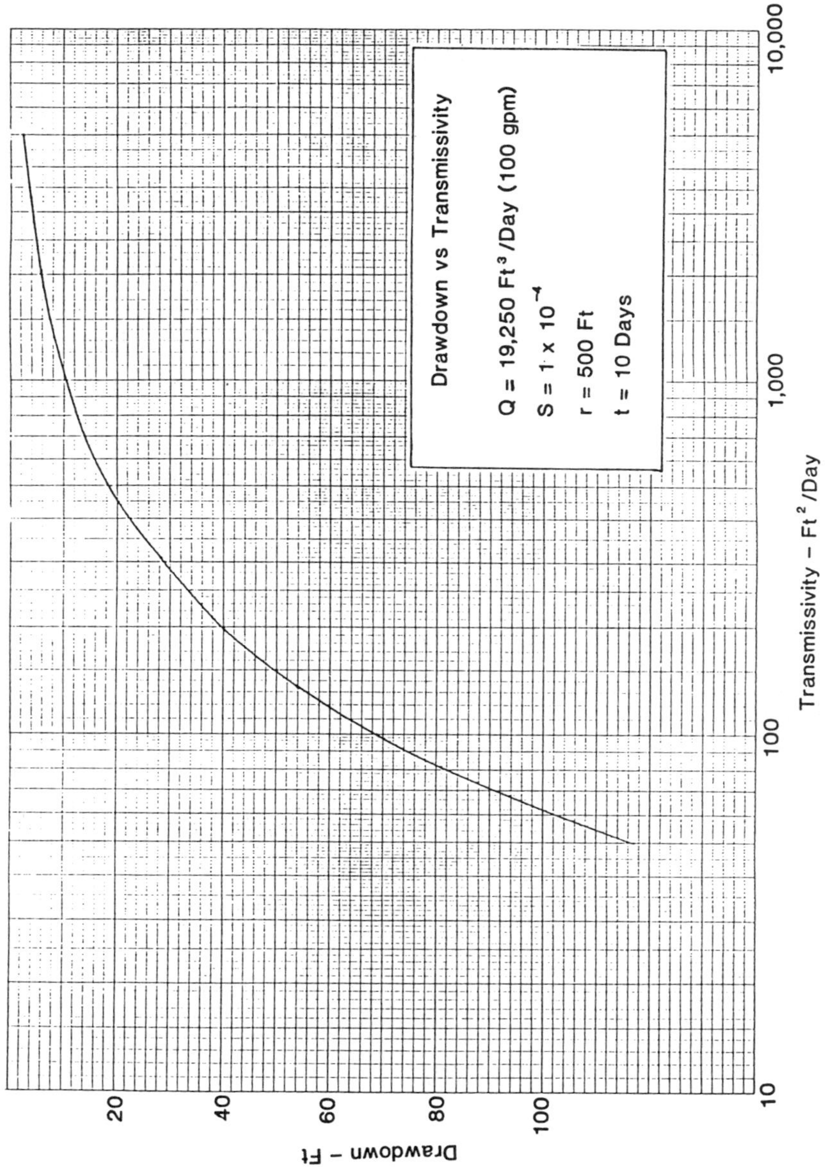

Drawdown vs. Transmissivity

Transient Flow to Linear Drain

The transient flow of ground water to a single linear drain in an infinite aquifer, with no recharge, can be determined from the following formulas (see accompanying sketch and table). The water level (h_x) at distance x from the drain is given by:

$$\frac{(h_o^2 - h_x^2)}{(h_o^2 - h_d^2)} = \text{erfc}\left(\frac{x/h_o}{2\sqrt{\beta}}\right)$$

where

$$\beta = \frac{T_a\, t}{S_y\, h_o^2}$$

$$T_a = \frac{K\,(h_o + h_d)}{2} \quad (= \text{average transmissivity})$$

erfc = complementary error function

S_y = specific yield

K = hydraulic conductivity (permeability)

t = time since flow started

The flow rate (Q) *per unit length* of drain after time period "t," with flow from *both sides* included:

$$\frac{Q\, h_o}{T_a\,(h_o - h_d)} = \frac{2}{\sqrt{\pi\,\beta}}$$

Volume (V) of flow *per unit length* of drain over time period "t," with flow from *both sides* included:

$$\frac{V}{S_y\, h_o\,(h_o - h_d)} = 4\sqrt{\frac{\beta}{\pi}}$$

For flow rate or volume for only one side, such as from bank storage along reservoir shoreline, use one-half of above calculations.

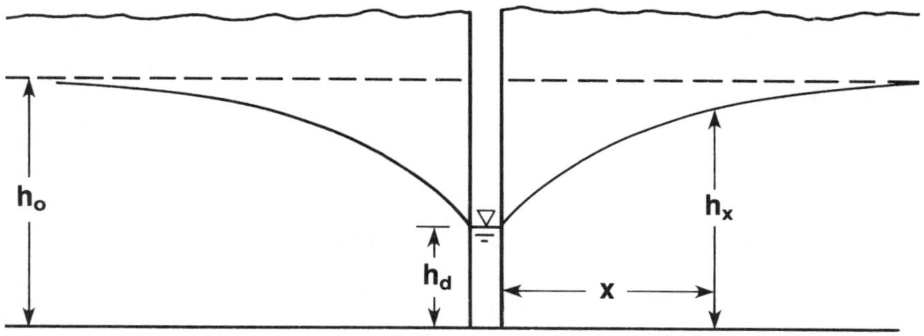

Flow to Linear Drain

Complementary Error Function (erfc)					
x	erfc (x)	x	erfc (x)	x	erfc (x)
0.00	1.000000	0.70	0.322199	1.80	0.010909
0.05	0.943628	0.75	0.288844	1.90	0.007210
0.10	0.887537	0.80	0.257899	2.00	0.004678
0.15	0.832004	0.85	0.229332	2.10	0.002979
0.20	0.777297	0.90	0.203092	2.20	0.001863
0.25	0.723674	0.95	0.179109	2.30	0.001143
0.30	0.671373	1.00	0.157299	2.40	0.000689
0.35	0.620618	1.10	0.119795	2.50	0.000407
0.40	0.571608	1.20	0.089686	2.60	0.000236
0.45	0.524518	1.30	0.065992	2.70	0.000134
0.50	0.479500	1.40	0.047715	2.80	0.000075
0.55	0.436677	1.50	0.033895	2.90	0.000041
0.60	0.396144	1.60	0.023652	3.00	0.000022
0.65	0.357971	1.70	0.016210		

Infiltration Galleries

Infiltration galleries, or horizontal screens installed typically adjacent to a stream or lake, can be a source of ground water when the aquifer may be too thin to allow installation of a typical vertical well. Flow can enter the gallery from one (or both) sides and from one (or both) ends, depending upon the type of installation and the geologic/hydrologic conditions.

Flow into *one side* of a gallery can be determined approximately from the equation:

$$Q = L\frac{T_a}{R}(H_2 - H_1)$$

where: Q = rate of inflow to one side
L = length of gallery
T_a = average transmissivity = $K\dfrac{(H_2 + H_1)}{2}$
K = hydraulic conductivity (permeability)
R = distance from line of recharge to gallery
H_1 and H_2 = saturated thicknesses (as shown)

(all above in *consistent* units)

Flow into *one end* of a gallery can be determined approximately from the equation:

$$Q = \frac{\pi T_a (H_2 - H_1)}{2.30 \log (R/r)}$$

where: r = radius of the gallery screen and the other notations as described above

(all above in *consistent* units)

The accompanying sketches illustrate these formulas, and the applicability to three typical conditions for field installation of infiltration galleries.

Design and Installation

Infiltration gallery screens are typically installed at a depth of from 5 to 25 feet below static water level. Unless the aquifer material is very coarse-grained, it is common to surround the screens with an artificial filter pack (similar to that for vertical wells). Screen slot width should be such that the entrance velocity is no greater than about 0.10 to 0.15 feet per second; screen diameter should be such that the velocity along the axis of the screen is not more than about 3.0 feet per second. The screen slot width should be small enough to retain all of the filter pack.

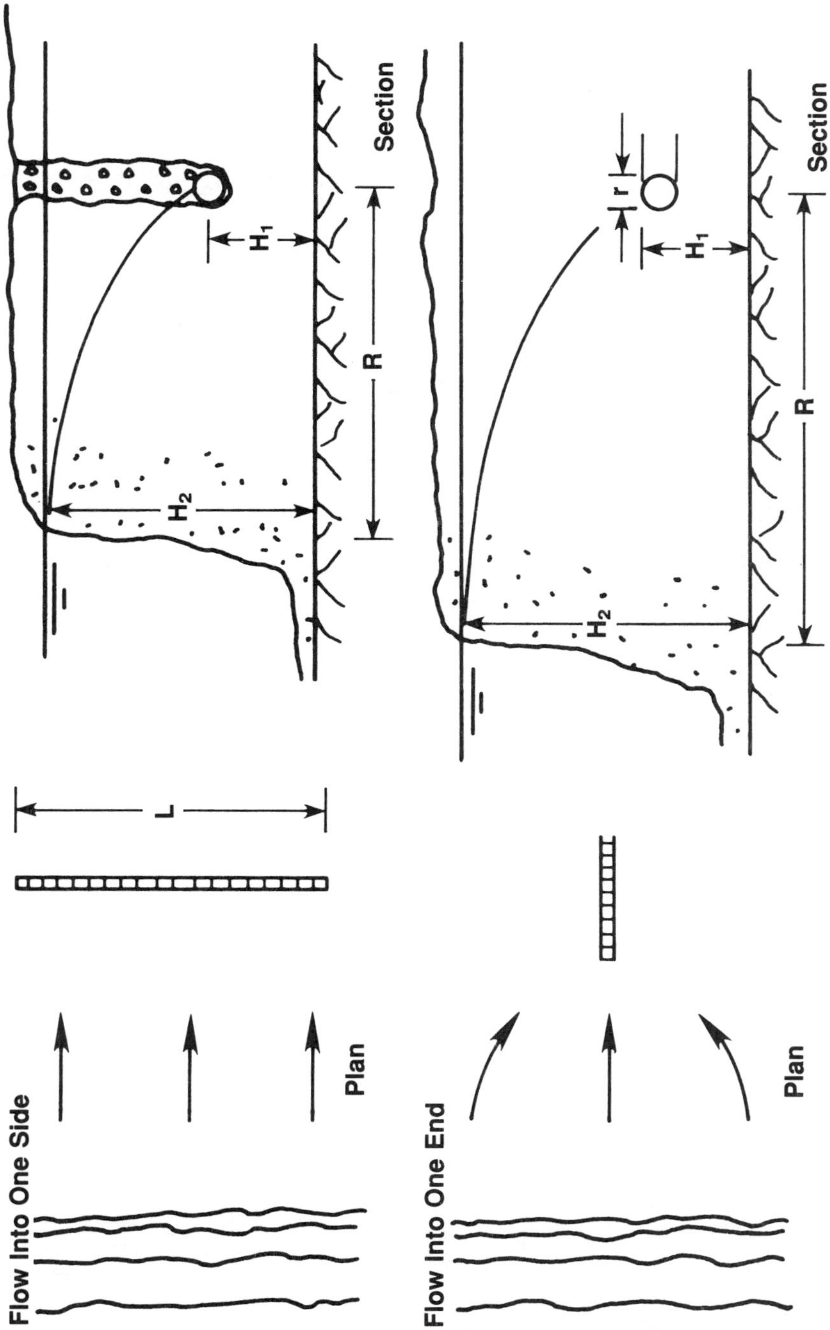

Infiltration Galleries

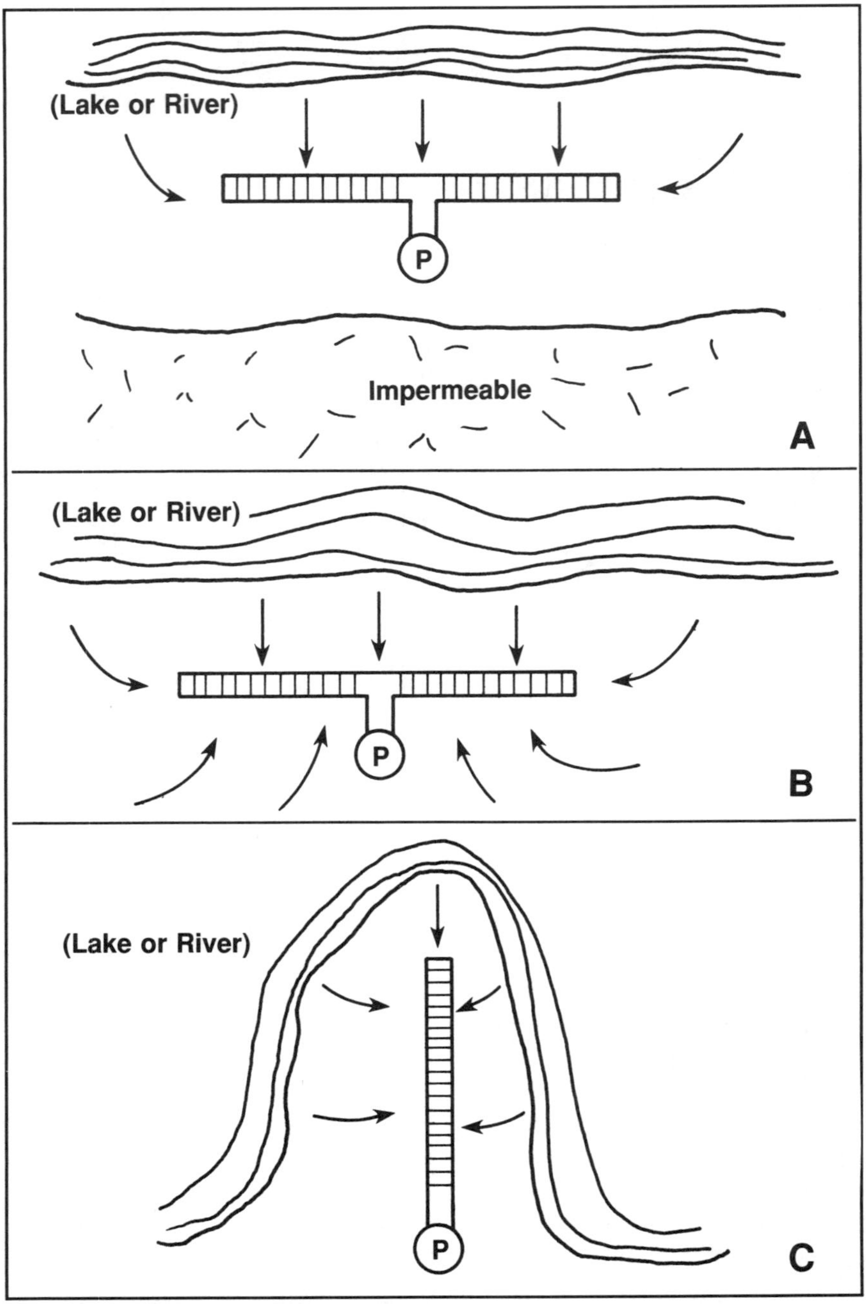

A – Flow into one side plus two ends
B – Flow into two sides plus two ends
C – Flow into two sides plus one end

Infiltration Galleries

Estimating Permeability and Transmissivity

A number of investigators have attempted to relate permeability to particle size. The Illinois State Water Survey made a study using field data to correlate particle-size distribution of aquifer materials (from sieve analyses) with permeability determined from pumping tests. This study established a fairly good correlation between the "effective grain size" (D_{10} , = 10% finer or 10% passing size). This relationship, which may be useful for estimating purposes when better data are not available, is shown on the accompanying graph.

A similar study, again using field data from numerous pumping tests, was undertaken to relate the transmissivity of sand and gravel aquifers to the specific capacity of wells. This study showed that these parameters, for most wells, fell within the range shown on the accompanying graph. This also may be useful for estimating purposes.

For preliminary estimating purposes, the following approximations may also be useful:

Confined Aquifers	**Unconfined Aquifers**
$T = Q/s \times 270$	$T = Q/s \times 200$

where: T = transmissivity in ft^2/day
Q/s = specific capacity in gpm/ft

or: $T = Q/s \times 120$ $\qquad$ $T = Q/s \times 90$

where: T = transmissivity in m^2/day
Q/s = specific capacity in l/s/m

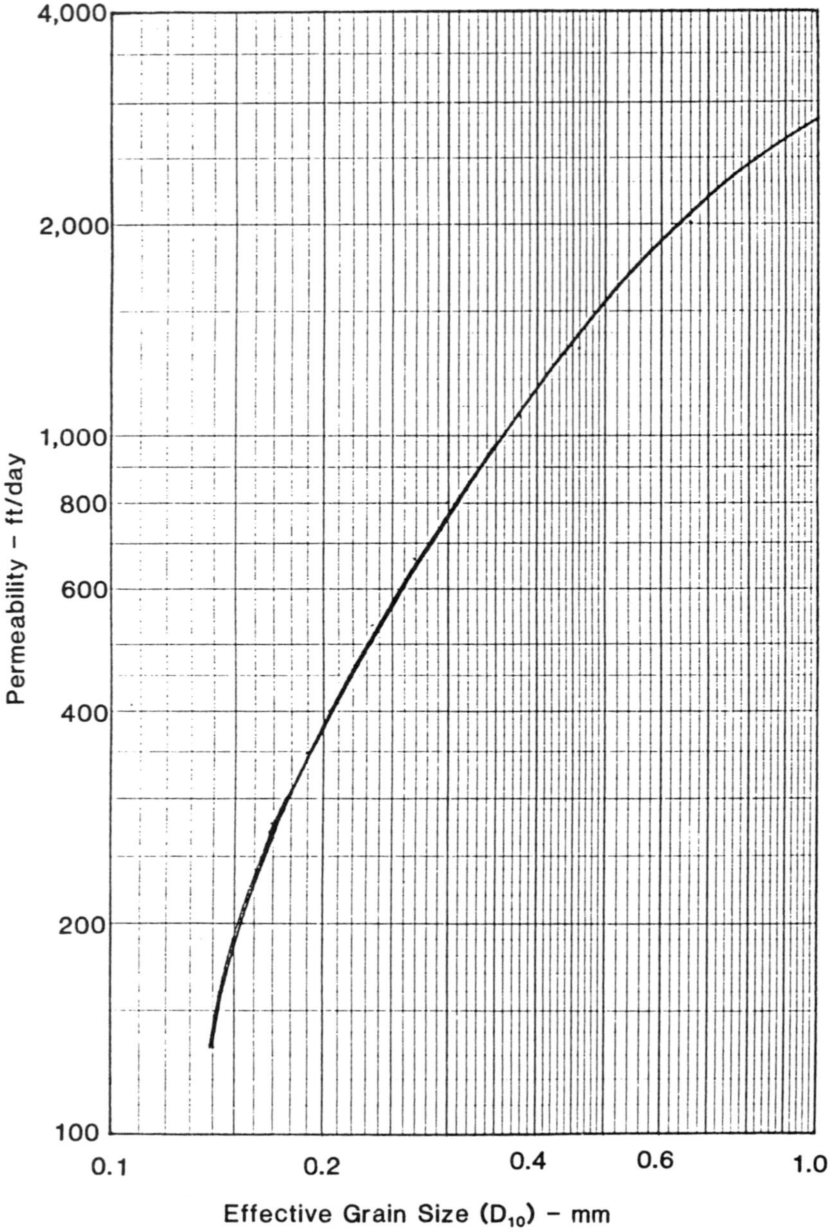

Estimating Permeability from Effective Grain Size

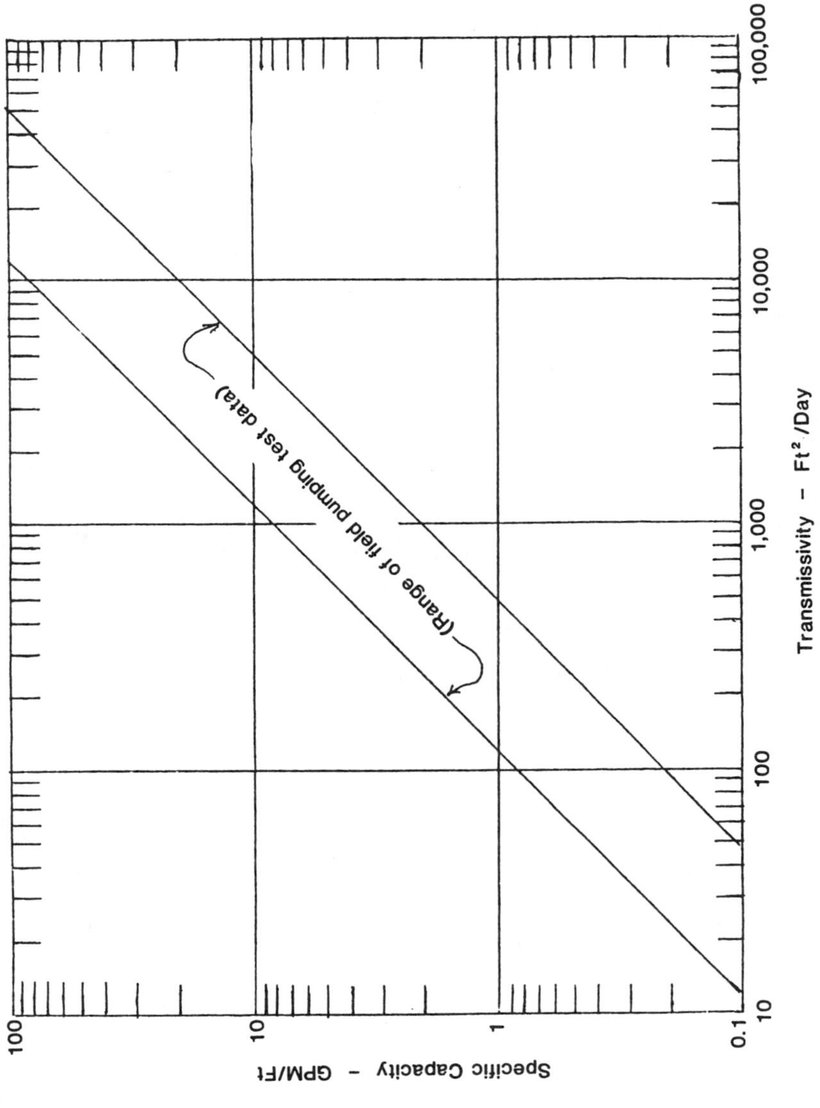

Specific Capacity vs Transmissivity from Field Tests

Theoretical Specific Capacity

Q/s (U.S. Gallons Per Minute/Foot) For Wells With 100% Efficiency

Transmissivity (gpd/ft)	Confined (Artesian) – S=0.0005				Unconfined (Water Table) – S=0.10			
	6-inch Well		12-inch Well		6-inch Well		12-inch Well	
	1 Day	10 Days	1 Day	10 Days	1 Day	10 Days	1 Day	10 Days
1,000	0.54	0.47	0.59	0.51	0.81	0.67	0.93	0.75
2,000	1.04	0.91	1.13	0.99	1.52	1.27	1.73	1.41
3,000	1.52	1.34	1.66	1.45	2.20	1.85	2.50	2.05
4,000	2.01	1.77	2.17	1.90	2.86	2.42	3.27	2.69
5,000	2.47	2.18	2.68	2.35	3.50	2.96	3.95	3.26
10,000	4.64	4.21	5.12	4.52	6.70	5.70	7.50	6.20
15,000	6.98	6.21	7.54	6.65	9.70	8.30	10.8	9.10
20,000	9.16	8.17	9.90	8.75	12.6	10.8	14.0	11.8
30,000	13.4	12.0	14.5	12.8	18.4	15.9	20.5	17.4
40,000	17.6	15.8	19.0	16.9	24.0	20.8	26.7	22.7
50,000	21.9	19.6	24.4	20.9	29.5	25.5	32.6	27.9
60,000	26.0	23.3	27.9	24.8	35.0	30.4	38.8	33.1
70,000	30.1	27.0	32.3	28.7	40.6	35.2	44.2	38.4
80,000	34.1	30.7	36.6	32.7	46.0	39.8	50.6	44.4
100,000	42.1	38.0	45.1	40.3	56.7	49.4	62.2	53.4
125,000	52.2	47.0	56.0	50.0	69.8	60.8	76.5	65.8
150,000	62.1	56.0	66.5	59.6	82.8	72.4	90.7	78.1
175,000	71.8	65.0	76.9	68.9	95.7	83.5	105	90.4
200,000	81.8	73.7	87.5	78.5	108	95.0	118	102

Pumping Tests

Introduction

Pumping tests are made on wells for a number of reasons, including determining aquifer characteristics, locating aquifer boundaries, selecting permanent pump discharge rate and depth of setting, predicting future drawdowns, estimating pumping levels at higher discharge rates, and determining well efficiency.

All pumping tests essentially include measurement of the discharge from the pumped well, measurement of water levels (or pressures) in the pumped well (and observation wells if available) during both the pumping and recovery periods, and recording the time of all measurements. Other observations can include periodic measurements of water temperature and quality (such as pH and specific conductivity), and collection of water samples for chemical analysis.

Some advance planning for any test is desirable in order to schedule the type and frequency of observations to be made, and the equipment and personnel that will be needed. The purpose for making the test should be a primary factor in the planning — determining the capacity and depth of setting for a domestic pump may only require a few hours of pumping with measurements of water level in the well and rate of discharge; estimating the ability of an aquifer to supply a large, long-term water supply for a municipality or industry may require a test of several days with a number of observation wells and frequent measurements at all points.

Records

Complete, accurate, and detailed records should be kept for the entire test. It is helpful to prepare forms for recording data in advance, with separate sheets for each location (pumping well, observation well A, observation well B, river stage, etc.). Each form should give basic information about that location: for the pumping well, include pump data (depth of setting, number and size of stages, length of airline if used, etc.), measuring point for water levels, method used for discharge measurements, etc.; for observations wells, include distance and direction from pumping well, measuring point for water levels, etc.

All forms should have separate columns for recording the time of all observations and for measurement of water levels; if for a pumping well it should also include columns for the discharge measurements (to show manometer readings, head on the weir, flowmeter readings, time to fill a container, or other applicable data). A separate column for "Remarks" should be provided to note such items as water temperature or quality, appearance (turbidity, sand, etc.), occurrence of rain or other climatic factors that may affect the test, time of collecting water samples, starting or stopping of the pump, changing of discharge rate, engine rpm if applicable, name of the observer, etc. *It is seldom possible to collect and record too much data!* All records should be permanently preserved.

Time Measurements

Each column for time observations should start with the day and calendar date. Observations should be shown to the nearest minute. It is probably easiest to record the actual clock time for each observation, which can then be converted later to "minutes since pumping started" or "minutes since pumping stopped." The same timepiece should be used throughout the test; if there are more than one observer, the watches should be synchronized at the start of the test. If an observation is not made at exactly the planned time, the *actual time of observation* should be the one recorded.

Water Level (or Pressure) Measurements

There are many ways to measure water levels or pressures during a pumping test. The principal ones include: airlines, electric sounders, transducers, wetted tapes, automatic float-type recorders, acoustic devices, water or mercury manometers, bourdon gages, and even timing how long it takes (with a stop watch) to hear a pebble that is dropped hit the water!

Airlines:

The airline is probably one of the most widely used methods to measure depth to water in a pumping well. The accuracy of measurement is governed by the accuracy of the pressure-measuring device; the accuracy is not sufficient for use in observation wells.

The operation of an airline is shown on the accompanying drawing. The airline itself is a small diameter metal or plastic tube installed below the maximum pumping level (usually down to or near the pump intake), fitted at the surface with a Schrader (tire) valve and pressure gage. In operation, air (or an inert gas) is pumped into the airline until a maximum pressure is reached on the gage — at which point the air will be bubbling out the bottom of the airline. The pressure at the gage is then equal to the distance from the water level in the well to the bottom of the airline and, by knowing the length of the airline, this distance can be subtracted from that length to obtain the depth to water below the gage. Bourdon gages are commonly used for the pressure measurement; if the gage reads in psi it must first be converted to feet by multiplying by 2.31. During falling water level conditions (drawdown) it is usually not necessary to keep adding air for each measurement because the expanding air continues to evacuate the line; during rising water levels (recovery) is is necessary to re-pressure the line for each observation. *CAUTION:* the source of air *must* have sufficient pressure to blow all the water out of the airline. A typical manual tire pump cannot usually develop more pressure than about 40 psi (or about 90 feet); for airlines submerged to greater depths it may be necessary to use a motor-driven compressor, or a tank or cylinder of compressed air or nitrogen, to obtain a true reading.

Electric Sounders:

Another common, and accurate, method of measuring water levels is the electric sounder. This is typically a two-conductor wire (or tape) on a reel with a weighted electrode on the lower end. When the electrode touches the water surface, it completes a battery-operated circuit and operates a signal or indicator (buzzer, meter, or light). The wire is marked at intervals of 5- or 10-feet and the depth to water calculated by measuring from the point that the indicator operates to the nearest calibration mark on the wire. The electrode is normally shielded in such a manner that falling or cascading water in the well will not give a false reading. In measuring very deep water levels, or where there is not much clearance between the pump and the casing, it may help to first install a small diameter tube (similar to an airline but with sufficient diameter to accommodate the electrode) and then measure inside the tube. This is also helpful if there is oil on the water surface in the well.

Transducers:

Electrical transducers are being used increasingly to measure water levels accurately, particularly where very frequent early-time measurements are needed, or where there are many observation wells (requiring extra personnel for the test). The transducers are pressure-sensing devices that record the depth of water over the instrument. Commercial equipment is now available that will record pressures at pre-determined time intervals and store the information for later down-loading into a personal computer data base.

Wetted Tapes:

Wetted tapes have been used for decades to measure water levels accurately. The method consists of rubbing chalk (such as carpenter's blue chalk) on the lower end of a tape, lowering the tape into the well until it enters the water, holding an even foot mark at the measuring point, then removing the tape and measuring the wetted depth (cut). The depth to water is then the "hold" minus the "cut." The method is good for observation well measurements. Disadvantages are the time it takes to chalk the tape, insert it, remove it, and make the wetted measurement for each observation. The method is also not satisfactory during drawdown or recovery , when water levels change rapidly, or where there is falling or cascading water.

Water Level Recorders:

Automatic water level recorders are widely used to continuously measure and record the water level in observation wells or on bodies of suface water such as drains, streams, lakes, or reservoirs. Observation wells must be of sufficient diameter to accommodate the float which usually requires a well with at least a 4-inch diameter.

Acoustic Methods:

Acoustic-type instruments are available that measure depths to water by the time it takes for a sound wave generated at the surface to travel down to the water surface in the well and echo back to the surface. These are particularly suited for measurement of very deep water levels. Another type of "acoustic instrument" is the "popper." This is a cylindrical weight with concave bottom that is attached to a cloth (or steel) measuring tape. When the weight just touches the water surface a sharp "pop" can be heard and the depth noted on the tape.

Manometers:

Where a test involves wells that have water levels above land surface (flowing artesian conditions), pressure measurements can be made accurately by using mercury manometers, and converting inches of mercury to feet of water (multiply inches of mercury by 1.133). If the water level is only a few feet above land surface, a standpipe or glass or plastic tube can be extended above ground and the level measured directly. Mercury manometers can also be used for airline pressure measurement.

Bourdon Gages:

Standard bourdon-type pressure gages, with circular dial, can be used to measure water levels (pressure) above land surface. The gage selected should be of the lowest range possible, for the maximum pressure condition, to improve accuracy. For maximum accuracy, these gages should be tested (calibrated) on a dead-weight test machine.

Flow Measurement

The pump rate of discharge can be measured in a number of different ways. The most common methods are discussed more fully in Chapter 6 "Flow Measurement" of this handbook.

Conducting the Test

It is desirable to measure water levels for several days prior to the test so that any regional trend can be observed and an appropriate correction applied to the measurements made during the test. It may also be desirable to pump the well briefly in advance of the test , either at one or several discharge rates, to establish the best rate that can be maintained throughout the test and to select the proper method of measuring discharge (selecting orifice diameter, weir crest length, etc.)

Tests are normally made with the pump operating at a constant discharge rate. Where the well to be tested flows at the surface, a "constant drawdown" test can be made by shutting in the well to obtain a static level (pressure) then opening the discharge fully and making measurements of the decline in discharge (flow) rate. For constant discharge tests, a valve should be installed on the pump discharge line so that the desired flow rate can be maintained. If the pump is engine-driven, a tachometer is useful to check the rotating speed of the pump shaft.

If the aquifer being tested is confined (artesian), water levels may have to be corrected to account for the effect of changes in barometric pressure or for tidal effects (ocean or earth tides). A recording barograph can be used at the site, if available, or periodic observations made with a barometer or barometric altimeter (or hourly barometric pressure readings from a nearby airport control tower or weather station). The attraction of the moon, and to a lesser extent the sun, causes earth tides that can produce twice daily (semi-diurnal) water-level fluctuations of as much as nearly one-tenth foot. The time of one cycle of earth tide is about 12 hours 25 minutes, with two cycles in the lunar day of about 24 hr 50 min. Ocean tides have approximately the same semi-diurnal cycle time although in some parts of the world there is only a once daily (diurnal) cycle. Local tide tables are published for coastal locations that give the time of high and low tides; data on time of moon passage are given in the annual Nautical Almanac published in Washington, D.C. by the United States Naval Observatory, or in London by Her Majesty's Stationery Office.

If the test is on a shallow unconfined aquifer, the water discharged by the pump during the test must be carried away a sufficient distance, preferably in pipe, so that it does not recirculate quickly to the aquifer. If there are nearby bodies of surface water that could influence the measurements in wells, those water levels should also be monitored. Temperature measurements are important when testing very deep or geothermal wells, because the density of water decreases with increasing temperature and measured water levels may need to be corrected for proper interpretation of the data. If contouring the water table or piezometric surface is important, the elevation of all measuring points should be determined by leveling.

When the pump is stopped, the recovery (rise) of water levels should be measured at all points in the same manner as when pumping started.

Measurement Frequency:

To the extent possible, water level measurements should be made so that there are at least eight observations during each log cycle of time. Such a schedule for measurement (for both drawdown and recovery) could be:

First 10 minutes:	at 1, 1.5, 2, 3, 4, 6, 8, and 10 minutes
From 10–100 minutes:	at 15, 20, 25, 30, 40, 60, 80, and 100 minutes
After 100 minutes:	at 1- or 2-hour intervals until completion

Duration of Test:

The duration of a pumping test will depend largely upon such factors as the purpose of the test, the character of the aquifer (confined or unconfined), the existence of boundaries or leaky aquifer conditions, the apparent adequacy of the data from plots made in the field, and available funds and personnel. Wells for domestic or livestock use do not normally need to be tested more than 2 to 4 hours. Wells for irrigation, industrial, and municipal use should be tested for at least 8 to 12 hours and often for 24 hours. Tests to determine aquifer properties carefully, to determine boundaries or leaky conditions, or to predict long-term conditions, may require days or even weeks.

Unconfined aquifer tests will require longer pumping than for confined aquifers where the effects of pumping spread rapidly. Water levels (or drawdowns) should be plotted in the field as the test is in progress. Pumping should continue until the water level plotted against the logarithm of time falls on a straight line (or, for three or more observation wells, the plot against logarithm of distance from the pumped well falls on a straight line). If boundaries or leaky aquifer conditions are present, the data plot will begin to depart (upward or downward) from the initial straight line, and pumping should continue. If estimates of transmissivity and storativity can be made, pumping should usually be continued until data from the most distant observation point provides for the factor "u" to be ≤ 0.01 ("u" is defined in the section "Useful Formulas" in this Chapter).

Measurement of recovery water levels should continue if possible, after the pump stops, for at least 1/4 to 1/2 of the time that the pump operated.

If the pump stops during a planned test, it should be restarted as quickly as possible and the exact times of stopping and restarting noted and recorded. If the interruption is not long, the test should be continued but for an additional time at least equal to the interruption. If the interruption is long (4 or more hours in a 24-hour test) the test should probably be abandoned and redone at a later time.

Other Types of Tests

Several other types of tests may be useful, depending upon field conditions and the objective for testing.

Step-Drawdown Tests:

In a step-drawdown test, the well is pumped at a low, constant rate until the water level has essentially stabilized, say 1 to 2 hours; then the discharge increased to a higher rate until the level again has essentially stabilized; then the procedure is repeated at successively higher pumping rates. At least three steps are normally done. Data from this type of test can be used to determine the well efficiency — determining what part of the total observed drawdown is caused by linear flow through the aquifer and what part is caused by turbulent flow within the aquifer adjacent to the well bore or as entrance loss.

Slug Tests:

These tests involve the rapid removal from (or introduction into) a well of a "slug" or small volume of water — followed by measurement of the recovering rise (or fall) of water level in the well. A slug can be removed from the well with a bailer, or by placing a solid cylinder beneath the water level, allowing the water level to stabilize, and then removing the cylinder rapidly. These tests are popular because no pump is required, they can be conducted in a short time, and no observation wells are needed. Data from slug tests can be used to analyze for aquifer paramaters in both confined and unconfined aquifers, but only determine conditions for a volume of material very close to the well.

Bailer Tests:

Bailer tests are used, usually during drilling operations, to get an approximate indication of the specific capacity at a particular well depth. The water level is allowed to stabilize prior to the test to establish a static water level. Water is then taken out, using a bailer of known volume, as rapidly as possible and the bailing continued until the depth at which the bailer is heard to strike the water has essentially stabilized. By keeping track of the number of full bailers taken out in a unit time (such as 10 or 20 minutes) the approximate rate of discharge can be calculated. The approximate drawdown for that rate is the difference between the static level and the "stabilized" depth at the end of bailing.

Interpretation and Analysis of Data

Suggested reference publications discussing the interpretation and analysis of pumping test data are given in the Introduction to this Chapter.

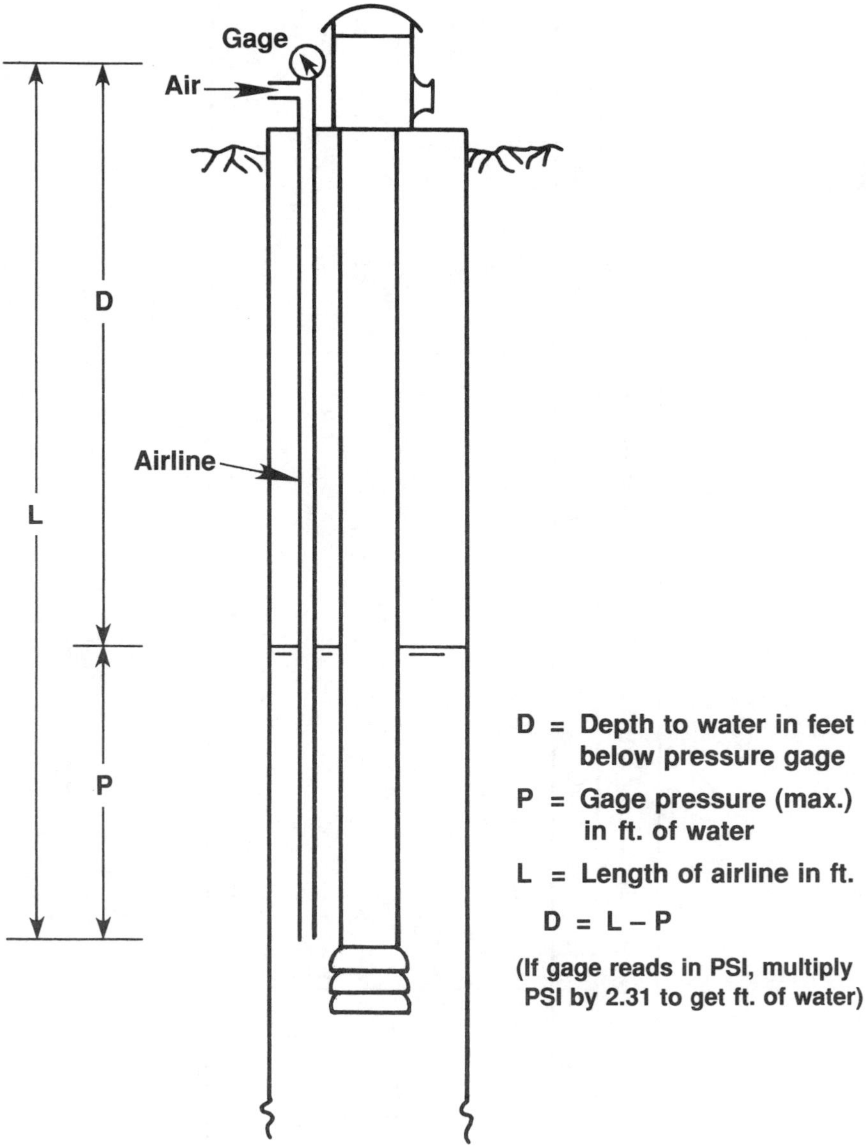
Gage
Air
Airline
D
L
P
D = Depth to water in feet below pressure gage
P = Gage pressure (max.) in ft. of water
L = Length of airline in ft.
D = L – P
(If gage reads in PSI, multiply PSI by 2.31 to get ft. of water)

Airline

Hydraulic Conductivity Conversions

	Ft/Yr	Ft/Day	Ft/Hr	Ft/Min	Ft/Sec	In/Day	In/Hr	Gal/Ft²/Day	M/Day	Cm/Hr
Ft/Yr =	ONE	2.74×10^{-3}	1.141×10^{-4}	1.903×10^{-6}	3.171×10^{-8}	3.287×10^{-2}	1.370×10^{-3}	2.049×10^{-2}	8.35×10^{-4}	3.479×10^{-3}
Ft/Day =	365	ONE	4.167×10^{-2}	6.945×10^{-4}	1.157×10^{-5}	12	0.50	7.4805	0.305	1.270
Ft/Hr =	8,760	24	ONE	1.667×10^{-2}	2.778×10^{-4}	288	12	179.5	7.32	30.48
Ft/Min =	525,600	1,440	60	ONE	1.667×10^{-2}	17,280	720.0	10,772	438.9	1,829
Ft/Sec =	31,536,000	86,400	3,600	60	ONE	1,036,800	43,200	646,315	26,335	109,728
In/Day =	30.42	8.333×10^{-2}	3.472×10^{-3}	5.787×10^{-5}	9.645×10^{-7}	ONE	4.166×10^{-2}	0.6234	2.54×10^{-2}	0.1058
In/Hr =	730	2.0	8.333×10^{-2}	1.389×10^{-3}	2.315×10^{-5}	24	ONE	14.96	0.61	2.540
Gal/Ft²/Day =	48.78	0.1337	5.569×10^{-3}	9.282×10^{-5}	1.547×10^{-6}	1.604	6.682×10^{-2}	ONE	4.07×10^{-2}	0.1697
M/Day =	1,198	3.28	0.1368	2.27×10^{-3}	3.78×10^{-5}	39.38	1.64	24.54	ONE	4.167
Cm/Hr =	287.4	0.7874	3.281×10^{-2}	5.469×10^{-4}	9.114×10^{-6}	9.449	0.3939	5.890	0.24	ONE

Temperature of Ground Water

The temperature of ground water at a depth of about 50 feet, unless influenced by nearby bodies of surface water (streams or lakes), is usually fairly constant throughout the year and is typically 2 to 3 degrees (F) higher than the mean annual air temperature at the site. Ground water temperatures will increase with greater depth, typically at about 1.5 degrees (F) for each additional 100 feet of depth.

Water from very shallow wells, or from wells (or galleries or collectors) that derive a large part of their supply by infiltration from surface water, will typically have temperatures that vary seasonally. This seasonal variation generally follows the pattern of air or surface water temperature changes — and usually lags by a month or two.

Capillary Rise of Water

Material	Grain Size		Capillary Rise	
	Millimeters	Inches	Centimeters	Feet
Gravel, fine	5 – 2	0.2 – 0.08	2.5	0.08
Sand, very coarse	2 – 1	0.08 – 0.04	6.5	0.21
Sand, coarse	1 – 0.5	0.04 – 0.02	14	0.46
Sand, medium	0.5 – 0.2	0.02 – 0.008	25	0.82
Sand, fine	0.2 – 0.1	0.008 – 0.004	43	1.4
Silt	0.1 – 0.05	0.004 – 0.002	106	3.5
Silt	0.05 – 0.02	0.002 – 0.0008	200	6.6

Numbering System for Wells and Springs

In areas using the Township and Range subdivision of the public lands, (particularly in the central and western United States), the U. S. Geological Survey has adopted a standard numbering system for wells and springs, with reference to the appropriate local standard base line and meridian.

The first two segments of the number designate the township and range. The third segment gives the number of the section, followed by three letters which give the 1/4-section (160-acre tract), the 1/4-1/4 section (40-acre tract), and the 1/4-1/4-1/4 section (10-acre tract), and lastly a number indicating the order in which the well was inventoried. Quarter sections are lettered A, B, C, and D (or sometimes a, b, c, d) in a counter-clockwise direction starting with the northeast quarter; within a quarter section the 40-acre and 10-acre tracts are lettered in the same fashion. For example: a well listed as 4S-1E-34BAD1 is in the SE 1/4, NE 1/4, NW 1/4, sec. 34, T. 4 S., R. 1 E. and is the first well listed for that tract (see accompanying diagram). If the location is known only to a 40-acre tract, only the first two letters are given. If it is a spring instead of a well, the letter “S” is added at the end.

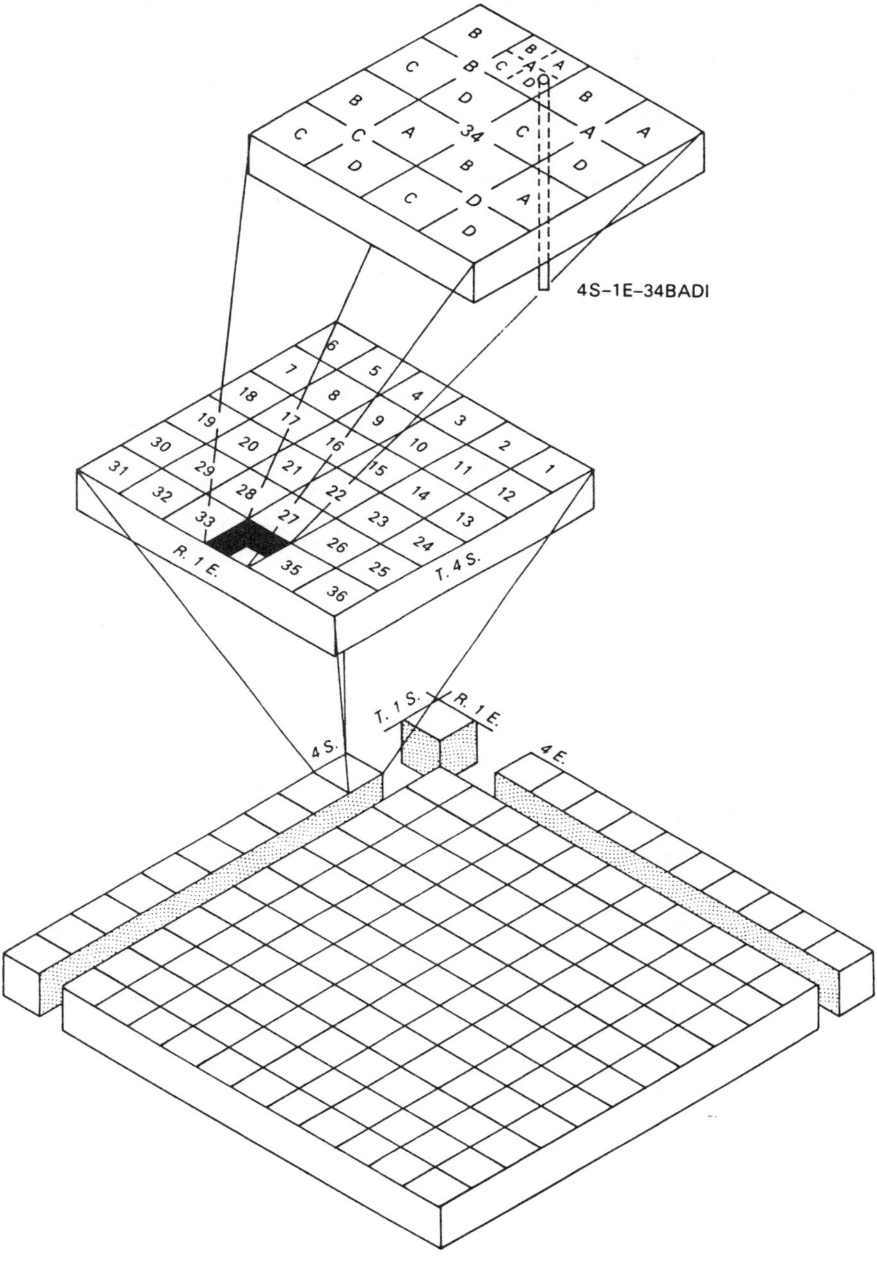

U.S.G.S. Numbering System for Wells and Springs

Relation of Well Diameter to Yield

The yield of a well is dependent upon a number of factors, including the permeability and thickness of the aquifer, the radius of influence (R), the drawdown when pumping, and the well diameter (r). The yield does not, however, vary directly with the well diameter — that is, if the diameter of the well is doubled the yield will not be twice as much. All other factors being equal, the yield varies according to the formula:

$$Q = \frac{C}{\log\frac{R}{r}}$$

where C is a constant related to the aquifer properties. The accompanying table shows the relative increase in yield for wells of different diameters for typical radius of influence associated with both unconfined (water table) and confined (artesian) conditions. For example, a 24-inch well in a confined aquifer, with all other conditions being the same, could be expected to yield only 1.13 times as much as an 8-inch well.

Selecting the diameter for a well must consider, however, such factors as the anticipated depth, type of construction, pump installation required, etc. Increasing the diameter of a well will also reduce the entrance velocity of water which can be important in reducing the possibility of pumping fine sand. In some localities with aquifers of very low permeability and slow inflow rates, large diameter dug wells (up to 20 ft or more) may be advantageous simply because of their large storage volumes.

Ratio of Well Yields From Wells of Different Diameters

Diameter of Well – Inches									
4	6	8	10	12	18	24	30	36	48
(Unconfined Aquifer – R = 500 Ft.)									
1.00	1.05	1.09	1.13	1.16	1.23	1.29	1.34	1.38	1.45
	1.00	1.04	1.07	1.10	1.17	1.22	1.27	1.31	1.38
		1.00	1.03	1.06	1.12	1.18	1.22	1.26	1.32
			1.00	1.03	1.09	1.14	1.18	1.22	1.28
				1.00	1.06	1.11	1.15	1.19	1.25
					1.00	1.05	1.09	1.12	1.18
						1.00	1.04	1.07	1.13
							1.00	1.03	1.09
								1.00	1.05
									1.00
(Confined Aquifer – R = 5,000 Ft.)									
1.00	1.04	1.07	1.10	1.12	1.17	1.21	1.24	1.27	1.32
	1.00	1.03	1.05	1.08	1.12	1.16	1.19	1.22	1.27
		1.00	1.02	1.04	1.09	1.13	1.16	1.19	1.23
			1.00	1.02	1.07	1.10	1.13	1.16	1.20
				1.00	1.05	1.08	1.11	1.14	1.18
					1.00	1.03	1.06	1.09	1.13
						1.00	1.03	1.05	1.09
							1.00	1.02	1.06
								1.00	1.04
									1.00

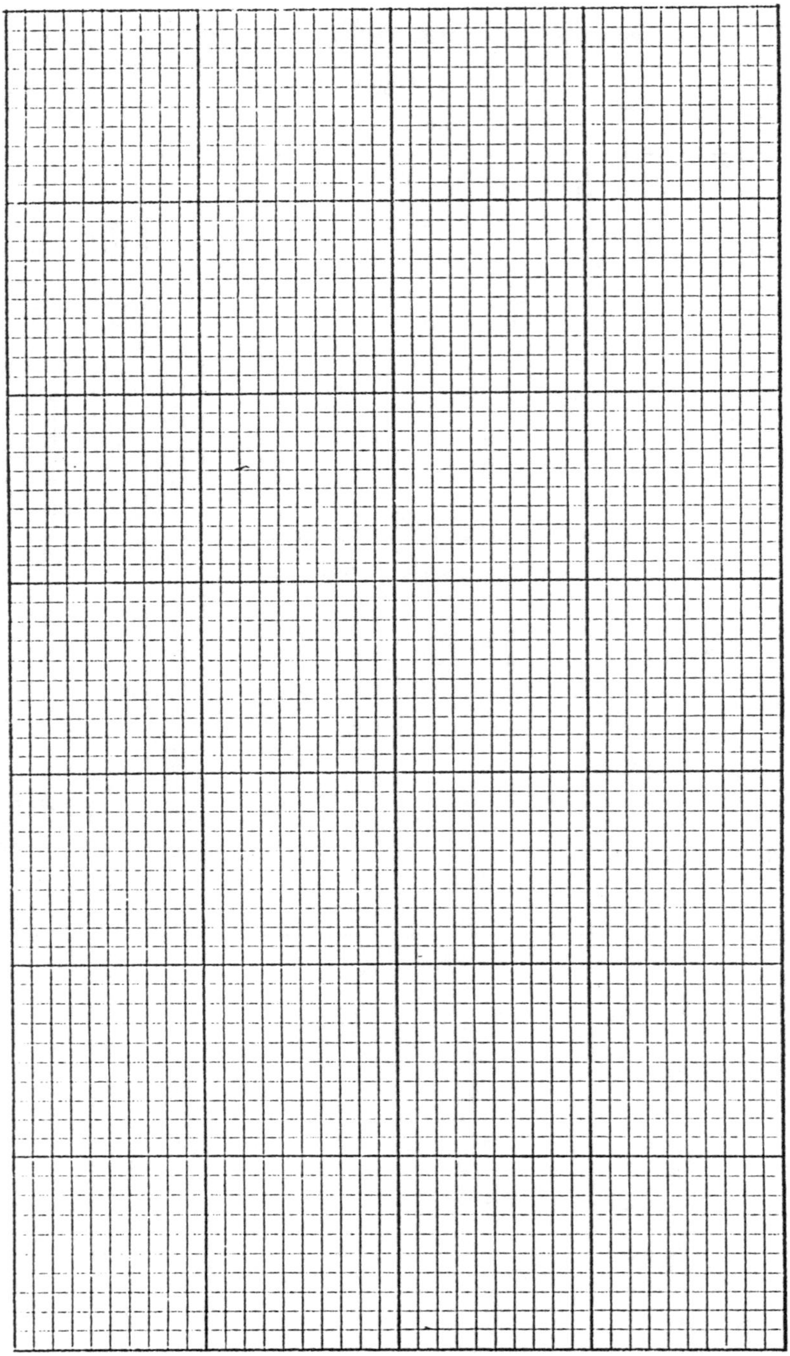

Rectangular Coordinate (In.)

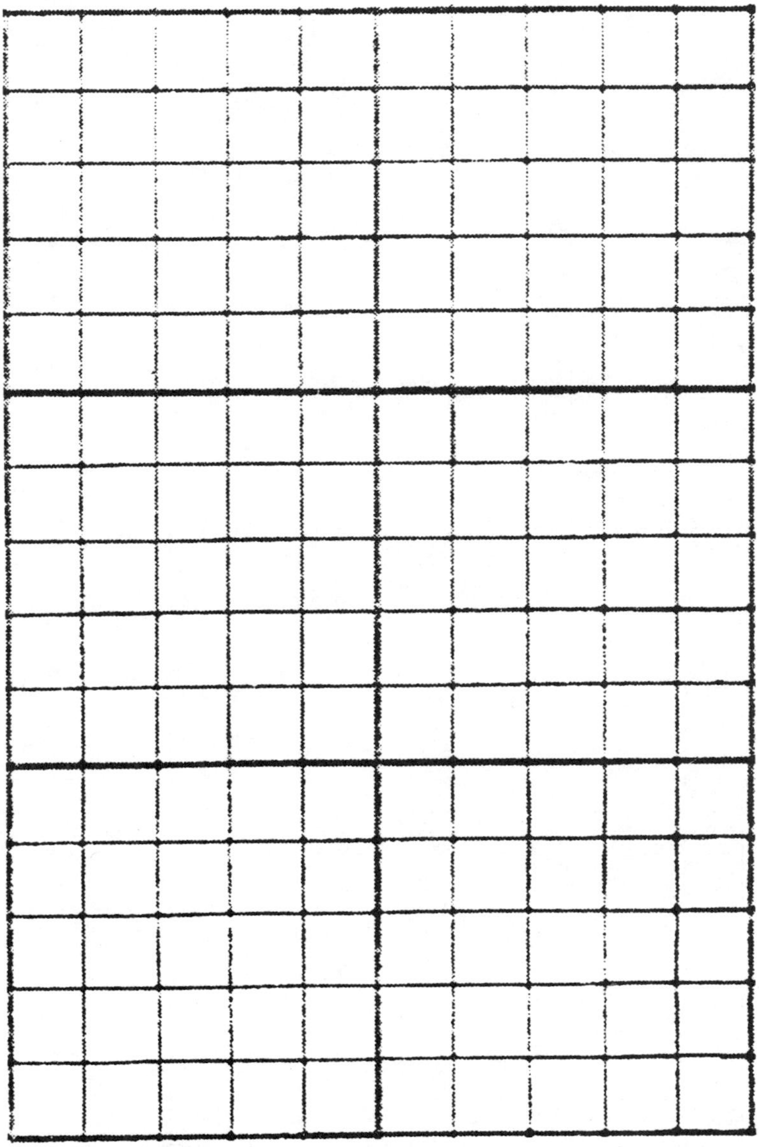

Rectangular Coordinates (Cm.)

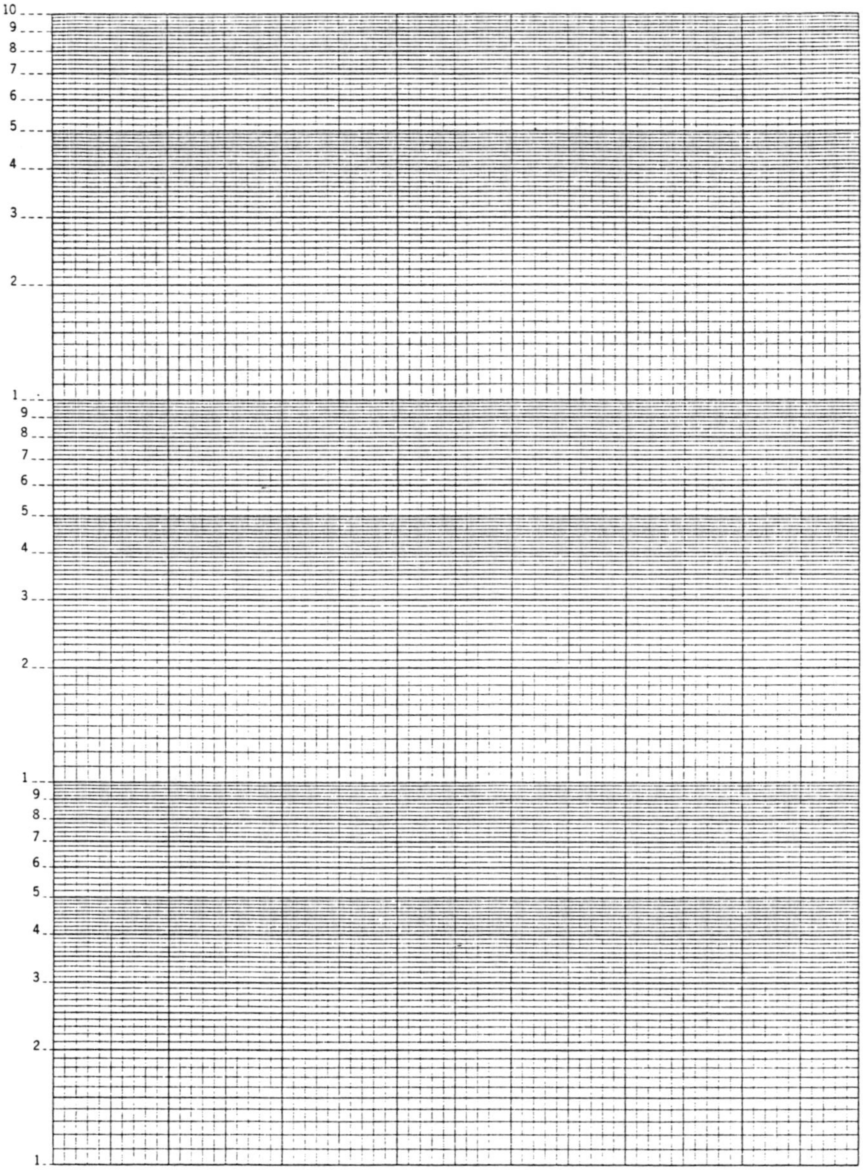

Semi-Log

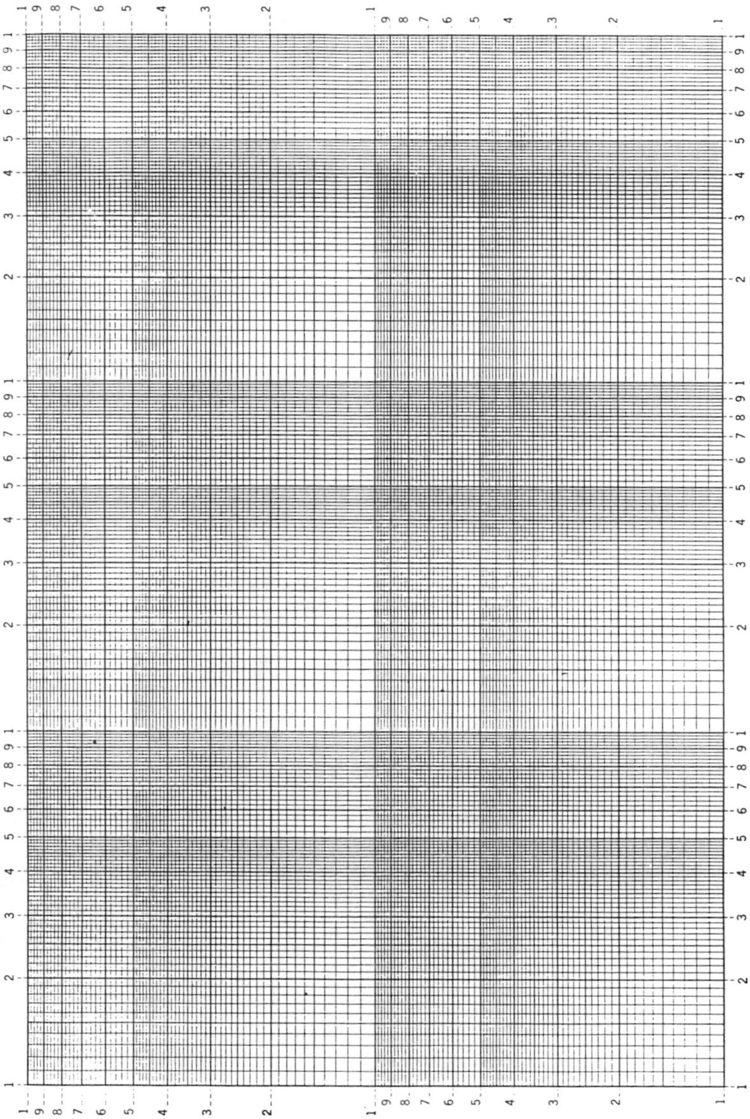

Log-Log

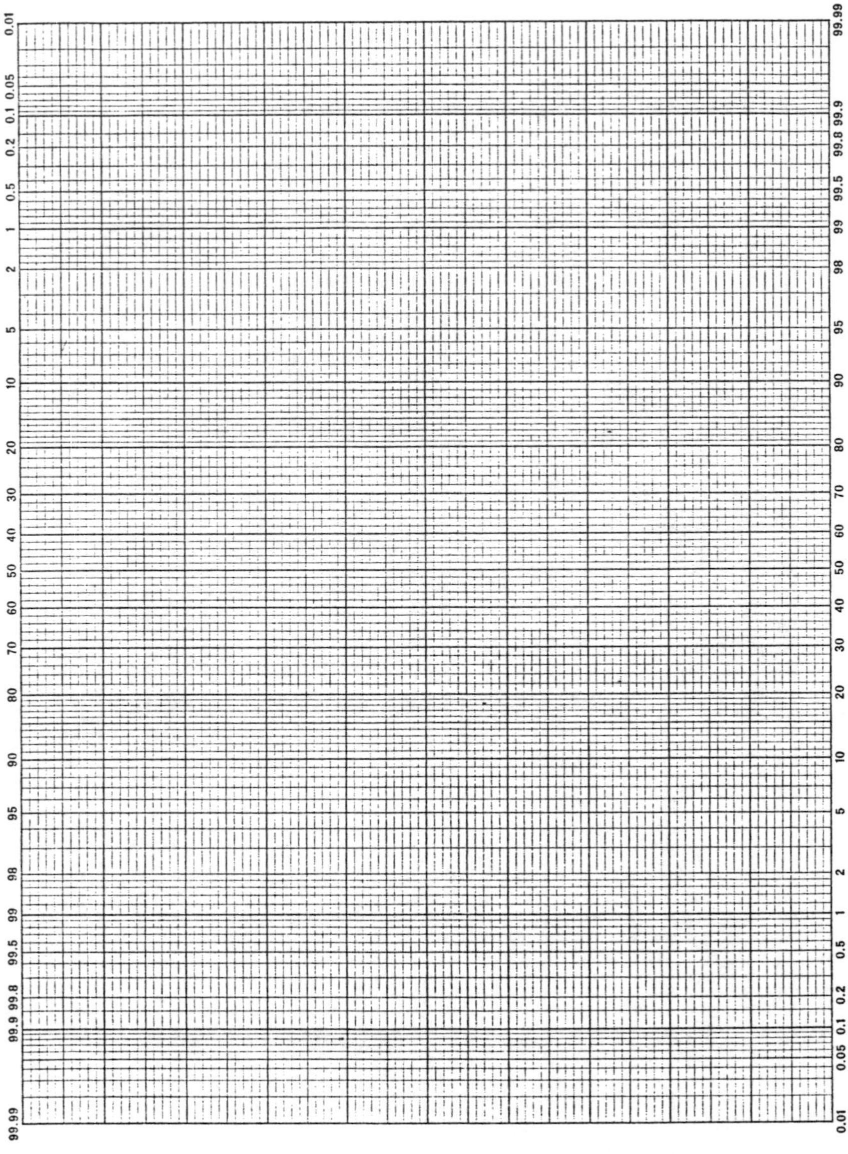

Arithmetic Probability

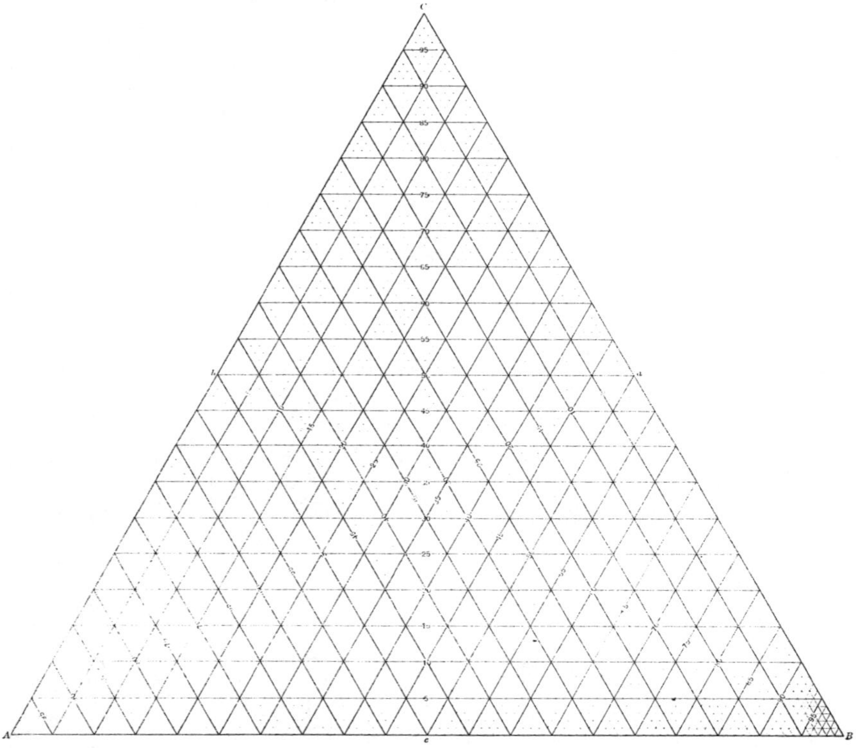

Triangular Coordinates

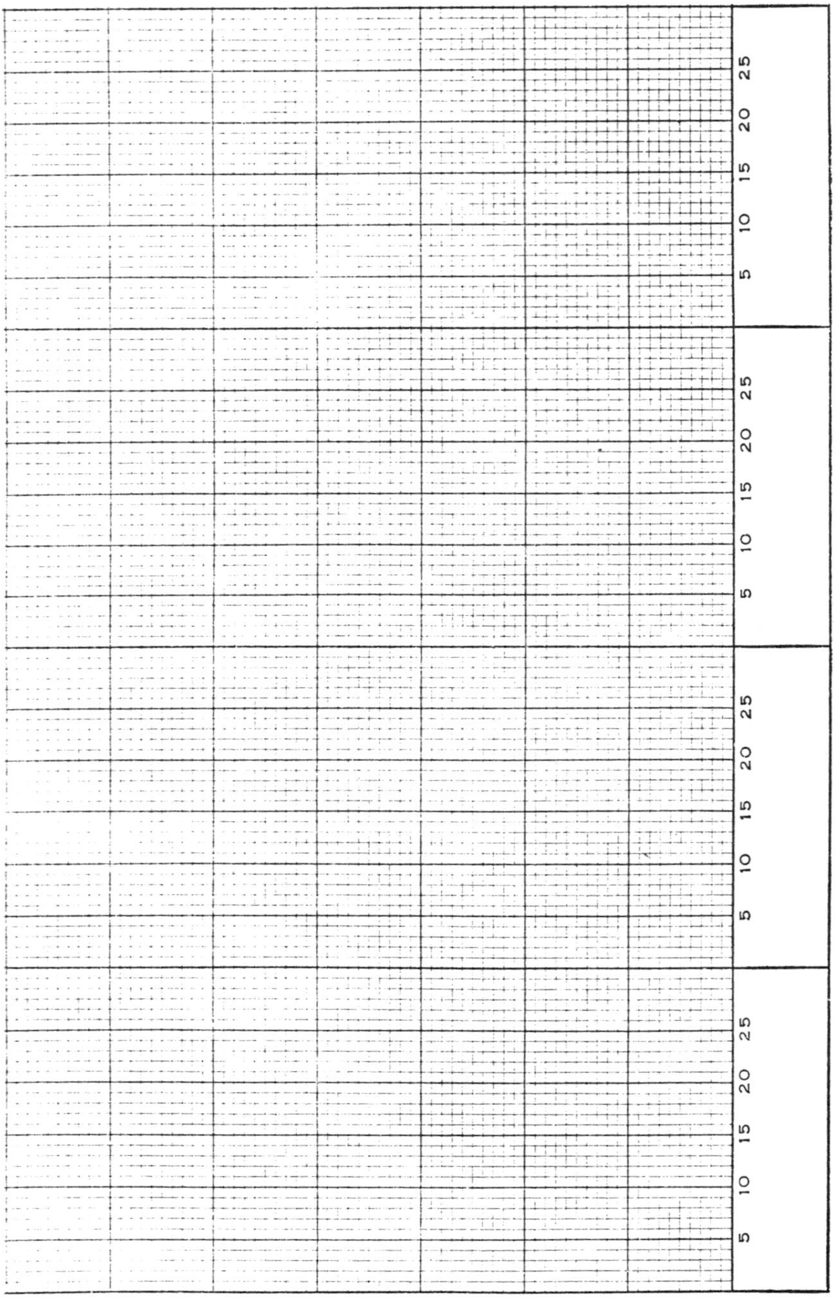

Months by Days

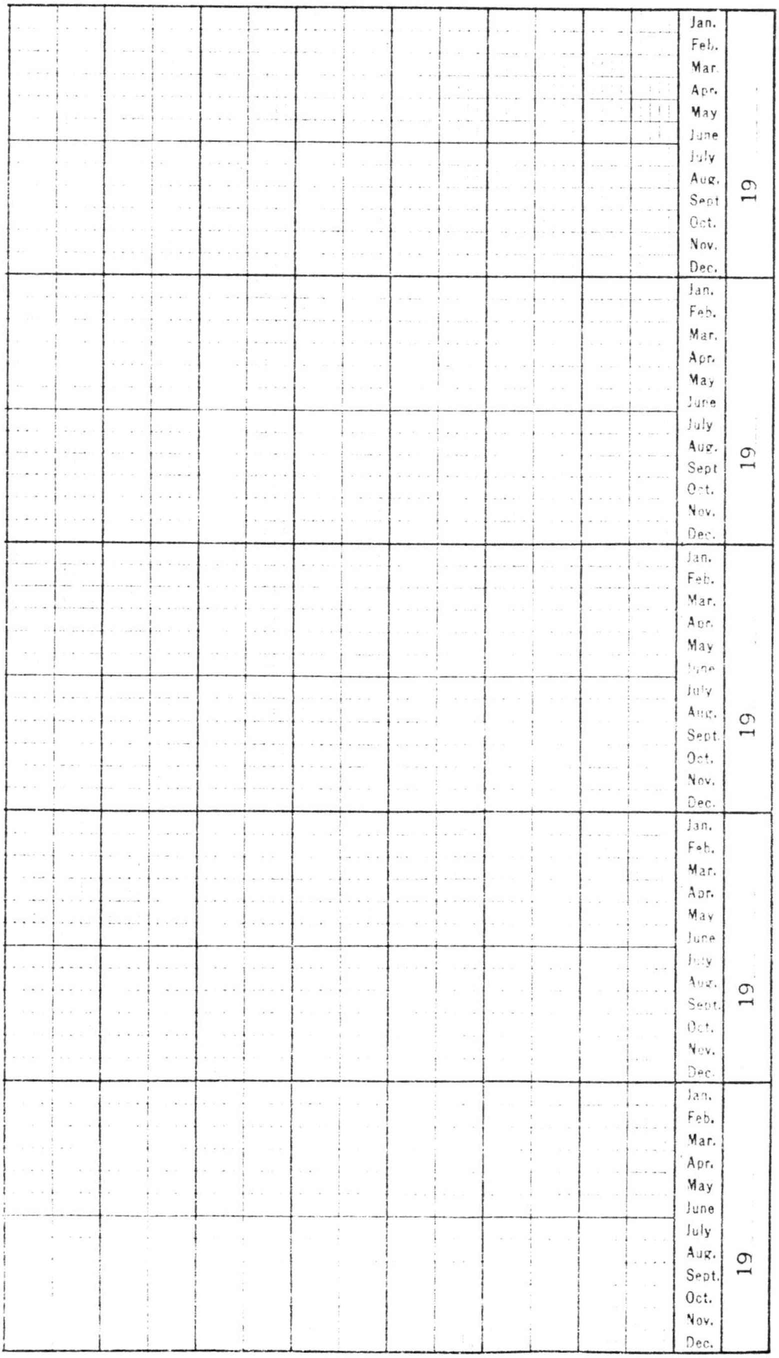

Years by Months

Chapter 12

Miscellaneous

Capacities of Horizontal Cylindrical Tanks

Depth Ratio	Fraction of Capacity	Depth Ratio	Fraction of Capacity	Depth Ratio	Fraction of Capacity
0.00	0.000000	0.40	0.373539	0.80	0.857654
0.02	0.004799	0.42	0.398525	0.82	0.877579
0.04	0.013480	0.44	0.423749	0.84	0.896766
0.06	0.024509	0.46	0.449132	0.86	0.914886
0.08	0.037501	0.48	0.464560	0.88	0.932021
0.10	0.052025	0.50	0.500000	0.90	0.947975
0.12	0.067979	0.52	0.525440	0.92	0.962499
0.14	0.085114	0.54	0.550868	0.94	0.975491
0.16	0.103234	0.56	0.576251	0.96	0.986520
0.18	0.122421	0.58	0.601475	0.98	0.995201
0.20	0.142346	0.60	0.626461	1.00	1.000000
0.22	0.163077	0.62	0.651333		
0.24	0.184466	0.64	0.675939		
0.26	0.206502	0.66	0.700238		
0.28	0.229195	0.68	0.724271		
0.30	0.252298	0.70	0.747702		
0.32	0.275729	0.72	0.770805		
0.34	0.299762	0.74	0.793498		
0.36	0.324061	0.76	0.815334		
0.38	0.348667	0.78	0.836923		

Example: Determine the depth ratio by dividing the liquid depth in the tank by the diameter of the tank; opposite this depth ratio the table gives the fraction of the full tank capacity that is occupied by the liquid.

Capacity of Vertical Cylindrical Tanks

Diam.	Capacity Per Ft. Depth		Diam.	Capacity Per Ft. Depth		Diam.	Capacity Per Ft. Depth	
Feet	(Gallons)	(Cu. Feet)	Feet	Gallons	(Cu. Feet)	Feet	(Gallons)	(Cu. Feet)
1	5.875	0.785	9	475.9	63.62	22	2,844	380.1
1.25	9.180	1.227	9.25	502.7	67.20	22.5	2,974	397.6
1.5	13.22	1.767	9.5	530.2	70.88	23	3,108	415.5
1.75	17.99	2.405	9.75	558.5	74.66	23.5	3,245	433.7
2	23.50	3.142	10	587.5	78.54	24	3,384	452.4
2.25	29.74	3.976	10.25	617.3	82.52	24.5	3,527	471.4
2.5	36.72	4.909	10.5	647.7	86.59	25	3,672	490.9
2.75	44.43	5.940	10.75	679.0	90.76	25.5	3,820	510.7
3	52.88	7.069	11	710.9	95.03	26	3,972	530.9
3.25	62.06	8.296	11.25	743.6	99.40	26.5	4,126	551.5
3.5	71.97	9.621	11.5	777.0	103.9	27	4,283	572.6
3.75	82.62	11.04	11.75	811.1	108.4	27.5	4,443	594.0
4	94.00	12.57	12	846.0	113.1	28	4,606	615.8
4.25	106.1	14.19	12.5	918.0	122.7	28.5	4,772	637.9
4.5	119.0	15.90	13	992.9	132.7	29	4,941	660.5
4.75	132.6	17.72	13.5	1,071	143.1	29.5	5,113	683.5
5	146.9	19.64	14	1,152	153.9	30	5,288	706.9
5.25	161.9	21.65	14.5	1,235	165.1	30.5	5,465	730.6
5.5	177.7	23.76	15	1,322	176.7	31	5,646	754.8
5.75	194.2	25.97	15.5	1,412	188.7	31.5	5,830	779.3
6	211.5	28.27	16	1,504	201.1	32	6,016	804.2
6.25	229.5	30.68	16.5	1,600	213.8	33	6,398	855.3
6.5	248.2	33.18	17	1,698	227.0	34	6,792	907.9
6.75	267.7	35.78	17.5	1,799	240.5	35	7,197	962.1
7	287.9	38.48	18	1,904	254.5	36	7,614	1,018
7.25	308.8	41.28	18.5	2,011	268.8	37	8,043	1,075
7.5	330.5	44.18	19	2,121	283.5	38	8,484	1,134
7.75	352.9	47.17	19.5	2,234	298.6	39	8,936	1,195
8	376.0	50.27	20	2,350	314.2	40	9,400	1,257
8.25	399.9	53.46	20.5	2,469	330.1	45	11,897	1,590
8.5	424.5	56.75	21	2,591	346.4	50	14,688	1,964
8.75	449.8	60.13	21.5	2,716	363.1	60	21,151	2,827

Geologic Time Scale

Era	Period	Epoch	Time (Million Yrs. Ago)
Cenozoic	Quaternary	Recent	0 – 0.1
		Pleistocene	0.1 – 1.6
	Tertiary	Pliocene	1.6 – 5.3
		Miocene	5.3 – 23.7
		Oligocene	23.7 – 36.6
		Eocene	36.6 – 57.8
		Paleocene	57.8 – 66.4
Mesozoic	Cretaceous		66.4 – 144
	Jurassic		144 – 208
	Triassic		208 – 245
Paleozoic	Permian		245 – 286
	Carboniferous	Pennsylvanian	286 – 320
		Mississippian	320 – 360
	Devonian		360 – 408
	Silurian		408 – 438
	Ordovician		438 – 505
	Cambrian		505 – 570
Archeozoic	Precambrian		570 – 4,600+

Weight of Various Rock Types

Rock Type	Pounds Per Cubic Foot		Cubic Feet Per Ton		Tons Per Cubic Yard	
	In Place	Broken	In Place	Broken	In Place	Broken
Andesite	181	97	11.1	20.6	2.44	1.31
Basalt	181	97	11.1	20.6	2.44	1.31
Diabase	187	94	10.6	21.3	2.52	1.27
Diorite	187	94	10.6	21.3	2.52	1.27
Gneiss	168	96	11.9	20.8	2.27	1.30
Granite	170	97	11.8	20.6	2.30	1.31
Limestone	168	96	11.9	20.8	2.27	1.30
Porphyry	170	97	11.8	20.6	2.30	1.31
Rhyolite	150	86	13.4	23.3	2.02	1.16
Quartzite	165	94	12.2	21.3	2.23	1.27
Sandstone	151	86	13.2	23.3	2.04	1.16
Schist	168	91	11.9	22.0	2.27	1.23
Shale	175	95	11.4	21.1	2.36	1.28
Slate	175	95	11.4	21.1	2.36	1.28

Unified Soil Classificiation

Division			Symbol	Description	% Fines (Minus #200)	Typical Dry Wt. (Lbs/CuFt)
Coarse-grained Soils	Gravel	Clean Gravel	G W	Well-graded gravel or sandy gravel	0 to 15	125-140
			G P	Poorly graded gravel or sandy gravel	0 to 15	120-130
		Gravel with Fines	G M	Silty gravel, non-plastic fines	15 to 50	130-145
			G C	Clayey gravel, plastic fines	15 to 50	120-140
	Sand	Clean Sand	S W	Well-graded sand or gravelly sand	0 to 15	110-130
			S P	Poorly graded sand or gravelly sand	0 to 15	105-120
		Sand with Fines	S M	Silty sand, non-plastic fines	15 to 50	120-135
			S C	Clayey sand, plastic fines	15 to 50	105-130
Fine-grained Soils	Silt and Clay	Low Plasticity Fines	M L	Very fine sand; silty or clayey fine sand; silt; clayey silt	More than 50	100-125
			C L	Gravelly, sandy, or silty clay; lean clay	More than 50	100-125
			O L	Organic silt; organic silty clay	More than 50	90-105
	Silt and Clay	High Plasticity Fines	M H	Inorganic silt; micaceous or diatomaceous fine sandy silts	More than 50	80-100
			C H	Highly plastic inorganic clay; fat clay	More than 50	90-110
			O H	Organic silt; organic clay	More than 50	80-105
Highly Organic Soils			P t	Peat; humus; highly organic soil	More than 50	—
				Well Graded = Poorly Sorted Poorly Graded = Well Sorted		

Soil Classification (Particle Size)

Designation		Particle Size Range: Inches	Particle Size Range: Millimeters
Boulders		More than 12	More than 305
Cobbles		3 – 12	76 – 305
Gravel	Coarse	0.75 – 3	19 – 76
	Fine	0.187 – 0.75	4.76 – 19
Sand	Coarse	0.079 – 0.187	2.00 – 4.76
	Medium	0.016 – 0.079	0.42 – 2.00
	Fine	0.003 – 0.079	0.074 – 0.42
Silt and Clay		Less than 0.003	Less than 0.074

Relative Mineral Hardness (Mohs Scale)

Scale No.	Mineral	Scale No.	Mineral
1	Talc	6	Orthoclase
2	Gypsum	7	Quartz
3	Calcite	8	Topaz
4	Fluorite	9	Corundum
5	Apatite	10	Diamond

Standard Penetration Test (SPT)

Standard penetration tests are made by driving a standard split-spoon sampler with a 140-pound hammer falling 30 inches, and counting the number of blows required to advance the sampler a distance of 12 inches (blows per foot).

Density (Sand and Gravel)				Consistency (Silt and Clay)			
	Spoon Diameter – Inches				Spoon Diameter – Inches		
	1.4	2.0	2.5		1.4	2.0	2.5
Description	(Blows/Ft)			Description	(Blows/Ft)		
Very Loose	0 – 4	0 – 5	0 – 7	Very Soft	0 – 2	0 – 2	0 – 2
Loose	4 – 10	5 – 12	7 – 18	Soft	2 – 4	2 – 4	2 – 4
Medium Dense	10 – 29	12 – 37	18 – 51	Medium Stiff	4 – 8	4 – 9	4 – 9
Dense	29 – 47	37 – 60	51 – 86	Stiff	8 – 15	9 – 17	9 – 18
Very Dense	47+	60+	86+	Very Stiff	15 – 30	17 – 39	18 - 42
				Hard	30 – 60	39 – 78	42 – 85
				Very Hard	60+	78+	85+

Typical Properties of Engineering Materials

Material	Ultimate Strength – psi			Yield Point	Modulus of Elasticity – psi		Weight
	Tension	Compression*	Shear	(Tension – psi)	Tension or Compression	Shear	(Lbs/Cu In)
Steel, forged-rolled:							
C, 0.10-0.20	60,000	39,000	48,000	39,000	30,000,000	12,000,000	0.28
C, 0.20-0.30	67,000	43,000	53,000	43,000	30,000,000	12,000,000	0.28
C, 0.30-0.40	70,000	46,000	56,000	46,000	30,000,000	12,000,000	0.28
C, 0.60-0.80	125,000	65,000	75,000	65,000	30,000,000	12,000,000	0.28
Nickel	115,000	—	92,000	—	30,000,000	12,000,000	0.28
Steel, cast:							
Low C	60,000	—	—	—	—	—	0.28
Medium C	70,000	—	—	—	—	—	0.28
High C	80,000	45,000	—	45,000	—	—	0.28
Wrought iron	48,000	25,000	38,000	25,000	27,000,000	—	0.28
Aluminum alloy:							
Structural, #350	16,000	5,000	11,000	5,000	10,000,000	3,750,000	0.10
Structural, #17ST	58,000	35,000	35,000	35,000	10,000,000	3,750,000	0.10
Brass:							
Annealed	54,000	18,000	—	18,000	—	—	0.30
Cold drawn	97,000	49,000	—	49,000	15,500,000	6,200,000	0.30
Concrete:							
1:2:4, 28-days	—	2,000	—	—	3,000,000	—	0.087

* - Yield point usually taken as ultimate compressive strength for ductile materials.

Weight of Materials

Substance	Lbs/Cu Ft	Substance	Lbs/Cu Ft
Bituminous		**Liquids (cont'd)**	
Asphalt	81	Kerosene, jet fuel	51
Coal, anthracite	97	Liquefied natural gas	34
Coal, bituminous	84	Oil, disel fuel	54
Coal, coke	75	Oil, lubricating	57
Coal, lignite	78	Oil, vegetable	58
Paraffin	56	Water (max density @ 4C)	62.428
Tar, bituminous	75	Water (at 100 deg C)	59.83
Building materials		Water, ice	56
Ashes, cinders	40-45	Water, sea	64
Brick, common	120	Water, snow, fresh fallen	8
Cement, portland, loose	90	**Metals**	
Cement, portland, set	183	Aluminum. cast, hammered	165
Concrete, 1:2:4	146	Brass, cast, rolled	534
Concrete, 1:3:6	156	Bronze, tin	509
Concrete, typical	150	Bronze, aluminum	481
Glass, common	156	Copper, cast, rolled	556
Glass, plate	161	Gold, cast, hammered	1205
Lime, gypsum, loose	53-64	Iron, cast, pig	450
Mortar, set	103	Iron, wrought	485
Excavated materials		Lead	710
Clay, damp, plastic	110	Magnesium, alloys	112
Clay, dry	63	Mercury	849
Clay & gravel, dry	100	Monel metal	556
Earth, dry, loose	76	Nickel	565
Earth, dry, packed	95	Platinum, cast, hammered	1330
Earth, moist, loose	78	Silver, cast, hammered	656
Earth, moist, packed	96	Steel, rolled	490
Riprap, limestone	80-85	Tin, cast, hammered	459
Riprap, sandstone	90	Zinc, cast, rolled	440
Sand & gravel, dry, loose	90-105	**Timber (seasoned)**	
Sand & gravel, dry, packed	100-120	Ash, white, red	40
Excavations in water		Cedar, white, red	22
Clay	80	Chestnut	41
Sand or gravel	60	Cypress	30
Soil	70	Elm, white	45
Stone riprap	65	Fir, Douglas spruce	32
Gases		Fir, eastern	25
Air	0.08072	Hemlock	29
Acetylene	0.07361	Hickory	49
Ammonia	0.04812	Locust	46
Butane	0.1619	Maple, hard	43
Carbon dioxide	0.12341	Maple, white	33
Carbon monoxide	0.07806	Oak, chestnut	54
Chlorine	0.2011	Oak, live	59
Helium	0.01115	Oak, red, black	41
Hydrogen	0.00561	Oak, white	46
Methane	0.04475	Pine, Oregon	32
Nitrogen	0.07808	Pine, red	30
Oxygen	0.08921	Pine, white	26
Liquids*		Pine, yellow, longleaf	44
Alcohol, 100%	49	Pine, yellow, shortleaf	38
Acid, hydrochloric, 40%	75	Poplar	30
Acid, nitric, 91%	112	Redwood, California	26
Acid, sulfuric, 87%	112	Spruce, white, black	27
Benzene	56	Walnut, black	38
Gasoline	41-43	Walnut, white	26

* – 1 Cu Ft = 7.48 US Gallons

Drill Sizes

Size	Diameter (Inches)	Area (Sq.In.)	Size	Diameter (Inches)	Area (Sq.In.)	Size	Diameter (Inches)	Area (Sq.In.)
80	0.0135	0.000143	45	0.0820	0.00528	10	0.1935	0.0294
79	0.0145	0.000165	44	0.0860	0.00580	9	0.1960	0.0302
78	0.0160	0.000201	43	0.0890	0.00622	8	0.1990	0.0311
77	0.0180	0.000254	42	0.0935	0.00686	7	0.2010	0.0316
76	0.0200	0.000314	41	0.0960	0.00723	6	0.2040	0.0327
75	0.0210	0.000346	40	0.0980	0.00755	5	0.2055	0.0332
74	0.0225	0.000398	39	0.0995	0.00777	4	0.2090	0.0343
73	0.0240	0.000453	38	0.1015	0.00810	3	0.2130	0.0356
72	0.0250	0.000491	37	0.1040	0.00850	2	0.2210	0.0384
71	0.0260	0.000531	36	0.1065	0.00893	1	0.2280	0.0409
70	0.0280	0.000615	35	0.1100	0.00950	A	0.234	0.0430
69	0.0293	0.000672	34	0.1110	0.00968	B	0.238	0.0440
68	0.0310	0.000755	33	0.1130	0.01002	C	0.242	0.0460
67	0.0320	0.000805	32	0.1160	0.01055	D	0.246	0.0475
66	0.0330	0.000856	31	0.1200	0.01130	E	0.250	0.0491
65	0.0350	0.000962	30	0.1285	0.01298	F	0.257	0.0519
64	0.0360	0.001018	29	0.1360	0.01452	G	0.261	0.0535
63	0.0370	0.001076	28	0.1405	0.01550	H	0.266	0.0556
62	0.0380	0.001134	27	0.1440	0.01629	I	0.272	0.0580
61	0.0390	0.001195	26	0.1470	0.01697	J	0.277	0.0601
60	0.0400	0.001257	25	0.1495	0.01755	K	0.281	0.0620
59	0.0410	0.001320	24	0.1520	0.01812	L	0.290	0.0660
58	0.0420	0.001385	23	0.1540	0.01863	M	0.295	0.0683
57	0.0430	0.00145	22	0.1570	0.01935	N	0.302	0.0716
56	0.0465	0.00170	21	0.1590	0.01985	O	0.316	0.0784
55	0.0520	0.00213	20	0.1610	0.02039	P	0.323	0.0820
54	0.0550	0.00238	19	0.1660	0.02162	Q	0.332	0.0866
53	0.0595	0.00278	18	0.1695	0.02258	R	0.339	0.0901
52	0.0635	0.00317	17	0.1730	0.02348	S	0.348	0.0950
51	0.0670	0.00352	16	0.1770	0.02461	T	0.358	0.1005
50	0.0700	0.00385	15	0.1800	0.02542	U	0.368	0.1063
49	0.0730	0.00419	14	0.1820	0.02603	V	0.377	0.1116
48	0.0760	0.00454	13	0.1850	0.02685	W	0.386	0.1170
47	0.0785	0.00484	12	0.1890	0.02805	X	0.397	0.1236
46	0.0810	0.00515	11	0.1910	0.02865	Y	0.404	0.1278
						Z	0.413	0.1340

Fractional Drill Sizes Also Available: By 1/64's from 1/64 to 1-3/4
By 1/32's from 1-25/32 to 2-1/4
By 1/16's from 2-5/16 to 3

Wood Screws

Number	Diameter Inches	Threads per Inch	Number	Diameter Inches	Threads per Inch
0	0.060	32	9	0.177	14
1	0.073	28	10	0.190	13
2	0.086	26	11	0.203	12
3	0.099	24	12	0.216	11
4	0.112	22	14	0.242	10
5	0.125	20	16	0.268	9
6	0.138	18	18	0.294	8
7	0.151	16	20	0.320	8
8	0.164	15	24	0.372	7

Standard (Common) Wire Nails

Size*	Length (Inches)	Number Per Pound	Size*	Length (Inches)	Number Per Pound
2d	1	876	10d	3	69
3d	1¼	568	12d	3¼	64
4d	1½	316	16d	3½	49
5d	1¾	271	20d	4	31
6d	2	181	30d	4½	24
7d	2¼	161	40d	5	18
8d	2½	106	50d	5½	14
9d	2¾	96	60d	6	11

*Sizes are shown as “penny”, abbreviated with the letter “d”. (An 8-penny nail is shown as 8d).

Concrete Proportioning

Mix Proportion			Material For One Cubic Yard Concrete		
Cement	Sand	Gravel	Cement (Sacks)	Sand (Cu. Yards)	Gravel* (Cu. Yards)
1	2	3	6.16	0.47	0.73
1	2	4	5.36	0.41	0.81
1	2	5	4.68	0.36	0.89
1	2½	4	4.96	0.47	0.75
1	2½	5	4.40	0.42	0.83
1	2½	6	3.92	0.37	0.89
1	3	5	4.12	0.47	0.78
1	3	6	3.68	0.42	0.84
1	3	7	3.36	0.38	0.89
1	4	6	3.32	0.51	0.77
1	4	7	3.08	0.47	0.81
1	4	8	2.84	0.43	0.86

* – ¾-inch minus

Concrete Required for Floors

Floor Thickness (Inches)	Concrete (Cu Yds) For 100 Sq. Ft.	1 : 2¼ : 3 Mix			1 : 2¾ : 4 Mix		
		Cement (Sacks)	Sand (Cu Yds)	Gravel (Cu Yds)	Cement (Sacks)	Sand (Cu Yds)	Gravel (Cu Yds)
3	0.92	5.8	0.48	0.65	4.6	0.48	0.68
3½	1.08	6.8	0.56	0.76	5.4	0.56	0.80
4	1.24	7.8	0.64	0.87	6.2	0.64	0.92
4½	1.39	8.7	0.72	0.98	7.0	0.72	1.03
5	1.56	9.8	0.80	1.10	7.8	0.80	1.16
6	1.85	11.5	0.96	1.30	9.3	0.96	1.37
7	2.16	13.4	1.12	1.52	10.8	1.12	1.59
8	2.46	15.4	1.27	1.73	12.3	1.27	1.83

Brick and Concrete Block Walls

Brick

Wall Thickness (Inches)	For 100 Square Feet of Wall		For 1000 Bricks
	No. of Bricks	Mortar (Cu Ft)	Mortar (Cu Ft)
8	1233	20	16.2
12	1850	32	17.3
16	2465	44	17.8

Standard size brick (2¼ × 3¾ × 8 inches)
Mortar joints ½-inch thick

Concrete Block

Standard blocks are 8 × 8 × 16 inches
One course of blocks, 100 ft long, requires 73 blocks
A wall 6 ft high requires about 9 courses of blocks
A wall 6 ft high and 100 ft long (600 sq ft) then requires about 9 × 73 = 657 blocks

Converting Linear Feet To Board Feet of Lumber

Nominal Size (Inches)	Length of Piece (In Feet)								
	8	10	12	14	16	18	20	22	24
1 × 4	2⅔	3⅓	4	4⅔	5⅓	6	6⅔	7⅓	8
1 × 6	4	5	6	7	8	9	10	11	12
1 × 8	5⅓	6⅔	8	9⅓	10⅔	12	13⅓	14⅔	16
1 × 10	6⅔	8⅓	10	11⅔	13⅓	15	16⅔	18⅓	20
1 × 12	8	10	12	14	16	18	20	22	24
2 × 4	5⅓	6⅔	8	9⅓	10⅔	12	13⅓	14⅔	16
2 × 6	8	10	12	14	16	18	20	22	24
2 × 8	10⅔	13⅓	16	18⅔	21⅓	24	26⅔	29⅓	32
2 × 10	13⅓	16⅔	20	23⅓	26⅔	30	33⅓	36⅔	40
2 × 12	16	20	24	28	32	36	40	44	48
4 × 4	10⅔	13⅓	16	18⅔	21⅓	24	26⅔	29⅓	32
4 × 6	16	20	24	28	32	36	40	44	48
4 × 8	21⅓	26⅔	32	37⅓	42⅔	48	53⅓	58⅔	64
4 × 10	26⅔	33⅓	40	46⅔	53⅓	60	66⅔	73⅓	80
4 × 12	32	40	48	56	64	72	80	88	96
4 × 14	37⅓	46⅔	56	65⅓	74⅔	84	93⅓	102⅔	112
6 × 6	24	30	36	42	48	54	60	66	72
6 × 8	32	40	48	56	64	72	80	88	96
6 × 10	40	50	60	70	80	90	100	110	120
6 × 12	48	60	72	84	96	108	120	132	144
6 × 14	56	70	84	98	112	126	140	154	168
6 × 16	64	80	96	112	128	144	160	176	192
8 × 8	42⅔	53⅓	64	74⅔	85⅓	96	106⅔	117⅓	128
8 × 10	53⅓	66⅔	80	93⅓	106⅔	120	133⅓	146⅔	160
8 × 12	64	80	96	112	128	144	160	176	192

Figures in table are given in board feet. A piece one inch thick, 12 inches wide, and one foot long constitutes one board foot. Lumber is usually priced by thousand board feet.

Standard Dimension of Surfaced Lumber (Inches)

Nominal	Surfaced (Actual) Size		Nominal	Surface (Actual) Size	
Size	Unseasoned	Dry	Size	Unseasoned	Dry
1 × 3	25/32 × 2 9/16	3/4 × 2 1/2	2 × 3	1 9/16 × 2 9/16	1 1/2 × 2 1/2
1 × 4	25/32 × 3 9/16	3/4 × 3 1/2	2 × 4	1 9/16 × 3 9/16	1 1/2 × 3 1/2
1 × 6	25/32 × 5 5/8	3/4 × 5 1/2	2 × 6	1 9/16 × 5 5/8	1 1/2 × 5 1/2
1 × 8	25/32 × 7 1/2	3/4 × 7 1/4	2 × 8	1 9/16 × 7 1/2	1 1/2 × 7 1/4
1 × 10	25/32 × 9 1/2	3/4 × 9 1/4	2 × 10	1 9/16 × 9 1/2	1 1/2 × 9 1/4
1 × 12	25/32 × 11 1/2	3/4 × 11 1/4	2 × 12	1 9/16 × 11 1/2	1 1/2 × 11 1/4

Steel Plates – Weight (Pounds/Square Foot)

Thickness (In.)	Weight (Lbs.)	Thickness (In.)	Weight (Lbs.)
3/16	7.65	2½	102.0
¼	10.20	2¾	112.2
5/16	12.75	3	122.4
⅜	15.30	3¼	132.6
7/16	17.86	3½	142.8
½	20.40	3¾	153.0
9/16	22.96	4	163.2
⅝	25.50	4¼	173.4
¾	30.60	4½	183.6
⅞	35.70	4¾	193.8
1	40.8	5	204.0
1⅛	45.9	5½	224.4
1¼	51.0	6	244.8
1⅜	56.1	6½	265.2
1½	61.2	7	285.6
1⅝	66.3	7½	306.0
1¾	71.4	8	326.4
2	81.6	9	367.2
2¼	91.8	10	408.0

U. S. Standard Gage for Iron and Steel Sheets

Gage Number	Thickness (Approx.) Inches Wrought Iron	Thickness (Approx.) Inches Steel	Thickness (Approx.) Millimeters Steel	Weight Pounds/ Square Foot	Weight Kilograms/ Square Meter
000	0.375	0.3587	9.111	15	73.24
00	0.34375	0.3288	8.352	13.75	67.13
0	0.3125	0.2989	7.592	12.5	61.03
1	0.28125	0.2690	6.833	11.25	54.93
2	0.265625	0.2541	6.454	10.625	51.88
3	0.25	0.2391	6.073	10	48.82
4	0.234375	0.2242	5.695	9.375	45.77
5	0.21875	0.2092	5.314	8.75	42.72
6	0.203125	0.1943	4.935	8.125	39.67
7	0.1875	0.1793	4.554	7.5	36.62
8	0.171875	0.1644	4.176	6.875	33.57
9	0.15625	0.1495	3.797	6.25	30.52
10	0.140625	0.1345	3.416	5.625	27.46
11	0.125	0.1196	3.038	5	24.41
12	0.109375	0.1046	2.657	4.375	21.36
13	0.09375	0.0897	2.278	3.75	18.31
14	0.078125	0.0747	1.897	3.125	15.26
15	0.0703125	0.0673	1.709	2.8125	13.73
16	0.0625	0.0598	1.519	2.5	12.21
17	0.05625	0.0538	1.367	2.25	10.99
18	0.05	0.0478	1.214	2	9.77
19	0.04375	0.0418	1.062	1.75	8.54
20	0.0375	0.0359	0.912	1.5	7.32

Gage numbers were established by Congress based on weight of wrought iron sheets (in ounces per square foot) rather than on thickness.

Corrugated Sheet Metal – (2½-Inch Corrugations)

U.S. Std. Gage	Thickness (inches)	Weight (lbs/sq.ft.)*		Maximum Span (Feet)	
		Galvanized	Black	Roofing	Siding
28	0.015	0.850	0.68	2 ft 9 in	3 ft 10 in
26	0.018	0.990	0.82	2 ft 9 in	3 ft 10 in
24	0.024	1.260	1.09	3 ft 9 in	4 ft 10 in
22	0.030	1.530	1.36	4 ft 9 in	5 ft 10 in
20	0.036	1.810	1.64	5 ft 9 in	5 ft 10 in
18	0.048	2.350	2.18	5 ft 9 in	5 ft 10 in
16	0.060	2.900	2.73	5 ft 9 in	5 ft 10 in
14	0.075	3.580	3.41	5 ft 9 in	5 ft 10 in
12	0.105	4.940	4.77	5 ft 9 in	5 ft 10 in

*Based on sheets 27 inches wide, lapped 3 inches

Reinforcing Steel Bars

Size	Dimensions	Cross-Sect. Area (Square Inches)	Weight (Pounds/Foot)	Feet Per Ton (2,000 Lbs)
No. 2	¼-In Round	0.0491	0.167	11985
	¼-In Square	0.0625	0.212	9413
No. 3	⅜-In Round	0.1104	0.375	5327
	⅜-In Square	0.1406	0.478	4183
No. 4	½-In Round	0.1963	0.668	2996
	½-In Square	0.2500	0.850	2353
No. 5	⅝-In Round	0.3068	1.043	1918
	⅝-In Square	0.3906	1.328	1506
No. 6	¾-In Round	0.4418	1.502	1332
	¾-In Square	0.5625	1.912	1046
No. 7	⅞-In Round	0.6013	2.044	978
	⅞-In Square	0.7656	2.603	768
No. 8	1-In Round	0.7854	2.670	749
	1-In Square	1.0000	3.400	588

Acceleration of Gravity - (g)

Latitude (degrees)	g (ft/sec/sec)	g (m/sec/sec)	Latitude (degrees)	g (ft/sec/sec)	g (m/sec/sec)
0 (Equator)	32.088	9.780	50	32.187	9.811
10	32.093	9.782	60	32.215	9.819
20	32.108	9.786	70	32.238	9.826
30	32.130	9.793	80	32.253	9.831
40	32.158	9.802	90 (Pole)	32.258	9.832
45	32.174	9.807			

Above table is for "g" at sea level
Corrections for altitude above sea level: -0.003 ft/sec/sec per 1,000 ft
-0.003 m/sec/sec per 1,000 m

Indicated Vacuum

P.S.I.A. (Absolute)	P.S.I.G. (Gauge)	Inches Of (Mercury)	P.S.I.A. (Absolute)	P.S.I.G. (Gauge)	Inches Of (Mercury)
0.00	−14.70000	29.92000	0.19342	−14.50658	29.5263
0.00002	−14.69998	29.91996	0.2902	−14.4098	29.3294
0.00004	−14.69996	29.91992	0.3868	−14.3132	29.1326
0.00006	−14.69994	29.91988	0.4816	−14.2184	28.9357
0.00008	−14.69992	29.91984	0.4913	−14.2087	28.92
0.00010	−14.69990	29.91980	0.5803	−14.1197	28.74
0.00019	−14.69981	29.91961	0.9433	−13.7567	28.00
0.00039	−14.69961	29.91921	2.417	−12.283	25.00
0.00058	−14.69942	29.91882	4.382	−10.318	21.00
0.00077	−14.69923	29.91843	5.856	−8.844	18.00
0.00097	−14.69903	29.91803	7.320	−7.380	15.00
0.00194	−14.69806	29.91606	8.804	−5.896	12.00
0.00387	−14.69613	29.91212	9.787	−4.913	10.00
0.00551	−14.69449	29.90818	10.770	−3.930	8.00
0.00774	−14.69226	29.90424	11.752	−2.948	6.00
0.00968	−14.69032	29.90030	12.735	−1.965	4.00
0.01934	−14.68066	29.88063	13.732	−0.968	2.00
0.03302	−14.66698	29.84126	14.209	−0.491	1.00
0.05803	−14.64197	29.80189	14.454	−0.246	0.50
0.07738	−14.62262	29.76252	14.700	0.00	0.00
0.09671	−14.60329	29.72315		(Atmospheric)	

Anti-Freeze Solutions

Freezing Point		Permanent Type (Ethylene Glycol)		Alcohol & Water
Degrees F	Degrees C	% By Volume	Sp.Gr.@ 60F	% Alcohol By Vol.
32	0.0	0.0	1.000	0.0
25	−3.9	12.5	1.019	9.1
20	−6.7	17.0	1.026	13.4
15	−9.4	21.6	1.032	17.0
10	−12.2	25.0	1.038	20.7
5	−15.0	29.0	1.043	24.0
Zero	−17.8	32.5	1.048	27.1
−5	−20.6	35.7	1.052	30.0
−10	−23.3	38.5	1.056	32.4
−20	−28.9	44.0	1.063	37.0
−30	−34.4	49.0	1.069	41.2
−40	−40.0	52.5	1.073	45.0
−50	−45.6	55.2	1.076	48.7

Typical Composition of Sea Water

Constituent	Concentration (mg/l)
Chloride	18,980
Sodium	10,560
Sulfate	2,560
Magnesium	1,272
Calcium	400
Potassium	380
Bicarbonate	142
Bromine	65
Strontium	13
Boron	4.6
Fluorine	1.4
Rubidium	0.2
Aluminum	0.16
Lithium	0.1
Barium	0.05
Iodine	0.05

(Specific Gravity = 1.025)

Typical Composition of Atmosphere (Dry Air)

Constituent	Percentage
Nitrogen	78.08%
Oxygen	20.95
Argon	0.93
Carbon Dioxide	0.03
Sub-total	99.99%
Neon	0.0018%
Helium	0.000524
Krypton	0.0001
Hydrogen	0.00005

Sand Settling Rates in Still Water

Particle Diameter		Settling Rate		Description
Millmeters	Inches	Millimeters/Sec.	Feet/Hour	
1.00	0.039	100	1181	Coarse Sand
0.80	0.031	83	980	"
0.60	0.024	63	744	"
0.50	0.020	53	626	Medium Sand
0.40	0.016	42	496	"
0.30	0.012	32	378	"
0.20	0.0079	21	248	Fine Sand
0.15	0.0059	15	177	"
0.10	0.0039	8	94.5	Very Fine Sand
0.08	0.0031	6	70.9	"
0.06	0.0024	3.8	44.9	"
0.05	0.0020	2.9	34.3	Silt
0.04	0.0016	2.1	24.8	"
0.03	0.0012	1.3	15.4	"
0.02	0.00079	0.62	7.32	"
0.015	0.00059	0.35	4.13	"
0.010	0.00039	0.154	1.82	Fine Silt, Clay
0.008	0.00031	0.098	1.16	"
0.006	0.00024	0.055	0.65	"
0.005	0.00020	0.0385	0.45	Clay
0.004	0.00016	0.0247	0.29	"
0.003	0.00012	0.0138	0.16	"
0.002	0.00008	0.0062	0.073	"
0.0015	0.000059	0.0035	0.041	"
0.0010	0.000039	0.00154	0.018	Fine Clay
0.0001	0.000004	0.0000154	0.00018	Colloidal

Air Bubbles – Rate of Rise in Still Water

Approximate Bubble Diameter		Rate Of Rise	
Inches	Millimeters	Feet Per Second	Centimeters Per Second
1/32	0.79	0.44	13.34
1/16	1.59	0.56	16.95
3/32	2.38	0.64	19.62
1/8	3.18	0.69	21.05
5/32	3.97	0.72	21.91
3/16	4.76	0.73	22.38
7/32	5.56	0.74	22.67
1/4	6.35	0.75	22.86

Approximate Relation of Particle Size to Channel Velocity

Particle Size		For Deposition (Velocities Below)		For Transportation (Velocities Between)		For Erosion (Velocities Above)	
Inches	Millimeters	Ft/Sec	Cm/Sec	Ft/Sec	Cm/Sec	Ft/Sec	Cm/Sec
3.937	100	7.4	225	7.4-10.2	225-310	10.2	310
1.969	50	5.9	180	5.9-8.9	180-270	8.9	270
0.394	10	2.3	70	2.3-3.9	70-120	3.9	120
0.197	5	1.2	36	1.2-2.3	36-70	2.3	70
0.039	1	0.25	7.5	0.25-0.82	7.5-25	0.82	25
0.020	0.5	0.12	3.8	0.12-0.66	3.8-20	0.66	20
0.004	0.1	0.025	0.75	0.025-0.69	0.75-21	0.69	21
0.0020	0.05	0.011	0.35	0.011-0.89	0.35-27	0.89	27
0.0004	0.01	0.002	0.075	0.002-2.0	0.075-60	2.0	60

Greek Alphabet

Α	α	*α*	Alpha	Ν	ν		Nu
Β	β	ϐ	Beta	Ξ	ξ		Xi
Γ	γ		Gamma	Ο	ο		Omicron
Δ	δ		Delta	Π	π	ϖ	Pi
Ε	ϵ		Epsilon	Ρ	ρ		Rho
Ζ	ζ		Zeta	Σ	σ	ς	Sigma
Η	η		Eta	Τ	τ		Tau
Θ	θ	ϑ	Theta	Υ	υ		Upsilon
Ι	ι		Iota	Φ	ϕ	φ	Phi
Κ	κ		Kappa	Χ	χ		Chi
Λ	λ		Lambda	Ψ	ψ		Psi
Μ	μ		Mu	Ω	ω		Omega

Map Scales in English and Metric Units

Map Scale	Per Inch				Per Centimeter				Per Mile		Per Kilometer	
1 to —	Miles	Kilometers	Feet	Meters	Miles	Kilometers	Feet	Meters	Inches	Centimeters	Inches	Centimeters
1,000,000	15.78	25.40	83,333	25,400	6.21	10.00	32,808	10,000	0.063	0.161	0.039	0.10
500,000	7.89	12.70	41,667	12,700	3.11	5.00	16,404	5,000	0.127	0.322	0.079	0.20
250,000	3.95	6.350	20,833	6,350	1.55	2.50	8,202	2,500	0.253	0.644	0.157	0.40
200,000	3.16	5.080	16,667	5,080	1.24	2.00	6,562	2,000	0.317	0.805	0.197	0.50
125,000	1.97	3.175	10,417	3,175	0.777	1.25	4,101	1,250	0.507	1.29	0.315	0.80
100,000	1.58	2.540	8,333	2,540	0.621	1.00	3,281	1,000	0.634	1.61	0.394	1.00
75,000	1.18	1.905	6,250	1,905	0.466	0.750	2,461	750.0	0.845	2.15	0.525	1.33
63,360	1.00	1.609	5,280	1,609	0.394	0.634	2,079	633.6	1.00	2.54	0.621	1.58
62,500	0.986	1.588	5,208	1,588	0.388	0.625	2,051	625.0	1.01	2.57	0.630	1.60
50,000	0.789	1.270	4,167	1,270	0.311	0.500	1,640	500.0	1.27	3.22	0.787	2.00
40,000	0.631	1.016	3,333	1,016	0.249	0.400	1,312	400.0	1.58	4.02	0.984	2.50
31,680	0.500	0.805	2,640	804.7	0.197	0.317	1,039	316.8	2.00	5.08	1.24	3.16
25,000	0.395	0.635	2,083	635.0	0.155	0.250	820.2	250.0	2.53	6.44	1.57	4.00
24,000	0.379	0.610	2,000	609.6	0.149	0.240	787.4	240.0	2.64	6.71	1.64	4.17
20,000	0.316	0.508	1,667	508.0	0.124	0.200	656.2	200.0	3.17	8.05	1.97	5.00
10,000	0.158	0.254	833.3	254.0	0.0621	0.100	328.1	100.0	6.34	16.1	3.94	10.0
5,000	0.079	0.127	416.7	127.0	0.0311	0.050	164.0	50.00	12.7	32.2	7.87	20.0
2,500	0.039	0.064	208.3	63.50	0.0155	0.025	82.02	25.00	25.3	64.4	15.7	40.0
1,200	0.019	0.030	100.0	30.48	0.0075	0.012	39.37	12.00	52.8	134.1	32.8	83.3

Length of A Degree of Latitude and Longitude

Latitude	Degree of Latitude		Degree of Longitude	
Degrees	Statute Miles	Feet	Statute Miles	Feet
0 (Equator)	68.703	362,753	69.172	365,226
5	68.709	362,782	68.910	363,845
10	68.724	362,864	68.128	359,715
15	68.750	363,002	66.830	352,864
20	68.785	363,186	65.026	343,337
25	68.828	363,412	62.729	331,207
30	68.878	363,675	59.955	316,562
35	68.934	363,970	56.725	299,508
40	68.993	364,281	53.063	280,171
45	69.054	364,603	48.994	258,691
50	69.114	364,925	44.551	235,230
55	69.174	365,240	39.765	209,961
60	69.229	365,531	34.674	183,077
65	69.280	365,801	29.314	154,780
70	69.324	366,030	23.729	125,289
75	69.359	366,217	17.959	94,826
80	69.385	366,355	12.051	63,629
85	69.402	366,440	6.049	31,939
90 (Polar)	69.407	366,467	0	0

Wind Chill Factor

Wind Speed (Miles/Hr)	Actual Temperature (Degrees Fahrenheit)														
	35	30	25	20	15	10	5	0	−5	−10	−15	−20	−25	−30	−35
5	33	27	21	16	12	7	0	−5	−10	−15	−21	−26	−31	−36	−42
10	22	16	10	3	−3	−9	−15	−22	−27	−34	−40	−46	−52	−58	−64
15	16	9	2	−5	−11	−18	−25	−31	−38	−45	−51	−58	−65	−72	−78
20	12	4	−3	−10	−17	−24	−31	−39	−46	−53	−60	−67	−74	−81	−88
25	8	1	−7	−15	−22	−29	−36	−44	−51	−59	−66	−74	−81	−88	−96
30	6	−2	−10	−18	−25	−33	−41	−49	−56	−64	−71	−79	−86	−93	−101
35	4	−4	−12	−20	−27	−35	−43	−52	−58	−67	−74	−82	−89	−97	−105
40	3	−5	−13	−21	−29	−37	−45	−53	−60	−69	−76	−84	−92	−100	−107
45	2	−6	−14	−22	−30	−38	−46	−54	−62	−70	−78	−85	−93	−102	−109

(Wind velocities greater than 45 MPH have little additional chilling effect)

Representative World Climatic Data

T = Average daily temperature (Max/Min); P = Average monthly precipitation
(Figures have been assembled from various sources and may not agree exactly with other published data)

United States			Jan	Feb	Mar	Apr	May	Jun	Jul	Aug	Sep	Oct	Nov	Dec
			Temperature (T) in degrees F; Precipitation (P) in inches; elevation in feet											
AK	Anchorage	T	21/4	28/9	33/13	44/28	54/36	62/44	65/50	64/47	57/39	42/28	30/15	20/6
	61.2N-149.9W; 85 ft	P	0.8	0.9	0.7	0.7	0.6	1.1	2.0	2.1	2.5	1.7	1.1	1.1
	Fairbanks	T	−1/−21	11/−10	23/−4	42/17	59/35	71/46	72/48	66/44	54/33	35/17	12/−5	1/−16
	64.8N-147.9W; 436 ft	P	0.5	0.5	0.4	0.3	0.6	1.3	1.8	1.9	1.1	0.7	0.7	0.7
	Juneau	T	27/16	34/22	37/25	47/31	55/38	61/44	64/47	63/47	56/42	47/37	38/28	32/22
	58.4N-134.6W; 12 ft	P	3.7	3.7	3.3	2.9	3.4	3.0	4.1	5.0	6.4	7.7	5.2	4.7
AL	Birmingham	T	57/36	58/38	67/47	76/50	82/62	89/69	93/71	92/70	87/67	79/52	65/44	56/38
	33.6N-86.8W; 620 ft	P	5.2	4.7	6.6	5.0	4.5	3.7	5.4	3.9	4.3	2.7	3.6	5.0
AR	Little Rock	T	50/34	54/36	63/44	72/53	79/61	82/69	90/72	90/71	84/65	74/54	61/43	52/36
	34.7N-92.3W; 286 ft	P	3.9	3.8	4.7	5.4	5.3	3.7	3.6	3.1	4.3	2.8	4.4	4.2
AZ	Phoenix	T	64/35	69/44	75/46	84/50	91/60	101/68	105/75	101/75	97/69	87/55	75/45	66/41
	33.4N-112.0W; 1117 ft	P	0.7	0.9	0.7	0.3	0.1	0.1	0.8	1.1	0.7	0.5	0.5	0.9
CA	Los Angeles	T	64/45	65/46	65/45	67/52	72/53	75/56	76/62	80/59	79/58	73/57	72/49	67/47
	33.9N-118.4W; 97 FT	P	2.7	2.9	1.8	1.1	0.1	0.1	0.0	0.0	0.2	0.4	1.1	2.4
	San Francisco	T	55/42	59/48	62/48	64/47	65/52	67/53	72/54	66/54	70/56	71/51	65/52	58/47
	37.6N-122.4W; 8 ft	P	4.0	3.5	2.7	1.3	0.5	0.1	0.0	0.0	0.2	0.7	1.6	4.1
CO	Denver	T	43/16	47/20	51/25	61/34	71/44	82/52	88/59	86/57	78/48	67/37	52/25	46/19
	39.7N-105.0W; 5280 ft	P	0.5	0.7	1.2	1.8	2.5	1.6	1.9	1.5	1.2	1.0	0.8	0.6
CT	Hartford	T	34/17	36/19	46/28	60/38	71/47	80/57	85/62	93/60	75/52	64/41	51/33	37/21
	41.8N-72.7W; 40 ft	P	3.5	3.2	4.2	4.0	3.4	3.4	3.1	4.0	3.9	3.5	4.1	4.2
D.C.	Washington	T	44/30	45/29	55/36	66/46	75/54	83/63	87/69	85/66	78/59	68/50	55/38	45/29
	38.9N-77.1W; 14ft	P	2.8	2.6	3.4	3.1	3.6	4.2	3.8	4.2	3.3	3.0	3.0	3.3

World Climate (Cont'd)

			Jan	Feb	Mar	Apr	May	Jun	Jul	Aug	Sep	Oct	Nov	Dec
FL	Miami	T	76/58	75/62	78/64	83/66	84/71	86/75	89/75	88/76	87/75	85/71	78/66	76/62
	25.8N-80.3W; 7 ft	P	2.1	2.1	1.9	3.1	6.5	9.2	6.0	7.0	8.1	7.1	2.7	1.9
GA	Atlanta	T	52/37	54/37	62/43	70/50	79/60	86/67	87/71	86/69	82/64	72/52	61/43	52/37
	33.7N-84.4W; 1010 ft	P	4.9	4.4	5.9	4.4	4.0	3.4	4.7	3.4	3.2	2.5	3.4	4.2
HI	Honolulu	T	76/69	76/67	77/67	78/68	80/70	81/72	82/73	83/74	83/74	82/72	80/70	78/69
	21.4N-157.9W; 7 ft	P	3.8	2.7	3.5	1.5	1.2	0.5	0.5	0.6	0.6	1.9	3.2	3.4
IA	Des Moines	T	29/11	33/15	46/27	59/38	71/51	80/61	87/65	84/63	76/55	66/46	48/30	34/18
	41.5N-93.7W; 938 ft	P	1.0	1.1	2.2	3.2	4.0	4.2	3.2	4.1	3.1	2.2	1.5	1.5
ID	Boise	T	36/22	43/27	52/33	63/37	71/45	80/51	91/59	88/56	76/47	65/38	50/31	40/23
	43.6N-116.2W; 2838 ft	P	1.6	1.1	1.0	1.2	1.2	1.0	0.3	0.4	0.6	0.8	1.3	1.3
IL	Chicago	T	33/19	34/20	43/29	57/41	66/51	76/60	84/67	79/65	74/58	63/47	47/34	36/23
	41.8N-87.8W; 607 ft	P	1.6	1.3	2.6	3.7	3.2	4.1	3.6	3.5	3.4	2.3	2.1	2.1
IN	Indianapolis	T	37/21	39/23	49/32	61/40	72/54	82/63	86/64	84/65	77/58	67/44	50/35	39/26
	39.7N-86.3W; 792 ft	P	2.7	2.5	3.6	3.7	3.7	4.0	4.3	3.5	2.7	2.5	3.0	3.0
KS	Dodge City	T	42/20	47/22	56/29	66/41	75/51	86/62	93/68	90/65	83/56	71/46	56/30	44/21
	37.8N-100.0W; 1582 ft	P	0.5	0.5	1.5	1.8	3.3	3.0	3.1	2.5	1.9	1.3	0.8	0.5
KY	Louisville	T	44/27	45/29	55/37	66/43	76/56	84/65	89/67	86/67	80/61	70/46	55/38	45/29
	38.2N-85.7W; 477 ft	P	3.4	3.2	4.7	4.1	4.2	3.6	4.1	3.3	3.6	2.6	3.5	3.5
LA	New Orleans	T	64/45	66/50	71/56	78/58	83/68	88/73	91/73	90/76	86/72	80/61	70/55	65/48
	30.0N-90.3W; 3 ft	P	5.0	5.2	4.7	4.5	5.1	4.6	6.7	6.0	5.9	2.7	4.1	5.3
MA	Boston	T	36/20	37/21	43/28	54/38	66/49	75/58	80/63	78/62	71/55	62/46	49/35	40/25
	42.4N-71.1W; 21 ft	P	4.0	3.7	4.1	3.7	3.5	2.9	2.7	3.7	3.4	3.4	4.2	4.9
MD	Baltimore	T	42/28	43/28	51/35	63/45	74/56	82/65	86/69	84/67	78/61	67/50	54/40	44/31
	39.3N-76.6W; 20 ft	P	3.0	3.0	3.7	3.4	3.4	3.8	3.9	4.6	3.5	3.1	3.1	3.4
ME	Portland	T	32/12	33/17	42/26	53/32	61/46	73/55	80/57	74/59	69/52	60/37	45/32	34/21
	43.7N-70.3W; 47 ft	P	3.8	3.6	4.0	3.9	3.3	3.1	2.8	2.8	3.3	3.8	4.7	4.5

World Climate (Cont'd)

			Jan	Feb	Mar	Apr	May	Jun	Jul	Aug	Sep	Oct	Nov	Dec
MI	Detroit	T	31/19	32/18	42/27	55/37	67/48	77/58	82/63	80/62	73/55	60/44	46/33	35/24
	42.3N-83.0W; 585 ft	P	1.9	1.7	2.5	3.2	2.8	3.4	3.1	3.2	2.3	2.1	2.3	2.5
MN	Minneapolis	T	22/2	25/8	38/22	56/33	68/48	77/58	84/61	80/61	72/52	61/37	40/26	27/12
	44.9N-93.2W; 834 ft	P	0.8	0.9	1.7	2.1	3.2	4.1	3.5	3.6	2.5	1.9	1.3	0.9
MO	Kansas City	T	40/23	41/24	53/34	66/46	74/56	83/65	92/71	87/68	80/60	72/49	53/36	41/26
	39.1N-94.6W; 742 ft	P	1.0	1.0	2.1	2.7	3.4	4.1	3.5	3.2	3.3	2.5	1.2	1.1
	St. Louis	T	40/24	43/26	54/36	65/47	75/57	84/66	88/71	87/69	80/62	68/50	54/38	43/28
	38.6N-90.2W; 455 ft	P	1.7	2.1	3.3	3.6	3.5	3.7	3.6	2.6	2.7	2.3	2.5	2.2
MS	Jackson	T	57/35	61/37	68/44	77/53	84/61	91/68	93/71	92/70	88/65	79/51	68/42	60/37
	32.3N-90.2W; 298 ft	P	5.0	4.9	5.9	5.9	4.8	2.9	4.4	3.7	3.6	2.6	4.2	5.4
MT	Billings	T	30/12	38/19	44/24	56/33	66/43	76/52	87/58	84/56	72/47	61/38	44/26	36/18
	45.8N-108.5W; 3120 ft	P	1.0	0.7	1.1	1.9	2.4	2.1	0.9	1.1	1.3	1.2	0.9	0.8
NC	Raleigh	T	50/29	53/30	61/38	72/47	79/55	85/63	88/67	87/67	82/60	72/48	62/38	53/31
	35.8N-78.6W; 365 ft	P	3.6	3.4	3.7	2.9	3.7	3.7	4.4	4.4	3.3	2.7	2.9	3.1
ND	Bismarck	T	18/−4	25/4	36/16	54/31	68/42	77/52	84/56	83/54	71/43	59/33	39/18	26/5
	46.8N-100.8W; 1647 ft	P	0.5	0.5	0.7	1.5	2.2	3.0	2.1	1.7	1.4	0.8	0.5	0.5
NE	Omaha	T	30/13	35/17	47/28	61/42	72/53	81/62	86/67	84/65	76/57	64/45	48/30	35/19
	41.3N-95.9W; 1040 ft	P	0.8	0.9	1.9	2.9	4.3	4.1	3.6	4.1	2.5	2.1	1.3	0.8
NH	Concord	T	31/9	33/11	42/22	57/32	69/41	78/52	83/56	80/55	72/46	61/36	47/27	34/15
	43.2N-71.5W; 290 ft	P	2.8	2.5	2.9	3.0	2.9	2.9	2.9	3.3	3.1	3.1	3.7	3.4
NM	Albuquerque	T	47/22	53/26	61/32	71/40	80/49	91/58	93/65	89/63	83/55	72/43	57/31	48/23
	35.1N-106.6W; 5311 ft	P	0.4	0.4	0.5	0.4	0.5	0.5	1.3	1.5	0.9	0.9	0.4	0.5
	Santa Fe	T	39/19	43/24	51/29	59/36	68/43	78/51	80/57	79/55	73/49	62/38	50/27	40/20
	35.7N-105.9W; 6950 ft	P	0.7	0.8	0.8	1.1	1.3	1.0	2.4	2.4	1.4	1.1	0.7	0.7
NV	Las Vegas	T	54/32	67/36	72/40	78/51	89/53	99/63	104/76	102/71	96/60	80/53	71/37	62/30
	36.1N-115.2W; 2162 ft	P	0.5	0.4	0.4	0.2	0.1	0.0	0.5	0.5	0.3	0.2	0.3	0.4
	Reno	T	45/16	51/24	56/26	65/31	72/38	82/43	89/46	89/45	80/38	69/29	56/23	46/19
	39.5N-119.8W; 4404 ft	P	1.2	1.0	0.7	0.5	0.5	0.4	0.3	0.2	0.2	0.5	0.6	1.1

World Climate (Cont'd)

			Jan	Feb	Mar	Apr	May	Jun	Jul	Aug	Sep	Oct	Nov	Dec
NY	New York City	T	40/27	39/25	46/32	60/43	69/55	78/61	85/68	80/66	78/59	66/50	51/37	41/29
	40.8N-74.0W; 132 ft	P	3.2	3.1	4.2	3.8	3.8	3.2	3.8	4.0	3.7	3.4	4.1	3.8
OH	Columbus	T	37/22	39/23	49/32	61/42	72/52	81/61	85/65	83/63	77/57	65/46	50/35	39/26
	40.0N-83.0W; 780 ft	P	2.8	2.2	3.2	3.4	3.8	4.0	4.0	3.7	2.8	1.9	2.6	2.6
OK	Oklahoma City	T	46/28	51/30	62/39	71/49	78/59	87/67	93/72	92/70	85/62	74/52	60/39	49/30
	35.4N-97.6W; 1285 ft	P	1.0	1.3	2.1	2.9	5.5	3.9	3.0	2.4	3.4	2.7	1.5	1.2
OR	Portland	T	44/33	48/36	54/40	62/42	66/47	73/53	79/56	77/56	71/52	63/45	53/41	46/37
	45.6N-122.6W; 21 ft	P	6.2	3.9	3.6	2.3	2.1	1.5	0.5	1.1	1.6	3.1	5.2	6.4
PA	Philadelphia	T	40/26	41/27	49/33	61/43	72/54	80/62	85/68	83/67	76/60	66/50	53/39	43/30
	39.9N-75.2W; 100 ft	P	3.2	2.8	3.9	3.5	3.2	3.9	3.9	4.1	3.4	2.8	3.3	3.5
	Pittsburgh	T	40/25	39/23	49/31	63/42	72/52	80/60	85/65	82/63	76/57	65/45	51/36	41/27
	40.5N-80.0W; 747 ft	P	2.9	2.4	3.6	3.3	3.5	3.3	3.8	3.3	2.8	2.5	2.3	2.6
RI	Providence	T	37/22	37/21	46/29	56/38	68/48	76/57	82/63	80/62	73/55	63/45	51/35	40/25
	41.8N-71.4W; 80 ft	P	4.1	3.7	4.3	4.0	3.5	2.8	3.0	4.0	3.5	3.8	4.2	4.5
SC	Charleston	T	58/43	59/44	66/50	73/57	80/66	86/73	88/75	87/75	83/71	75/61	66/51	59/44
	32.8N-79.9W; 9 ft	P	3.3	3.4	4.4	2.6	4.4	6.5	7.3	6.5	4.9	2.9	2.2	3.1
SD	Pierre	T	25/5	32/11	42/21	58/34	70/45	81/56	89/61	88/60	77/48	64/37	45/23	32/12
	44.4N-100.3W; 1480 ft	P	0.5	0.7	0.9	1.9	2.7	3.8	2.2	2.0	1.3	1.1	0.5	0.6
TN	Memphis	T	48/31	53/34	61/42	73/52	81/61	88/69	92/73	90/71	84/64	75/51	61/41	52/34
	35.1N-90.1W; 275 ft	P	4.6	4.3	5.4	5.8	5.1	3.6	4.0	3.7	3.6	2.4	4.2	4.9
	Nashville	T	49/31	50/33	60/40	71/48	78/58	87/67	91/70	88/68	82/62	74/49	59/40	49/33
	36.1N-86.7W; 590 ft	P	4.5	4.0	5.6	4.8	4.6	3.7	3.8	3.4	3.7	2.6	3.5	4.6
TX	Dallas-Ft. Worth	T	55/36	60/40	67/45	75/56	82/63	90/72	94/75	94/74	88/68	78/56	66/47	57/37
	32.8N-97.0W; 550 ft	P	1.7	1.9	2.4	3.6	4.3	2.6	2.0	1.8	3.3	2.5	1.8	1.7
	El Paso	T	56/30	62/37	69/43	78/49	86/59	94/67	95/69	91/68	86/62	79/50	67/40	58/33
	31.8N-106.4W; 3918 ft	P	0.5	0.4	0.4	0.3	0.4	0.7	1.3	1.2	1.1	0.9	0.3	0.5
	Houston	T	62/44	65/46	72/53	78/60	84/65	90/72	92/74	93/73	88/70	81/60	71/52	63/46
	29.8N-95.4W; 40 ft	P	3.2	3.3	2.7	4.2	4.7	4.0	3.3	3.7	4.9	3.7	3.4	3.7

World Climate (Cont'd)

			Jan	Feb	Mar	Apr	May	Jun	Jul	Aug	Sep	Oct	Nov	Dec
UT	Salt Lake City	T	37/18	41/24	52/31	63/36	73/45	82/53	94/60	90/59	79/49	65/38	49/27	40/22
	40.8N-112.0W; 4220 ft	P	1.4	1.3	1.7	2.2	1.5	1.0	0.7	0.9	0.9	1.1	1.2	1.4
VA	Norfolk	T	49/34	50/34	58/40	66/49	76/58	83/66	87/72	85/70	80/65	70/55	60/45	51/36
	36.9N-76.3W; 10 ft	P	3.7	3.3	3.9	2.9	3.8	3.5	5.2	5.3	4.4	3.4	2.9	3.2
WA	Seattle	T	44/33	48/37	52/38	58/40	65/47	70/52	76/54	74/54	67/52	60/44	51/41	47/38
	47.5N-122.3W; 400 ft	P	6.0	4.2	3.6	2.4	1.6	1.4	0.7	1.3	2.0	3.4	5.6	6.3
	Spokane	T	31/19	39/24	49/31	59/36	68/45	76/51	86/55	83/53	72/47	60/38	44/31	36/26
	47.6N-117.5W; 2356 ft	P	2.5	1.6	1.4	1.1	1.4	1.2	0.5	0.7	0.7	1.1	2.1	2.5
WI	Madison	T	25/7	30/11	41/22	58/34	70/44	79/54	83/58	81/56	72/48	61/38	44/26	31/14
	43.1N-89.4W; 860 ft	P	1.1	1.0	2.2	3.1	3.3	3.9	3.8	3.8	3.1	2.2	1.8	1.5
WV	Charleston	T	50/26	51/27	61/35	70/42	79/52	85/61	89/64	87/63	84/58	72/45	61/35	51/28
	38.4N-81.6W; 601 ft	P	3.8	3.5	4.3	3.7	4.0	4.4	4.2	4.5	3.0	2.9	3.1	3.4
WY	Cheyenne	T	37/14	38/16	44/23	56/30	63/39	76/50	85/55	81/53	72/45	63/32	47/23	39/18
	41.2N-104.8W; 6126 ft	P	0.5	0.6	1.2	1.9	2.5	2.1	1.8	1.4	1.1	0.8	0.6	0.5

Canada — **Temperature (T) in degrees C; Precipitation (P) in millimeters; elevation in meters**

		Jan	Feb	Mar	Apr	May	Jun	Jul	Aug	Sep	Oct	Nov	Dec
Edmonton	T	−9/−19	−6/−17	1/−10	11/−2	18/4	21/7	23/9	22/7	17/4	11/−2	1/−10	−6/−15
53.6N-113.5W; 677 m	P	23	18	18	25	48	81	84	61	33	20	23	23
Ottawa	T	−5−16	−6/−16	1/−8	11/−1	19/7	23/12	27/14	25/12	20/9	13/3	4/−3	−4/−13
45.3N-75.7W; 114 m	P	75	56	71	68	64	89	86	65	81	74	77	66
Quebec	T	−8/−17	−7/−15	−1/−9	7/−3	16/5	22/11	24/14	23/11	18/8	13/3	2/−4	−6/−13
46.8N-71.4W; 73 m	P	88	69	76	60	79	94	103	102	92	86	81	80
Saskatoon	T	−13/−23	−11/−22	−3/−14	9/−2	18/3	22/8	25/11	24/9	17/2	11/−3	−1/−11	−9/-20
52.1N-106.6W; 515 m	P	23	15	18	17	36	67	61	48	37	23	13	16

World Climate (Cont'd)

		Jan	Feb	Mar	Apr	May	Jun	Jul	Aug	Sep	Oct	Nov	Dec
Toronto	T	−1/−9	0/−8	3/−5	10/0	17/8	23/12	26/15	25/13	21/11	13/3	6/−1	1/−5
43.7N-79.4W; 116 m	P	67	60	66	64	75	69	76	68	74	61	70	66
Vancouver	T	5/0	7/0	10/3	14/4	18/8	21/10	23/12	23/12	18/8	14/7	9/4	6/2
49.3N-123.1W; 39 m	P	218	147	127	84	71	64	30	43	91	147	211	224
Winnipeg	T	−14/−25	−11/−22	−3/−14	9/−3	18/5	23/10	26/13	24/10	18/6	11/−2	−1/−11	−9/−19
49.9N-97.2W; 239 m	P	23	23	30	36	58	79	79	64	58	38	28	23
Mexico		**Temperature (T) in degrees C; Precipitation (P) in millimeters; elevation in meters**											
Guaymas	T	23/13	24/14	26/17	29/18	30/21	34/24	34/26	35/27	35/25	31/22	28/18	24/13
28.0N-110.9W; 18 m	P	13	5	5	3	0	3	43	69	53	18	8	20
Merida	T	28/17	30/18	37/19	41/20	40/22	33/23	33/22	33/22	32/22	31/22	29/20	28/18
21.0N-89.6W; 22 m	P	30	23	18	20	81	142	132	142	173	97	33	33
Mexico City	T	19/6	21/7	24/8	25/11	26/12	25/13	23/12	23/12	23/11	21/10	20/8	9/6
19.4N-99.1W; 2238 m	P	5	8	13	18	48	104	114	109	104	41	13	8
Monterrey	T	20/9	22/11	24/15	29/17	31/19	33/23	33/22	32/22	30/21	27/18	22/13	18/11
25.7N-100.3W; 528 m	P	15	18	20	33	33	76	58	61	132	76	38	20
Caribbean		**Temperature (T) in degrees C; Precipitation (P) in millimeters; elevation in meters**											
Havana	T	26/18	27/18	27/19	29/21	30/22	31/23	32/23	32/24	31/24	29/23	27/22	26/19
23.1N-82.4W; 24 m	P	71	46	47	58	119	168	124	135	151	173	80	58
Kingston	T	30/19	30/19	30/20	30/21	31/22	32/23	32/23	32/23	31/23	31/23	31/22	31/20
18.0N-76.8W; 34 m	P	23	15	24	30	103	89	40	90	99	180	76	36
Nassau	T	25/18	26/18	26/19	27/21	29/22	30/23	31/24	32/24	31/24	29/23	28/21	26/19
25.1N-77.4W; 4 m	P	36	39	36	64	115	163	147	138	175	165	73	33
Port au Prince	T	31/19	31/20	32/21	32/22	32/23	33/23	34/23	34/23	33/22	32/22	31/22	31/21
18.6N-72.3W; 37 m	P	33	59	86	160	230	102	74	149	175	170	88	33
San Juan	T	27/21	27/20	27/21	28/22	29/23	29/23	29/24	29/24	30/24	29/24	29/23	28/22
18.4N-66.0W; 4 m	P	119	74	55	94	182	145	160	179	173	149	165	135

World Climate (Cont'd)

		Jan	Feb	Mar	Apr	May	Jun	Jul	Aug	Sep	Oct	Nov	Dec
Central America		Temperature (T) in degrees C; Precipitation (P) in millimeters; elevation in meters											
Balboa Heights	T	31/22	32/22	32/22	31/22	30/23	31/23	31/23	30/23	30/23	29/23	30/23	31/22
09.0N-79.6W; 36 m	P	26	10	18	76	203	213	183	201	208	255	259	122
Guatemala City	T	23/12	25/12	27/13	28/14	29/15	27/16	26/16	26/16	25/16	24/16	23/15	22/13
14.6N-90.5W; 1480 m	P	9	3	13	34	152	274	207	198	230	173	25	8
Managua	T	31/20	31/21	33/22	34/23	34/23	32/23	31/23	31/22	31/22	31/22	31/21	31/20
12.1N-86.3W; 55 m	P	6	2	5	7	77	293	134	130	185	241	60	6
San Jose	T	24/13	24/14	26/15	26/16	27/17	26/17	25/17	26/17	26/16	26/16	25/16	25/14
09.9N-84.1W; 1146 m	P	16	5	20	46	229	240	211	241	307	300	145	41
San Salvador	T	32/16	33/17	34/17	34/18	34/19	33/19	32/18	32/19	32/19	31/18	31/17	32/16
13.7N-89.2W; 682 m	P	8	5	10	46	196	328	290	297	310	241	41	12
Tegucigalpa	T	25/14	28/14	29/15	30/16	30/18	29/18	27/18	28/17	28/17	27/16	26/16	25/15
14.1N-87.2W; 1004 m	P	13	3	2	26	180	172	70	75	152	87	39	14
South America		Temperature (T) in degrees C; Precipitation (P) in millimeters; elevation in meters											
Bogota	T	19/9	20/9	20/10	19/11	19/11	18/11	18/10	18/10	19/9	19/10	19/10	19/9
04.7N-74.1W; 2547 m	P	58	68	102	147	116	61	51	56	62	160	119	65
Buenos Aires	T	29/17	28/16	26/16	23/12	18/8	14/5	15/6	16/6	18/9	21/10	24/14	28/16
34.6S-58.5W; 27 m	P	78	71	110	89	76	62	56	61	78	86	84	98
Caracas	T	24/13	25/13	26/14	27/15	27/17	26/17	26/16	26/16	27/16	27/16	26/16	25/14
10.5N-66.9W; 1042 m	P	24	10	15	36	79	102	108	109	107	111	94	46
La Paz	T	17/6	17/6	18/5	18/4	18/3	17/2	17/1	17/2	18/3	19/4	19/5	18/6
16.5S-68.1W; 3659 m	P	114	108	66	33	14	8	10	13	28	40	48	94
Lima	T	28/19	29/19	28/19	27/17	24/16	20/15	19/14	19/13	20/14	22/15	23/16	26/18
12.1S-77.1W; 120 m	P	3	0	0	1	5	5	8	9	8	3	3	0

World Climate (Cont'd)

		Jan	Feb	Mar	Apr	May	Jun	Jul	Aug	Sep	Oct	Nov	Dec
Quito	T	22/8	22/8	21/8	21/8	21/7	22/7	22/7	23/7	23/7	22/7	22/7	22/8
00.1S-78.5W; 2812 m	P	99	110	142	176	137	43	21	30	69	110	97	79
Rio de Janeiro	T	28/23	29/23	28/22	27/21	25/20	24/18	24/17	24/18	24/19	25/19	26/21	28/22
22.9S-43.2W; 61 m	P	124	122	130	109	79	53	40	43	66	77	104	137
Santiago	T	29/12	28/11	27/9	23/7	19/5	14/3	15/4	17/4	19/6	22/7	25/9	28/11
33.5S-70.7W; 520 m	P	2	3	5	13	66	84	76	56	30	16	8	5
Europe		Temperature (T) in degrees C; Precipitation (P) in millimeters; elevation in meters											
Amsterdam	T	4/1	5/1	8/3	11/6	16/10	18/13	21/15	20/15	18/13	13/9	8/5	5/2
52.4N-04.9E; 2 m	P	51	36	33	41	46	48	66	69	72	71	66	56
Athens	T	13/6	15/7	16/8	20/12	25/16	30/21	33/23	33/22	29/19	24/16	19/12	15/7
38.0N-23.7E; 107 m	P	56	41	36	22	20	15	5	11	15	43	70	71
Belgrade	T	3/−3	5/−3	12/2	18/7	23/12	26/14	29/16	28/16	24/13	18/8	11/4	4/−1
44.8N-20.5E; 138 m	P	41	33	41	56	66	71	48	64	43	69	46	48
Berlin	T	2/−3	3/−3	8/0	13/4	19/8	22/13	24/14	23/13	20/9	13/6	7/2	3/−1
52.5N-13.3E; 57 m	P	48	33	38	43	48	58	79	56	48	43	43	48
Brussels	T	4/−1	7/0	10/2	14/5	18/9	21/11	23/12	22/11	21/11	15/7	9/3	6/0
50.9N-04.4E; 100 m	P	68	62	53	59	55	76	95	79	63	84	76	88
Budapest	T	2/−3	4/−2	11/2	17/7	22/11	26/14	28/16	27/15	23/17	16/7	8/3	3/−1
47.5N-19.0E; 120 m	P	38	38	43	51	69	66	51	48	46	53	61	51
Copenhagen	T	2/−2	2/−2	5/−1	10/3	16/7	19/11	22/13	21/12	17/9	12/6	6/2	3/0
55.7N-12.6E; 13 m	P	41	33	30	43	43	53	56	81	48	53	56	53
Dublin	T	8/1	9/2	10/3	13/4	15/7	18/9	20/11	19/11	17/8	14/6	10/4	8/3
53.4N-06.4W; 47 m	P	69	56	51	48	56	51	71	76	72	69	69	66
Frankfurt	T	3/−2	6/−1	9/2	14/5	19/9	22/12	24/13	23/13	19/11	13/6	7/2	4/−1
50.1N-08.7E; 103 m	P	43	33	41	38	51	64	71	66	48	56	51	51

World Climate (Cont'd)

		Jan	Feb	Mar	Apr	May	Jun	Jul	Aug	Sep	Oct	Nov	Dec
Geneva	T	4/−2	9/−1	11/2	14/5	19/9	23/13	25/14	24/14	21/11	14/7	8/3	4/−1
46.2N-06.2E; 405 m	P	48	46	56	64	76	79	74	91	91	97	79	61
Helsinki	T	−3/−8	−3/−9	0/−6	6/−1	13/5	17/9	22/14	19/13	14/8	7/3	3/−1	−1/−6
60.2N-25.0E; 9 m	P	56	43	43	43	48	51	58	84	71	74	69	61
Lisbon	T	13/8	14/8	16/9	18/11	21/13	24/16	26/17	27/18	24/17	21/14	17/11	14/8
38.7N-09.1W; 95 m	P	84	81	79	61	43	18	5	5	36	79	107	91
London	T	6/2	7/2	9/2	13/4	17/7	20/11	22/12	21/12	18/9	13/7	9/4	7/2
51.5N-00.00; 45 m	P	51	38	36	46	46	41	51	56	46	58	64	51
Madrid	T	8/1	11/2	14/4	18/7	22/10	27/14	31/17	30/17	25/13	19/9	12/4	9/2
40.4N-03.7W; 667 m	P	28	43	43	43	38	30	10	8	30	48	56	41
Marseille	T	10/2	12/3	15/5	18/8	22/11	26/15	29/17	28/16	25/15	20/10	15/6	11/3
43.3N-05.4E; 75 m	P	48	38	46	51	48	25	15	23	66	94	79	56
Moscow	T	−6/−13	−5/−12	0/−8	8/−1	18/7	23/11	24/13	22/11	16/6	8/1	−1/−5	−5/−11
55.8N-37.7E; 154 m	P	38	36	28	48	56	74	76	74	48	69	43	41
Oslo	T	−1/−7	0/−7	4/−4	10/1	17/6	21/11	23/13	21/12	16/7	9/3	3/−2	−1/−4
59.9N-10.7E; 94 m	P	43	33	36	41	46	61	74	97	64	74	58	58
Paris	T	6/1	7/1	12/3	16/6	19/9	23/13	24/14	24/14	21/12	16/8	10/4	7/2
48.8N-02.5E; 50 m	P	38	33	38	43	51	53	53	51	51	56	51	48
Prague	T	1/−4	3/−2	7/1	13/4	18/9	22/13	23/14	23/14	18/11	12/7	5/2	1/−2
50.1N-14.4E; 202 m	P	23	20	28	38	61	71	66	56	43	30	30	23
Rome	T	11/5	13/6	15/7	19/10	23/14	28/17	30/19	30/20	26/17	22/13	16/10	13/6
41.8N-12.6E; 115 m	P	84	74	54	51	48	18	10	18	68	109	112	104
Stockholm	T	−1/−5	−1/−6	3/−3	7/0	14/6	18/9	21/13	19/12	14/8	9/4	3/−1	1/−3
59.4N-18.1E; 45 m	P	38	28	28	38	41	48	71	79	53	53	48	48
Vienna	T	1/−3	3/−2	8/1	14/5	19/10	22/13	24/15	23/14	19/11	13/7	7/2	3/−1
48.3N-16.4E; 202 m	P	38	36	46	51	71	69	76	69	51	51	48	46
Warsaw	T	−1/−6	0/−5	5/−2	12/3	19/9	22/12	24/13	23/13	18/9	12/5	4/0	0/−4
52.2N-21.0E; 90 m	P	30	28	33	38	48	66	76	76	48	43	36	36

World Climate (Cont'd)

		Jan	Feb	Mar	Apr	May	Jun	Jul	Aug	Sep	Oct	Nov	Dec
Australia-New Zealand		Temperature (T) in degrees C; Precipitation (P) in millimeters; elevation in meters											
Adelaide	T	30/16	30/17	27/15	23/12	19/10	16/8	15/7	17/8	19/10	23/11	26/13	28/15
35.0S-138.5E; 6 m	P	20	18	25	46	69	76	66	66	53	43	28	25
Auckland	T	23/16	23/16	22/15	19/13	17/11	14/9	13/8	14/8	16/9	17/11	16/12	21/14
37.0S-174.8E; 7 m	P	79	94	81	97	127	137	145	117	102	102	89	79
Brisbane	T	29/21	28/20	28/19	26/16	23/13	21/10	20/9	22/10	24/13	27/16	28/18	29/19
27.4S-135.1E; 5 m	P	163	160	145	94	71	66	56	48	48	64	94	127
Christchurch	T	21/12	21/11	19/10	17/7	13/4	11/2	10/2	11/2	14/4	17/7	19/8	21/11
43.5S-172.5E; 36 m	P	56	43	48	48	66	66	69	48	46	43	48	56
Melbourne	T	25/14	26/14	24/13	20/11	17/8	15/7	13/6	15/6	17/8	19/9	22/11	24/12
37.8S-145.0E; 35 m	P	48	46	56	58	53	53	48	48	58	66	58	58
Perth	T	29/17	29/17	28/16	24/14	21/12	18/10	17/9	18/10	19/10	21/12	24/14	27/16
31.9S-116.0E; 20 m	P	8	10	20	43	130	180	170	145	86	56	20	13
Sydney	T	26/18	26/18	24/17	22/14	19/11	16/9	16/8	17/9	19/11	22/13	23/16	25/17
33.9S-151.0E; 19 m	P	89	102	127	135	127	117	117	76	74	71	74	74
Wellington	T	20/13	21/13	19/12	17/11	14/8	13/7	12/6	13/6	14/8	16/9	17/10	19/12
41.3S-174.8E; 127 m	P	81	81	81	97	117	117	137	117	97	102	89	89
Africa		Temperature (T) in degrees C; Precipitation (P) in millimeters; elevation in meters											
Addis Ababa	T	24/6	24/8	25/9	25/10	25/10	23/9	21/10	21/10	22/9	24/7	23/6	23/5
09.3N-38.8E; 2451 m	P	13	38	66	86	86	137	279	300	191	20	15	5
Algiers	T	15/9	16/10	18/11	20/13	23/15	26/18	29/21	29/22	27/20	23/17	19/13	16/10
36.8N-03.1E; 59 m	P	112	84	74	41	46	15	0	5	41	79	130	137
Cairo	T	18/8	21/9	24/11	28/11	33/17	35/20	36/22	35/22	32/20	30/18	26/14	20/10
29.9N-31.3E; 116 m	P	5	5	5	3	3	0	0	0	0	0	3	5

World Climate (Cont'd)

		Jan	Feb	Mar	Apr	May	Jun	Jul	Aug	Sep	Oct	Nov	Dec
Cape Town	T	26/16	26/16	25/14	22/12	19/9	18/5	17/7	18/8	18/9	21/11	23/13	24/14
33.9S-18.5E; 17 m	P	15	8	18	48	79	84	89	66	43	30	18	10
Dakar	T	26/18	27/17	27/18	27/18	29/20	31/23	31/24	31/24	32/24	32/24	30/23	27/19
14.7N-17.5W; 40 m	P	0	0	0	0	0	18	89	254	132	38	3	8
Dar-es-Salaam	T	31/25	31/25	31/24	30/23	29/22	29/20	28/19	28/19	28/19	29/21	30/22	31/24
06.8S-39.3E; 14 m	P	66	66	130	290	188	33	30	25	30	41	74	91
Johannesburg	T	26/14	25/14	24/12	22/10	19/6	17/4	17/4	20/6	23/9	24/12	25/13	26/14
26.2S-28.0E; 1750 m	P	114	109	89	39	25	8	7	8	23	56	107	125
Khartoum	T	32/15	34/16	38/19	41/23	42/25	41/26	39/25	37/24	39/25	40/24	36/20	33/17
15.6N-32.6E; 390 m	P	0	0	0	0	3	8	53	71	18	5	0	0
Kinshasha	T	31/21	31/22	32/22	32/22	31/21	29/19	27/18	29/19	31/20	31/21	31/22	30/21
04.3S-15.3E; 325 m	P	135	145	196	196	157	8	3	3	30	119	221	142
Monrovia	T	30/23	29/23	30/23	31/23	30/22	27/23	27/22	27/23	27/22	28/22	29/23	30/23
06.3N-10.8W; 23 m	P	5	3	112	297	340	917	615	472	759	640	208	74
Nairobi	T	25/12	26/13	25/14	24/14	22/13	21/12	21/11	21/11	24/11	24/13	23/13	23/13
01.3S-36.8E; 1820 m	P	38	64	124	211	157	46	15	23	30	53	109	86
Rabat	T	17/8	18/8	20/9	22/11	23/14	26/16	28/17	28/18	27/17	25/15	21/12	18/9
34.0N-06.8W; 65 m	P	66	64	66	43	28	8	0	0	10	48	84	86
Tripoli	T	16/8	17/9	19/11	22/14	25/16	27/19	29/21	30/22	29/22	27/18	23/14	18/9
32.9N-13.2E; 22 m	P	81	46	28	10	5	3	0	0	10	41	66	94
Tunis	T	14/6	16/7	18/8	21/11	24/13	29/17	32/20	33/21	31/19	25/15	20/11	16/7
36.8N-10.2E; 66 m	P	64	51	41	36	18	8	3	8	33	51	48	61

World Climate (Cont'd)

Asia-Pacific		Jan	Feb	Mar	Apr	May	Jun	Jul	Aug	Sep	Oct	Nov	Dec
		Temperature (T) in degrees C; Precipitation (P) in millimeters; elevation in meters											
Amman	T	12/4	13/4	16/6	23/9	28/14	31/16	32/18	32/18	31/17	27/14	21/10	15/6
32.0N-36.0E; 777 m	P	69	74	30	15	5	0	0	0	0	5	33	46
Ankara	T	4/−4	6/−3	11/−1	17/5	23/9	26/12	30/15	31/14	26/11	21/7	14/3	6/−2
40.0N-32.9E; 861 m	P	33	30	33	33	48	25	13	10	18	23	30	48
Baghdad	T	16/4	18/6	22/9	29/14	36/19	41/23	43/24	43/24	40/20	33/16	25/11	18/6
33.3N-44.4E; 34 m	P	23	25	28	13	3	0	0	0	0	3	20	25
Bangkok	T	32/20	33/22	34/24	34/25	34/25	33/24	32/24	32/24	32/24	31/24	31/22	31/20
13.7N-100.5E; 16 m	P	5	28	28	58	132	152	175	234	356	251	46	3
Beijing	T	1/−10	4/−7	11/−1	21/8	27/14	31/18	31/21	30/20	26/13	20/6	9/−2	3/−8
39.9N-116.5E; 183 m	P	5	6	8	18	36	78	245	140	58	16	12	3
Beirut	T	17/11	17/11	19/12	22/14	26/18	28/21	31/23	32/23	30/23	27/21	23/16	18/13
33.9N-35.5E; 34 m	P	191	157	94	56	18	3	0	0	5	51	132	185
Bombay	T	28/19	28/20	30/22	32/24	33/27	32/26	29/25	29/24	29/24	32/24	32/23	31/21
19.1N-72.9E; 8 m	P	3	3	3	0	18	485	617	340	264	64	13	3
Calcutta	T	27/13	29/15	34/22	36/24	36/25	33/26	32/26	32/26	32/25	32/24	29/18	26/13
22.5N-88.3E; 6 m	P	10	30	36	43	140	297	325	328	251	114	20	5
Damascus	T	12/2	14/4	18/6	24/9	29/13	33/16	36/18	37/18	33/16	27/12	19/8	13/4
33.5N-36.3E; 720 m	P	43	43	8	13	3	0	0	0	18	10	41	41
Dhaka	T	25/12	28/14	33/16	35/23	34/25	32/26	31/26	31/26	31/25	31/24	29/18	26/13
23.8N-90.4E; 7 m	P	8	30	61	137	244	315	330	338	249	135	2	5
Hong Kong	T	18/13	17/13	19/16	24/19	28/23	29/26	31/26	31/26	29/25	27/23	23/18	20/15
22.3N-114.2E; 33 m	P	33	46	74	137	292	394	381	361	257	114	43	38
Istanbul	T	7/2	8/3	11/4	16/7	20/12	25/16	27/18	27/19	24/16	19/12	15/9	11/5
41.0N-28.8E; 18 m	P	94	58	66	48	36	33	43	38	58	97	104	124
Jakarta	T	29/23	29/23	30/23	31/24	31/24	31/23	31/23	31/23	31/23	31/23	30/23	29/23
06.2S-106.8E; 8 m	P	300	300	211	147	114	97	64	43	66	112	142	203

World Climate (Cont'd)

		Jan	Feb	Mar	Apr	May	Jun	Jul	Aug	Sep	Oct	Nov	Dec
Jerusalem	T	13/5	13/6	18/8	23/10	27/14	29/16	31/17	31/18	29/17	27/15	21/12	15/7
31.8N-35.2E; 809 m	P	132	132	64	28	3	0	0	0	0	13	71	86
Kabul	T	2/−8	4/−6	12/1	19/6	26/11	31/13	33/16	33/15	29/11	23/6	17/1	8/−3
34.5N-69.2E; 1816 m	P	33	38	91	84	23	5	3	3	0	10	15	15
Karachi	T	25/13	26/14	29/19	32/23	33/26	33/28	33/27	31/26	31/25	33/22	31/18	27/14
24.8N-67.0E; 4 m	P	13	10	8	3	3	18	81	41	13	3	3	5
Kathmandu	T	22/−5	24/−1	26/2	29/6	32/11	35/16	31/18	30/19	31/14	28/9	26/2	23/−3
27.7N-85.4E; 1348 m	P	15	41	23	58	122	246	373	345	155	38	8	3
Manila	T	30/21	31/21	33/22	34/23	34/24	33/24	31/24	31/24	31/24	31/23	32/22	30/21
14.5N-121.0E; 15 m	P	23	13	18	33	130	254	432	422	356	193	145	66
New Delhi	T	21/7	24/9	31/14	36/20	40/26	38/28	36/27	34/26	34/24	34/18	29/11	23/8
28.6N-77.2E; 212 m	P	23	18	13	8	13	74	180	173	117	10	3	10
Riyadh	T	21/8	23/10	28/13	32/18	38/22	42/25	42/26	41/24	39/22	34/16	29/13	21/9
24.7N-46.7E; 591 m	P	3	20	23	25	10	0	0	0	0	0	0	0
Seoul	T	0/−9	3/−7	8/−2	17/5	22/11	27/16	29/21	31/22	26/15	19/7	11/0	3/−7
37.5N-126.9E; 10 m	P	30	20	38	76	81	130	376	267	119	41	46	25
Shanghai	T	8/1	9/1	13/4	19/10	25/16	28/19	32/23	32/23	28/18	23/14	17/7	12/2
31.2N-121.4E; 5 m	P	48	61	84	91	97	178	147	140	132	74	53	38
Singapore	T	30/23	31/23	31/23	31/24	31/24	31/24	31/24	31/24	31/24	31/23	31/23	30/23
01.3N-103.8E; 10 m	P	251	173	193	188	173	173	170	196	178	208	254	257
Taipei	T	19/12	18/12	22/14	25/17	28/21	32/23	33/24	33/24	31/23	27/19	24/17	21/14
25.1N-121.5E; 6 m	P	97	135	109	135	175	224	224	221	208	140	107	74
Tehran	T	7/−7	10/−4	15/4	22/9	28/14	34/19	37/22	36/22	32/18	24/12	17/6	11/−3
35.7N-51.3E; 1200 m	P	46	38	46	36	13	3	3	3	3	8	20	30
Tokyo	T	8/−2	9/−1	12/2	17/8	22/12	24/17	28/21	30/22	26/19	21/13	16/6	11/1
35.7N-139.8E; 6 m	P	48	74	107	135	147	165	142	152	234	208	97	56

Number of Each Day of the Year

Day of Month	Jan	Feb	Mar	Apr	May	Jun	Jul	Aug	Sep	Oct	Nov	Dec	Day of Month
1	1	32	60	91	121	152	182	213	244	274	305	335	1
2	2	33	61	92	122	153	183	214	245	275	306	336	2
3	3	34	62	93	123	154	184	215	246	276	307	337	3
4	4	35	63	94	124	155	185	216	247	277	308	338	4
5	5	36	64	95	125	156	186	217	248	278	309	339	5
6	6	37	65	96	126	157	187	218	249	279	310	340	6
7	7	38	66	97	127	158	188	219	250	280	311	341	7
8	8	39	67	98	128	159	189	220	251	281	312	342	8
9	9	40	68	99	129	160	190	221	252	282	313	343	9
10	10	41	69	100	130	161	191	222	253	283	314	344	10
11	11	42	70	101	131	162	192	223	254	284	315	345	11
12	12	43	71	102	132	163	193	224	255	285	316	346	12
13	13	44	72	103	133	164	194	225	256	286	317	347	13
14	14	45	73	104	134	165	195	226	257	287	318	348	14
15	15	46	74	105	135	166	196	227	258	288	319	349	15
16	16	47	75	106	136	167	197	228	259	289	320	350	16
17	17	48	76	107	137	168	198	229	260	290	321	351	17
18	18	49	77	108	138	169	199	230	261	291	322	352	18
19	19	50	78	109	139	170	200	231	262	292	323	353	19
20	20	51	79	110	140	171	201	232	263	293	324	354	20
21	21	52	80	111	141	172	202	233	264	294	325	355	21
22	22	53	81	112	142	173	203	234	265	295	326	356	22
23	23	54	82	113	143	174	204	235	266	296	327	357	23
24	24	55	83	114	144	175	205	236	267	297	328	358	24
25	25	56	84	115	145	176	206	237	268	298	329	359	25
26	26	57	85	116	146	177	207	238	269	299	330	360	26
27	27	58	86	117	147	178	208	239	270	300	331	361	27
28	28	59	87	118	148	179	209	240	271	301	332	362	28
29	29	*	88	119	149	180	210	241	272	302	333	363	29
30	30		89	120	150	181	211	242	273	303	334	364	30
31	31		90		151		212	243		304		365	31

* In leap years (divided evenly by 4), after February 29 add 1 to numbers in table

Length of Days
Time from Sunrise to Sunset (Hours:Minutes) At Beginning of Month

Latitude	Jan	Feb	Mar	Apr	May	June	July	Aug	Sept	Oct	Nov	Dec
N 60	06:05	08:00	10:23	13:23	16:04	18:18	18:43	16:55	14:04	11:23	08:42	06:28
50	08:11	09:20	10:54	12:58	14:45	16:04	16:17	15:17	13:26	11:36	09:48	08:24
45	08:52	09:48	11:06	12:50	14:19	15:22	15:33	14:44	13:13	11:40	10:10	09:02
40	09:24	10:10	11:16	12:42	13:56	14:50	14:58	14:18	13:02	11:44	10:29	09:33
N 35	09:52	10:30	11:24	12:36	13:38	14:22	14:29	13:56	12:53	11:48	10:45	09:59
30	10:16	10:47	11:31	12:31	13:22	13:57	14:03	13:36	12:44	11:50	10:59	10:22
20	10:57	11:16	11:44	12:22	12:54	13:16	13:19	13:02	12:31	11:56	11:25	11:00
10	11:34	11:42	11:55	12:14	12:30	12:40	12:42	12:34	12:17	12:01	11:46	11:31
Equator	12:07	12:07	12:07	12:07	12:07	12:07	12:07	12:07	12:07	12:07	12:07	12:07
S 10	12:42	12:32	12:17	11:59	11:44	11:35	11:33	11:41	11:55	12:11	12:28	12:40
20	13:18	12:59	12:31	11:53	11:21	11:00	10:56	11:13	11:44	12:17	12:51	13:15
30	14:02	13:29	12:45	11:45	10:55	10:20	10:15	10:41	11:31	12:25	13:16	13:56
40	14:56	14:09	13:02	11:35	10:22	09:31	09:22	10:01	11:20	12:31	13:50	14:47

With clear skies, normal outdoor activity can be conducted during twilight before sunrise and after sunset. Each of the two daily twilight periods is about 20 minutes at the equator, about 30 minutes at latitudes of 30 degrees, and about 40 minutes at latitudes of 50 degrees.

World Standard Time

Showing hours fast (+) or slow(−) with respect to Greenwich Mean Time (GMT); and time at noon (1200 hrs) GMT

Locality	Hrs + or − of GMT	Time at 1200 GMT
Afghanistan	+4½	1630
Alaska		
Juneau, Ketch'kn	−8	0400
Central	−10	0200
Algeria	0	1200
Angola	+1	1300
Argentina	−3	0900
Australia		
Western	+8	2000
North-South	+9½	2130
NSW,Queensland, & Victoria	+10	2200
Austria	+1	1300
Azores Isl.	−1	1100
Bahamas	−5	0700
Bahrain	+3	1500
Bangladesh	+6	1800
Belgium	+1	1300
Belize	−6	0600
Bermuda	−4	0800
Bolivia	−4	0800
Botswana	+2	1400
Brazil	−3	0900
Eastern		
Amaz,Matto Gr	−4	0800
British Isles	0	1200
Brunei	+8	2000
Bulgaria	+2	1400
Burma	+6½	1830
Cambodia	+7	1900
Cameroon	+1	1300
Canada	−4	0800
W.Queb, E Ont.	−5	0700
W Ont, Manitoba	−6	0600
Alb, Saskatch	−7	0500
B.C.	−8	0400
Cent. African Rep.	+1	1300
Chad	+1	1300
Chile	−4	0800
China	+8	2000
Colombia	−5	0700
Costa Rica	−6	0600
Cuba	−5	0700
Cyprus	+2	1400
Czechoslovakia	+1	1300
Denmark	+1	1300
Dominican Repub.	−4	0800
Ecuador	−5	0700
Egypt	+2	1400
El Salvador	−6	0600
Ethiopia	+3	1500
Fiji Isl.	+12	2400
Finland	+2	1400
France	+1	1300
French Guiana	−3	0900
Germany	+1	1300
Ghana	0	1200
Greece	+2	1400
Guam	+10	2200
Guatemala	−6	0600
Haiti	−5	0700
Hawaii	−10	0200
Honduras	−6	0600
Hong Kong	+8	2000
Hungary	+1	1300
Iceland	0	1200
India	+5½	1730
Indonesia		
Java, Sumatra	+7	1900
Bomeo, Celebes	+8	2000
W. Irian	+9	2100
Iran	+3½	1530
Iraq	+3	1500
Ireland	0	1200
Israel	+2	1400
Italy	+1	1300
Ivory Coast	0	1200
Jamaica	−5	0700
Jordan	+2	1400
Kenya	+3	1500
Kuwait	+3	1500
Laos	+7	1900
Lebanon	+2	1400
Lesotho	+2	1400
Liberia	0	1200
Libya	+2	1400
Luxembourg	+1	1300

World Standard Time (Cont'd)

Locality	Hrs + or − of GMT	Time at 1200 GMT
Madagascar	+3	1500
Malawi	+2	1400
Malaysia	+7½	1930
Mali	0	1200
Manchuria	+8	2000
Mauritania	0	1200
Mexico		
Central, East	−6	0600
Sonora, Sinaloa		
Nayarit, S Baja	−7	0500
N Baja	−8	0400
Mongolia	+8	2000
Morocco	0	1200
Mozambique	+2	1400
Namibia	+2	1400
Nepal	+5¾	1745
Netherlands	+1	1300
New Zealand	+12	2400
Nicaragua	−6	0600
Niger	+1	1300
Nigeria	+1	1300
Norway	+1	1300
Oman	+4	1600
Pakistan	+5	1700
Panama	−5	0700
Papua New Guinea	+10	2200
Paraguay	−4	0800
Peru	−5	0700
Phillippines	+8	2000
Poland	+1	1300
Portugal	0	1200
Puerto Rico	−4	0800
Qatar	+3	1500
Romania	+2	1400
Rwanda	+2	1400
Saudi Arabia	+3	1500
Senegal	0	1200
Sicily	+1	1300
Sierra Leone	0	1200
Singapore	+7½	1930
South Africa	+2	1400
Spain	+1	1300
Sri Lanka	+5½	1730
Sudan	+2	1400
Swaziland	+2	1400
Sweden	+1	1300
Switzerland	+1	1300
Syria	+2	1400
Taiwan	+8	2000
Tanzania	+3	1500
Thailand	+7	1900
Togo	0	1200
Trindad, Tobago	−4	0800
Tunisia	+1	1300
Turkey	+3	1500
Uganda	+3	1500
United Arab Emir.	+4	1600
United States		
Eastern	−5	0700
Central	−6	0600
Mountain	−7	0500
Pacific	−8	0400
Uruguay	−3	0900
(U.S.S.R.) Commonwealth of States		
Estonia, Latvia, Lithuania, Ukraine Eur. Russia	+3	1500
Georgia, Armenia Azerbaijan	+4	1600
Kazakh, Uzbek, Turkman	+5	1700
Kirghiz, W. Siberia	+6	1800
Buryat, E Siberia	+8	2000
Kuril, Sakhlin Is	+10	2200
Venezuela	−4	0800
Vietnam	+7	1900
Western Sahara	0	1200
West Indies	−4	0800
Yemen	3	1500
Yugoslavia	1	1300
Zaire	2	1400
Zambia	2	1400
Zimbabwe	2	1400

Perpetual Calendar

The following table gives the Calendar Number to use for each year listed.

Year	Calendar Number	Year	Calendar Number	Year	Calendar Number	Year	Calendar Number
1971	6	1991	3	2011	7	2031	4
1972	14	1992	11	2012	8	2032	12
1973	2	1993	6	2013	3	2033	7
1974	3	1994	7	2014	4	2034	1
1975	4	1995	1	2015	5	2035	2
1976	12	1996	9	2016	13	2036	10
1977	7	1997	4	2017	1	2037	5
1978	1	1998	5	2018	2	2038	6
1979	2	1999	6	2019	3	2039	7
1980	10	2000	14	2020	11	2040	8
1981	5	2001	2	2021	6	2041	3
1982	6	2002	3	2022	7	2042	4
1983	7	2003	4	2023	1	2043	5
1984	8	2004	12	2024	9	2044	13
1985	3	2005	7	2025	4	2045	1
1986	4	2006	1	2026	5	2046	2
1987	5	2007	2	2027	6	2047	3
1988	13	2008	10	2028	14	2048	11
1989	1	2009	5	2029	2	2049	6
1990	2	2010	6	2030	3	2050	7

Calendar No. One

January

S	M	T	W	T	F	S
1	2	3	4	5	6	7
8	9	10	11	12	13	14
15	16	17	18	19	20	21
22	23	24	25	26	27	28
29	30	31				

February

S	M	T	W	T	F	S
			1	2	3	4
5	6	7	8	9	10	11
12	13	14	15	16	17	18
19	20	21	22	23	24	25
26	27	28				

March

S	M	T	W	T	F	S
			1	2	3	4
5	6	7	8	9	10	11
12	13	14	15	16	17	18
19	20	21	22	23	24	25
26	27	28	29	30	31	

April

S	M	T	W	T	F	S
						1
2	3	4	5	6	7	8
9	10	11	12	13	14	15
16	17	18	19	20	21	22
23	24	25	26	27	28	29
30						

May

S	M	T	W	T	F	S
	1	2	3	4	5	6
7	8	9	10	11	12	13
14	15	16	17	18	19	20
21	22	23	24	25	26	27
28	29	30	31			

June

S	M	T	W	T	F	S
				1	2	3
4	5	6	7	8	9	10
11	12	13	14	15	16	17
18	19	20	21	22	23	24
25	26	27	28	29	30	

July

S	M	T	W	T	F	S
						1
2	3	4	5	6	7	8
9	10	11	12	13	14	15
16	17	18	19	20	21	22
23	24	25	26	27	28	29
30	31					

August

S	M	T	W	T	F	S
		1	2	3	4	5
6	7	8	9	10	11	12
13	14	15	16	17	18	19
20	21	22	23	24	25	26
27	28	29	30	31		

September

S	M	T	W	T	F	S
					1	2
3	4	5	6	7	8	9
10	11	12	13	14	15	16
17	18	19	20	21	22	23
24	25	26	27	28	29	30

October

S	M	T	W	T	F	S
1	2	3	4	5	6	7
8	9	10	11	12	13	14
15	16	17	18	19	20	21
22	23	24	25	26	27	28
29	30	31				

November

S	M	T	W	T	F	S
			1	2	3	4
5	6	7	8	9	10	11
12	13	14	15	16	17	18
19	20	21	22	23	24	25
26	27	28	29	30		

December

S	M	T	W	T	F	S
					1	2
3	4	5	6	7	8	9
10	11	12	13	14	15	16
17	18	19	20	21	22	23
24	25	26	27	28	29	30
31						

Calendar No. Two

January

S	M	T	W	T	F	S
	1	2	3	4	5	6
7	8	9	10	11	12	13
14	15	16	17	18	19	20
21	22	23	24	25	26	27
28	29	30	31			

February

S	M	T	W	T	F	S
				1	2	3
4	5	6	7	8	9	10
11	12	13	14	15	16	17
18	19	20	21	22	23	24
25	26	27	28			

March

S	M	T	W	T	F	S
				1	2	3
4	5	6	7	8	9	10
11	12	13	14	15	16	17
18	19	20	21	22	23	24
25	26	27	28	29	30	31

April

S	M	T	W	T	F	S
1	2	3	4	5	6	7
8	9	10	11	12	13	14
15	16	17	18	19	20	21
22	23	24	25	26	27	28
29	30					

May

S	M	T	W	T	F	S
		1	2	3	4	5
6	7	8	9	10	11	12
13	14	15	16	17	18	19
20	21	22	23	24	25	26
27	28	29	30	31		

June

S	M	T	W	T	F	S
					1	2
3	4	5	6	7	8	9
10	11	12	13	14	15	16
17	18	19	20	21	22	23
24	25	26	27	28	29	30

July

S	M	T	W	T	F	S
1	2	3	4	5	6	7
8	9	10	11	12	13	14
15	16	17	18	19	20	21
22	23	24	25	26	27	28
29	30	31				

August

S	M	T	W	T	F	S
			1	2	3	4
5	6	7	8	9	10	11
12	13	14	15	16	17	18
19	20	21	22	23	24	25
26	27	28	29	30	31	

September

S	M	T	W	T	F	S
						1
2	3	4	5	6	7	8
9	10	11	12	13	14	15
16	17	18	19	20	21	22
23	24	25	26	27	28	29
30						

October

S	M	T	W	T	F	S
	1	2	3	4	5	6
7	8	9	10	11	12	13
14	15	16	17	18	19	20
21	22	23	24	25	26	27
28	29	30	31			

November

S	M	T	W	T	F	S
				1	2	3
4	5	6	7	8	9	10
11	12	13	14	15	16	17
18	19	20	21	22	23	24
25	26	27	28	29	30	

December

S	M	T	W	T	F	S
						1
2	3	4	5	6	7	8
9	10	11	12	13	14	15
16	17	18	19	20	21	22
23	24	25	26	27	28	29
30	31					

Calendar No. Three

January

S	M	T	W	T	F	S
		1	2	3	4	5
6	7	8	9	10	11	12
13	14	15	16	17	18	19
20	21	22	23	24	25	26
27	28	29	30	31		

February

S	M	T	W	T	F	S
					1	2
3	4	5	6	7	8	9
10	11	12	13	14	15	16
17	18	19	20	21	22	23
24	25	26	27	28		

March

S	M	T	W	T	F	S
					1	2
3	4	5	6	7	8	9
10	11	12	13	14	15	16
17	18	19	20	21	22	23
24	25	26	27	28	29	30
31						

April

S	M	T	W	T	F	S
	1	2	3	4	5	6
7	8	9	10	11	12	13
14	15	16	17	18	19	20
21	22	23	24	25	26	27
28	29	30				

May

S	M	T	W	T	F	S
7			1	2	3	4
5	6	7	8	9	10	11
12	13	14	15	16	17	18
19	20	21	22	23	24	25
26	27	28	29	30	31	

June

S	M	T	W	T	F	S
						1
2	3	4	5	6	7	8
9	10	11	12	13	14	15
16	17	18	19	20	21	22
23	24	25	26	27	28	29
30						

July

S	M	T	W	T	F	S
	1	2	3	4	5	6
7	8	9	10	11	12	13
14	15	16	17	18	19	20
21	22	23	24	25	26	27
28	29	30	31			

August

S	M	T	W	T	F	S
				1	2	3
4	5	6	7	8	9	10
11	12	13	14	15	16	17
18	19	20	21	22	23	24
25	26	27	28	29	30	31

September

S	M	T	W	T	F	S
1	2	3	4	5	6	7
8	9	10	11	12	13	14
15	16	17	18	19	20	21
22	23	24	25	26	27	28
29	30					

October

S	M	T	W	T	F	S
		1	2	3	4	5
6	7	8	9	10	11	12
13	14	15	16	17	18	19
20	21	22	23	24	25	26
27	28	29	30	31		

November

S	M	T	W	T	F	S
					1	2
3	4	5	6	7	8	9
10	11	12	13	14	15	16
17	18	19	20	21	22	23
24	25	26	27	28	29	30

December

S	M	T	W	T	F	S
1	2	3	4	5	6	7
8	9	10	11	12	13	14
15	16	17	18	19	20	21
22	23	24	25	26	27	28
29	30	31				

Calendar No. Four

January

S	M	T	W	T	F	S
			1	2	3	4
5	6	7	8	9	10	11
12	13	14	15	16	17	18
19	20	21	22	23	24	25
26	27	28	29	30	31	

February

S	M	T	W	T	F	S
						1
2	3	4	5	6	7	8
9	10	11	12	13	14	15
16	17	18	19	20	21	22
23	24	25	26	27	28	

March

S	M	T	W	T	F	S
						1
2	3	4	5	6	7	8
9	10	11	12	13	14	15
16	17	18	19	20	21	22
23	24	25	26	27	28	29
30	31					

April

S	M	T	W	T	F	S
		1	2	3	4	5
6	7	8	9	10	11	12
13	14	15	16	17	18	19
20	21	22	23	24	25	26
27	28	29	30			

May

S	M	T	W	T	F	S
				1	2	3
4	5	6	7	8	9	10
11	12	13	14	15	16	17
18	19	20	21	22	23	24
25	26	27	28	29	30	31

June

S	M	T	W	T	F	S
1	2	3	4	5	6	7
8	9	10	11	12	13	14
15	16	17	18	19	20	21
22	23	24	25	26	27	28
29	30					

July

S	M	T	W	T	F	S
		1	2	3	4	5
6	7	8	9	10	11	12
13	14	15	16	17	18	19
20	21	22	23	24	25	26
27	28	29	30	31		

August

S	M	T	W	T	F	S
					1	2
3	4	5	6	7	8	9
10	11	12	13	14	15	16
17	18	19	20	21	22	23
24	25	26	27	28	29	30
31						

September

S	M	T	W	T	F	S
	1	2	3	4	5	6
7	8	9	10	11	12	13
14	15	16	17	18	19	20
21	22	23	24	25	26	27
28	29	30				

October

S	M	T	W	T	F	S
			1	2	3	4
5	6	7	8	9	10	11
12	13	14	15	16	17	18
19	20	21	22	23	24	25
26	27	28	29	30	31	

November

S	M	T	W	T	F	S
						1
2	3	4	5	6	7	8
9	10	11	12	13	14	15
16	17	18	19	20	21	22
23	24	25	26	27	28	29
30						

December

S	M	T	W	T	F	S
	1	2	3	4	5	6
7	8	9	10	11	12	13
14	15	16	17	18	19	20
21	22	23	24	25	26	27
28	29	30	31			

Calendar No. Five

January

S	M	T	W	T	F	S
				1	2	3
4	5	6	7	8	9	10
11	12	13	14	15	16	17
18	19	20	21	22	23	24
25	26	27	28	29	30	31

February

S	M	T	W	T	F	S
1	2	3	4	5	6	7
8	9	10	11	12	13	14
15	16	17	18	19	20	21
22	23	24	25	26	27	28

March

S	M	T	W	T	F	S
1	2	3	4	5	6	7
8	9	10	11	12	13	14
15	16	17	18	19	20	21
22	23	24	25	26	27	28
29	30	31				

April

S	M	T	W	T	F	S
			1	2	3	4
5	6	7	8	9	10	11
12	13	14	15	16	17	18
19	20	21	22	23	24	25
26	27	28	29	30		

May

S	M	T	W	T	F	S
					1	2
3	4	5	6	7	8	9
10	11	12	13	14	15	16
17	18	19	20	21	22	23
24	25	26	27	28	29	30
31						

June

S	M	T	W	T	F	S
	1	2	3	4	5	6
7	8	9	10	11	12	13
14	15	16	17	18	19	20
21	22	23	24	25	26	27
28	29	30				

July

S	M	T	W	T	F	S
			1	2	3	4
5	6	7	8	9	10	11
12	13	14	15	16	17	18
19	20	21	22	23	24	25
26	27	28	29	30	31	

August

S	M	T	W	T	F	S
						1
2	3	4	5	6	7	8
9	10	11	12	13	14	15
16	17	18	19	20	21	22
23	24	25	26	27	28	29
30	31					

September

S	M	T	W	T	F	S
		1	2	3	4	5
6	7	8	9	10	11	12
13	14	15	16	17	18	19
20	21	22	23	24	25	26
27	28	29	30			

October

S	M	T	W	T	F	S
				1	2	3
4	5	6	7	8	9	10
11	12	13	14	15	16	17
18	19	20	21	22	23	24
25	26	27	28	29	30	31

November

S	M	T	W	T	F	S
1	2	3	4	5	6	7
8	9	10	11	12	13	14
15	16	17	18	19	20	21
22	23	24	25	26	27	28
29	30					

December

S	M	T	W	T	F	S
		1	2	3	4	5
6	7	8	9	10	11	12
13	14	15	16	17	18	19
20	21	22	23	24	25	26
27	28	29	30	31		

Calendar No. Six

January

S	M	T	W	T	F	S
					1	2
3	4	5	6	7	8	9
10	11	12	13	14	15	16
17	18	19	20	21	22	23
24	25	26	27	28	29	30
31						

February

S	M	T	W	T	F	S
	1	2	3	4	5	6
7	8	9	10	11	12	13
14	15	16	17	18	19	20
21	22	23	24	25	26	27
28						

March

S	M	T	W	T	F	S
	1	2	3	4	5	6
7	8	9	10	11	12	13
14	15	16	17	18	19	20
21	22	23	24	25	26	27
28	29	30	31			

April

S	M	T	W	T	F	S
				1	2	3
4	5	6	7	8	9	10
11	12	13	14	15	16	17
18	19	20	21	22	23	24
25	26	27	28	29	30	

May

S	M	T	W	T	F	S
						1
2	3	4	5	6	7	8
9	10	11	12	13	14	15
16	17	18	19	20	21	22
23	24	25	26	27	28	29
30	31					

June

S	M	T	W	T	F	S
		1	2	3	4	5
6	7	8	9	10	11	12
13	14	15	16	17	18	19
20	21	22	23	24	25	26
27	28	29	30			

July

S	M	T	W	T	F	S
				1	2	3
4	5	6	7	8	9	10
11	12	13	14	15	16	17
18	19	20	21	22	23	24
25	26	27	28	29	30	31

August

S	M	T	W	T	F	S
1	2	3	4	5	6	7
8	9	10	11	12	13	14
15	16	17	18	19	20	21
22	23	24	25	26	27	28
29	30	31				

September

S	M	T	W	T	F	S
			1	2	3	4
5	6	7	8	9	10	11
12	13	14	15	16	17	18
19	20	21	22	23	24	25
26	27	28	29	30		

October

S	M	T	W	T	F	S
					1	2
3	4	5	6	7	8	9
10	11	12	13	14	15	16
17	18	19	20	21	22	23
24	25	26	27	28	29	30
31						

November

S	M	T	W	T	F	S
	1	2	3	4	5	6
7	8	9	10	11	12	13
14	15	16	17	18	19	20
21	22	23	24	25	26	27
28	29	30				

December

S	M	T	W	T	F	S
			1	2	3	4
5	6	7	8	9	10	11
12	13	14	15	16	17	18
19	20	21	22	23	24	25
26	27	28	29	30	31	

Calendar No. Seven

January

S	M	T	W	T	F	S
						1
2	3	4	5	6	7	8
9	10	11	12	13	14	15
16	17	18	19	20	21	22
23	24	25	26	27	28	29
30	31					

February

S	M	T	W	T	F	S
		1	2	3	4	5
6	7	8	9	10	11	12
13	14	15	16	17	18	19
20	21	22	23	24	25	26
27	28					

March

S	M	T	W	T	F	S
		1	2	3	4	5
6	7	8	9	10	11	12
13	14	15	16	17	18	19
20	21	22	23	24	25	26
27	28	29	30	31		

April

S	M	T	W	T	F	S
					1	2
3	4	5	6	7	8	9
10	11	12	13	14	15	16
17	18	19	20	21	22	23
24	25	26	27	28	29	30

May

S	M	T	W	T	F	S
1	2	3	4	5	6	7
8	9	10	11	12	13	14
15	16	17	18	19	20	21
22	23	24	25	26	27	28
29	30	31				
30	31					

June

S	M	T	W	T	F	S
			1	2	3	4
5	6	7	8	9	10	11
12	13	14	15	16	17	18
19	20	21	22	23	24	25
26	27	28	29	30		

July

S	M	T	W	T	F	S
					1	2
3	4	5	6	7	8	9
10	11	12	13	14	15	16
17	18	19	20	21	22	23
24	25	26	27	28	29	30
31						

August

S	M	T	W	T	F	S
	1	2	3	4	5	6
7	8	9	10	11	12	13
14	15	16	17	18	19	20
21	22	23	24	25	26	27
28	29	30	31			

September

S	M	T	W	T	F	S
				1	2	3
4	5	6	7	8	9	10
11	12	13	14	15	16	17
18	19	20	21	22	23	24
25	26	27	28	29	30	

October

S	M	T	W	T	F	S
						1
2	3	4	5	6	7	8
9	10	11	12	13	14	15
16	17	18	19	20	21	22
23	24	25	26	27	28	29
30	31					

November

S	M	T	W	T	F	S
		1	2	3	4	5
6	7	8	9	10	11	12
13	14	15	16	17	18	19
20	21	22	23	24	25	26
27	28	29	30			

December

S	M	T	W	T	F	S
				1	2	3
4	5	6	7	8	9	10
11	12	13	14	15	16	17
18	19	20	21	22	23	24
25	26	27	28	29	30	31

Calendar No. Eight

January

S	M	T	W	T	F	S
1	2	3	4	5	6	7
8	9	10	11	12	13	14
15	16	17	18	19	20	21
22	23	24	25	26	27	28
29	30	31				

February

S	M	T	W	T	F	S
			1	2	3	4
5	6	7	8	9	10	11
12	13	14	15	16	17	18
19	20	21	22	23	24	25
26	27	28	29			

March

S	M	T	W	T	F	S
				1	2	3
4	5	6	7	8	9	10
11	12	13	14	15	16	17
18	19	20	21	22	23	24
25	26	27	28	29	30	31

April

S	M	T	W	T	F	S
1	2	3	4	5	6	7
8	9	10	11	12	13	14
15	16	17	18	19	20	21
22	23	24	25	26	27	28
29	30					

May

S	M	T	W	T	F	S
		1	2	3	4	5
6	7	8	9	10	11	12
13	14	15	16	17	18	19
20	21	22	23	24	25	26
27	28	29	30	31		

June

S	M	T	W	T	F	S
					1	2
3	4	5	6	7	8	9
10	11	12	13	14	15	16
17	18	19	20	21	22	23
24	25	26	27	28	29	30

July

S	M	T	W	T	F	S
1	2	3	4	5	6	7
8	9	10	11	12	13	14
15	16	17	18	19	20	21
22	23	24	25	26	27	28
29	30	31				

August

S	M	T	W	T	F	S
			1	2	3	4
5	6	7	8	9	10	11
12	13	14	15	16	17	18
19	20	21	22	23	24	25
26	27	28	29	30	31	

September

S	M	T	W	T	F	S
						1
2	3	4	5	6	7	8
9	10	11	12	13	14	15
16	17	18	19	20	21	22
23	24	25	26	27	28	29
30						

October

S	M	T	W	T	F	S
	1	2	3	4	5	6
7	8	9	10	11	12	13
14	15	16	17	18	19	20
21	22	23	24	25	26	27
28	29	30	31			

November

S	M	T	W	T	F	S
				1	2	3
4	5	6	7	8	9	10
11	12	13	14	15	16	17
18	19	20	21	22	23	24
25	26	27	28	29	30	

December

S	M	T	W	T	F	S
						1
2	3	4	5	6	7	8
9	10	11	12	13	14	15
16	17	18	19	20	21	22
23	24	25	26	27	28	29
30	31					

Calendar No. Nine

January

S	M	T	W	T	F	S
	1	2	3	4	5	6
7	8	9	10	11	12	13
14	15	16	17	18	19	20
21	22	23	24	25	26	27
28	29	30	31			

February

S	M	T	W	T	F	S
				1	2	3
4	5	6	7	8	9	10
11	12	13	14	15	16	17
18	19	20	21	22	23	24
25	26	27	28	29		

March

S	M	T	W	T	F	S
					1	2
3	4	5	6	7	8	9
10	11	12	13	14	15	16
17	18	19	20	21	22	23
24	25	26	27	28	29	30
31						

April

S	M	T	W	T	F	S
	1	2	3	4	5	6
7	8	9	10	11	12	13
14	15	16	17	18	19	20
21	22	23	24	25	26	27
28	29	30				

May

S	M	T	W	T	F	S
			1	2	3	4
5	6	7	8	9	10	11
12	13	14	15	16	17	18
19	20	21	22	23	24	25
26	27	28	29	30	31	

June

S	M	T	W	T	F	S
						1
2	3	4	5	6	7	8
9	10	11	12	13	14	15
16	17	18	19	20	21	22
23	24	25	26	27	28	29
30						

July

S	M	T	W	T	F	S
	1	2	3	4	5	6
7	8	9	10	11	12	13
14	15	16	17	18	19	20
21	22	23	24	25	26	27
28	29	30	31			

August

S	M	T	W	T	F	S
				1	2	3
4	5	6	7	8	9	10
11	12	13	14	15	16	17
18	19	20	21	22	23	24
25	26	27	28	29	30	31

September

S	M	T	W	T	F	S
1	2	3	4	5	6	7
8	9	10	11	12	13	14
15	16	17	18	19	20	21
22	23	24	25	26	27	28
29	30					

October

S	M	T	W	T	F	S
		1	2	3	4	5
6	7	8	9	10	11	12
13	14	15	16	17	18	19
20	21	22	23	24	25	26
27	28	29	30	31		

November

S	M	T	W	T	F	S
					1	2
3	4	5	6	7	8	9
10	11	12	13	14	15	16
17	18	19	20	21	22	23
24	25	26	27	28	29	30

December

S	M	T	W	T	F	S
1	2	3	4	5	6	7
8	9	10	11	12	13	14
15	16	17	18	19	20	21
22	23	24	25	26	27	28
29	30	31				

Calendar No. Ten

January

S	M	T	W	T	F	S
		1	2	3	4	5
6	7	8	9	10	11	12
13	14	15	16	17	18	19
20	21	22	23	24	25	26
27	28	29	30	31		

February

S	M	T	W	T	F	S
					1	2
3	4	5	6	7	8	9
10	11	12	13	14	15	16
17	18	19	20	21	22	23
24	25	26	27	28	29	

March

S	M	T	W	T	F	S
						1
2	3	4	5	6	7	8
9	10	11	12	13	14	15
16	17	18	19	20	21	22
23	24	25	26	27	28	29
30	31					

April

S	M	T	W	T	F	S
		1	2	3	4	5
6	7	8	9	10	11	12
13	14	15	16	17	18	19
20	21	22	23	24	25	26
27	28	29	30			

May

S	M	T	W	T	F	S
				1	2	3
4	5	6	7	8	9	10
11	12	13	14	15	16	17
18	19	20	21	22	23	24
25	26	27	28	29	30	31

June

S	M	T	W	T	F	S
1	2	3	4	5	6	7
8	9	10	11	12	13	14
15	16	17	18	19	20	21
22	23	24	25	26	27	28
29	30					

July

S	M	T	W	T	F	S
		1	2	3	4	5
6	7	8	9	10	11	12
13	14	15	16	17	18	19
20	21	22	23	24	25	26
27	28	29	30	31		

August

S	M	T	W	T	F	S
					1	2
3	4	5	6	7	8	9
10	11	12	13	14	15	16
17	18	19	20	21	22	23
24	25	26	27	28	29	30
31						

September

S	M	T	W	T	F	S
	1	2	3	4	5	6
7	8	9	10	11	12	13
14	15	16	17	18	19	20
21	22	23	24	25	26	27
28	29	30				

October

S	M	T	W	T	F	S
			1	2	3	4
5	6	7	8	9	10	11
12	13	14	15	16	17	18
19	20	21	22	23	24	25
26	27	28	29	30	31	

November

S	M	T	W	T	F	S
						1
2	3	4	5	6	7	8
9	10	11	12	13	14	15
16	17	18	19	20	21	22
23	24	25	26	27	28	29
30						

December

S	M	T	W	T	F	S
	1	2	3	4	5	6
7	8	9	10	11	12	13
14	15	16	17	18	19	20
21	22	23	24	25	26	27
28	29	30	31			

Calendar No. Eleven

January

S	M	T	W	T	F	S
			1	2	3	4
5	6	7	8	9	10	11
12	13	14	15	16	17	18
19	20	21	22	23	24	25
26	27	28	29	30	31	

February

S	M	T	W	T	F	S
						1
2	3	4	5	6	7	8
9	10	11	12	13	14	15
16	17	18	19	20	21	22
23	24	25	26	27	28	29

March

S	M	T	W	T	F	S
1	2	3	4	5	6	7
8	9	10	11	12	13	14
15	16	17	18	19	20	21
22	23	24	25	26	27	28
29	30	31				

April

S	M	T	W	T	F	S
			1	2	3	4
5	6	7	8	9	10	11
12	13	14	15	16	17	18
19	20	21	22	23	24	25
26	27	28	29	30		

May

S	M	T	W	T	F	S
					1	2
3	4	5	6	7	8	9
10	11	12	13	14	15	16
17	18	19	20	21	22	23
24	25	26	27	28	29	30
31						

June

S	M	T	W	T	F	S
	1	2	3	4	5	6
7	8	9	10	11	12	13
14	15	16	17	18	19	20
21	22	23	24	25	26	27
28	29	30				

July

S	M	T	W	T	F	S
			1	2	3	4
5	6	7	8	9	10	11
12	13	14	15	16	17	18
19	20	21	22	23	24	25
26	27	28	29	30	31	

August

S	M	T	W	T	F	S
						1
2	3	4	5	6	7	8
9	10	11	12	13	14	15
16	17	18	19	20	21	22
23	24	25	26	27	28	29
30	31					

September

S	M	T	W	T	F	S
		1	2	3	4	5
6	7	8	9	10	11	12
13	14	15	16	17	18	19
20	21	22	23	24	25	26
27	28	29	30			

October

S	M	T	W	T	F	S
				1	2	3
4	5	6	7	8	9	10
11	12	13	14	15	16	17
18	19	20	21	22	23	24
25	26	27	28	29	30	31

November

S	M	T	W	T	F	S
1	2	3	4	5	6	7
8	9	10	11	12	13	14
15	16	17	18	19	20	21
22	23	24	25	26	27	28
29	30					

December

S	M	T	W	T	F	S
		1	2	3	4	5
6	7	8	9	10	11	12
13	14	15	16	17	18	19
20	21	22	23	24	25	26
27	28	29	30	31		

Calendar No. Twelve

January

S	M	T	W	T	F	S
				1	2	3
4	5	6	7	8	9	10
11	12	13	14	15	16	17
18	19	20	21	22	23	24
25	26	27	28	29	30	31

February

S	M	T	W	T	F	S
1	2	3	4	5	6	7
8	9	10	11	12	13	14
15	16	17	18	19	20	21
22	23	24	25	26	27	28
29						

March

S	M	T	W	T	F	S
	1	2	3	4	5	6
7	8	9	10	11	12	13
14	15	16	17	18	19	20
21	22	23	24	25	26	27
28	29	30	31			

April

S	M	T	W	T	F	S
				1	2	3
4	5	6	7	8	9	10
11	12	13	14	15	16	17
18	19	20	21	22	23	24
25	26	27	28	29	30	

May

S	M	T	W	T	F	S
						1
2	3	4	5	6	7	8
9	10	11	12	13	14	15
16	17	18	19	20	21	22
23	24	25	26	27	28	29
30	31					

June

S	M	T	W	T	F	S
		1	2	3	4	5
6	7	8	9	10	11	12
13	14	15	16	17	18	19
20	21	22	23	24	25	26
27	28	29	30			

July

S	M	T	W	T	F	S
				1	2	3
4	5	6	7	8	9	10
11	12	13	14	15	16	17
18	19	20	21	22	23	24
25	26	27	28	29	30	31

August

S	M	T	W	T	F	S
1	2	3	4	5	6	7
8	9	10	11	12	13	14
15	16	17	18	19	20	21
22	23	24	25	26	27	28
29	30	31				

September

S	M	T	W	T	F	S
			1	2	3	4
5	6	7	8	9	10	11
12	13	14	15	16	17	18
19	20	21	22	23	24	25
26	27	28	29	30		

October

S	M	T	W	T	F	S
					1	2
3	4	5	6	7	8	9
10	11	12	13	14	15	16
17	18	19	20	21	22	23
24	25	26	27	28	29	30
31						

November

S	M	T	W	T	F	S
	1	2	3	4	5	6
7	8	9	10	11	12	13
14	15	16	17	18	19	20
21	22	23	24	25	26	27
28	29	30				

December

S	M	T	W	T	F	S
			1	2	3	4
5	6	7	8	9	10	11
12	13	14	15	16	17	18
19	20	21	22	23	24	25
26	27	28	29	30	31	

Calendar No. Thirteen

January

S	M	T	W	T	F	S
					1	2
3	4	5	6	7	8	9
10	11	12	13	14	15	16
17	18	19	20	21	22	23
24	25	26	27	28	29	30
31						

February

S	M	T	W	T	F	S
	1	2	3	4	5	6
7	8	9	10	11	12	13
14	15	16	17	18	19	20
21	22	23	24	25	26	27
28	29					

March

S	M	T	W	T	F	S
		1	2	3	4	5
6	7	8	9	10	11	12
13	14	15	16	17	18	19
20	21	22	23	24	25	26
27	28	29	30	31		

April

S	M	T	W	T	F	S
					1	2
3	4	5	6	7	8	9
10	11	12	13	14	15	16
17	18	19	20	21	22	23
24	25	26	27	28	29	30

May

S	M	T	W	T	F	S
1	2	3	4	5	6	7
8	9	10	11	12	13	14
15	16	17	18	19	20	21
22	23	24	25	26	27	28
29	30	31				

June

S	M	T	W	T	F	S
			1	2	3	4
5	6	7	8	9	10	11
12	13	14	15	16	17	18
19	20	21	22	23	24	25
26	27	28	29	30		

July

S	M	T	W	T	F	S
					1	2
3	4	5	6	7	8	9
10	11	12	13	14	15	16
17	18	19	20	21	22	23
24	25	26	27	28	29	30
31						

August

S	M	T	W	T	F	S
	1	2	3	4	5	6
7	8	9	10	11	12	13
14	15	16	17	18	19	20
21	22	23	24	25	26	27
28	29	30	31			

September

S	M	T	W	T	F	S
				1	2	3
4	5	6	7	8	9	10
11	12	13	14	15	16	17
18	19	20	21	22	23	24
25	26	27	28	29	30	

October

S	M	T	W	T	F	S
						1
2	3	4	5	6	7	8
9	10	11	12	13	14	15
16	17	18	19	20	21	22
23	24	25	26	27	28	29
30	31					

November

S	M	T	W	T	F	S
		1	2	3	4	5
6	7	8	9	10	11	12
13	14	15	16	17	18	19
20	21	22	23	24	25	26
27	28	29	30			

December

S	M	T	W	T	F	S
				1	2	3
4	5	6	7	8	9	10
11	12	13	14	15	16	17
18	19	20	21	22	23	24
25	26	27	28	29	30	31

Calendar No. Fourteen

January

S	M	T	W	T	F	S
						1
2	3	4	5	6	7	8
9	10	11	12	13	14	15
16	17	18	19	20	21	22
23	24	25	26	27	28	29
30	31					

February

S	M	T	W	T	F	S
		1	2	3	4	5
6	7	8	9	10	11	12
13	14	15	16	17	18	19
20	21	22	23	24	25	26
27	28	29				

March

S	M	T	W	T	F	S
			1	2	3	4
5	6	7	8	9	10	11
12	13	14	15	16	17	18
19	20	21	22	23	24	25
26	27	28	29	30	31	

April

S	M	T	W	T	F	S
						1
2	3	4	5	6	7	8
9	10	11	12	13	14	15
16	17	18	19	20	21	22
23	24	25	26	27	28	29
30						

May

S	M	T	W	T	F	S
	1	2	3	4	5	6
7	8	9	10	11	12	13
14	15	16	17	18	19	20
21	22	23	24	25	26	27
28	29	30	31			

June

S	M	T	W	T	F	S
				1	2	3
4	5	6	7	8	9	10
11	12	13	14	15	16	17
18	19	20	21	22	23	24
25	26	27	28	29	30	

July

S	M	T	W	T	F	S
						1
2	3	4	5	6	7	8
9	10	11	12	13	14	15
16	17	18	19	20	21	22
23	24	25	26	27	28	29
30	31					

August

S	M	T	W	T	F	S
		1	2	3	4	5
6	7	8	9	10	11	12
13	14	15	16	17	18	19
20	21	22	23	24	25	26
27	28	29	30	31		

September

S	M	T	W	T	F	S
					1	2
3	4	5	6	7	8	9
10	11	12	13	14	15	16
17	18	19	20	21	22	23
24	25	26	27	28	29	30

October

S	M	T	W	T	F	S
1	2	3	4	5	6	7
8	9	10	11	12	13	14
15	16	17	18	19	20	21
22	23	24	25	26	27	28
29	30	31				

November

S	M	T	W	T	F	S
			1	2	3	4
5	6	7	8	9	10	11
12	13	14	15	16	17	18
19	20	21	22	23	24	25
26	27	28	29	30		

December

S	M	T	W	T	F	S
					1	2
3	4	5	6	7	8	9
10	11	12	13	14	15	16
17	18	19	20	21	22	23
24	25	26	27	28	29	30
31						

Index

	Page No.
A.P.I. casing	108
drill pipe	80
tubing	110
Acceleration of gravity	354
Acid, hydrochloric	95
sulfamic	95
Additives, drilling fluid	71
Air bubbles, rise	358
Air hose friction	76
Air pipe friction	75
Air volume for annulus velocities	77
Air-lift pumping	167
Airlines	312
Alkalinity	280
Alphabet, Greek	359
Aluminum pipe, friction loss	253
Aluminum siphon tubes, discharge	256
Annulus velocity, air volume	77
Annulus, volume of	65
Anti-freeze solutions	355
Area conversion table	3
Areas and volumes	45
Asbestos-cement pipe	120
Ascending mud velocities	68
Atmosphere, composition	356
Atmospheric pressure	133
Atomic weights	284
Bailer tests	317
Belt drives	179
Bentonite cement slurry	74
Bits, cable tool	84
rotary tricone roller	81
Blaney-Criddle formula	259
Boiling point of water	133
Boron	271
Breaking strength of wire ropes	88
Brick walls	348
Cable tool bits	84
Cable tools, dimensions	83
Calendar, perpetual	380
California pipe method for flow measurement	219
Canal seepage loss	257
Capacity of drill collars	79
of holes	64
of rope drums	85
Capillary rise	321
Capital recovery factor	44
Casing stretch	111
Casing, A.P.I.	108
Cement slurry, bentonite	74
neat	73
Centrifugal pump capacity	166
Chains, safe loads on	86
Channel velocity, erosion and deposition	358
Chemical elements	284
Chlorination, material required	289
wells and piping	288
Cipolletti weir table	201
Circles, area and circumference	47
Circular arc for radius one	59
Circular conduits, hydraulic elements of	146
Climatic data, world	363
Complementary error function	303
Compounding and discounting formulas	36
Compounding factor for 1	40
Compounding factor for one per annum	41
Concrete block walls	348
floors	347
irrigation pipe	121
proportioning	347
Conductors, copper, capacity	233
Consumptive use, Blaney-Criddle formula	259
conversion to flow	252
irrigation	258
pan evaporation	260
pan evaporation coefficients	264
Conversion factors	7
Conversion table, area	3
decimal equivalents of fractions	32
decimal time	33
degrees to radians	35
discharge-runoff	128

	Page No.
feet of water to psi	125
flow	6
flow rates	127
hydraulic conductivity	320
inches and fractions to decimal feet	30
inches and fractions to centimers	31
inches of mercury to feet of water	29
length	3
minutes and seconds to decimal degrees	34
mud weights	69
power	6
specific gravity	135
temperature	28
volume	4
water analysis	284
water pressure	125
weight	5
Cooper-Jacob formula	295
Coordinate method for pipe flow	215
Copper conductors, capacity	233
Copper wire properties	232
Core drilling, diamond	82
Corrosion, galvanic series	286
Ryznar stability index	287
Corrugated sheet metal	353
Critical moisture for irrigation	266
Crops, boron tolerance	274
growing season	265
salt tolerance	271
Current meter	195
Cylinder pump discharge	183
Days of year, number	376
Days, length of	377
Daytime hours, monthly percentage of annual	262
Decimal degree conversions	34
Decimal equivalents of fractions	32
Decimal numbers, square root of	142
Decimal time conversions	33
Degrees to radians conversion	35
Density and viscosity of water	132

	Page No.
Deposition and erosion, channel velocity	358
Development of wells	90
Diamond core drilling	82
Discharge-runoff conversion	128
Discount factor	39
Disinfection, wells and piping	288
Displacement of drill collars	79
of mud pumps	67
Dissolved oxygen, saturation	286
Drill collars, capacity of	79
displacement of	79
weight of	78
Drill pipe, A.P.I.	80
stretch	111
Drill sizes	345
Drilling fluid, additives	71
ascending mud velocities	68
mud weight conversions	69
salt-water	71
solid contents effect on mud weight	70
weight adjustment	72
Drinking water standards	281
Drum calibration, 55-gallon	214
Ductile iron pipe	119
Effective grain size	308
Effective precipitation	260
average monthly	263
Efficiency, motors	228
pumps	161
Electrical conductivity	279
Electrical formulas	225
Engine operation	177
Engineering materials, properties	343
Equivalent pipe carrying capacities	149
Erfc	303
Erosion and deposition, channel velocity	358
Error function, complementary	303
Fiber ropes, safe working loads	87
Filter pack, design	241
gradation	248
Fitting and valves, friction losses	151
Five-halves power of numbers	58

	Page No.
Flanges, steel pipe	113
Flat belt drives	179
horsepower rating	180
Flow conversion table	6
from open pipes	215
in parallel pipes	144
measurement	191
meters	194
rate conversion	127
rate from filling containers	213
through hose	152
through nozzles	153
through pipe orifices	208
Flumes	193
(Parshall) table	204
(Trapezoidal) table	206
Formula, Blaney-Criddle	259
Cooper-Jacob	295
Hazen-Williams	136
Manning	137
Theis	294
Thiem	295
Formulas, areas and volumes	45
compounding and discounting	36
ground water flow	294
triangles	46
Fractions, decimal equivalents of	32
Freezing of water in pipes	156
Friction loss, air in hose	76
air in pipe	75
aluminum pipe	253
fittings and valves	151
pipe flow	136
pipe sizes for reasonable	148
pump column	164
sprinkler laterals	254
Fuel requirement for pumping	177
Full-load current of motors	227
Galleries, infiltration	304
Galvanic series	286
Gauge, iron and steel sheets	352
wire	231
Gear drives, right-angle	178
Generation vs. pumping	155
Geologic time scale	339
Grain size, effective	308
Graph sheet, log-log	329
months by days	332

	Page No.
probability	330
rectangular coordinates	326
semi-logarithmic	328
sieve analyses	246
triangular coordinates	331
years by months	333
Gravel pack, design	241
Gravity	354
Greek alphabet	359
Ground water flow formulas	294
Growing season, crops	265
Hardness	278
Hazen-Williams formula	136
friction loss graph	138
table of 0.54 powers	139
table of 0.63 powers	141
Holes, capacity of	64
Hooks, steel, safe loads on	86
Horizontal cylindrical tanks	337
Horizontal pipe flow	215
Horsepower loss, pump shaft	163
Horsepower rating, flat belts	180
V-belts	180
Horsepower, theoretical to lift water	175
Hose, flow of water through	152
Hydraulic conductivity conversion table	320
Hydraulic elements of circular conduits	146
Hydraulic jet	154
Hydraulic jetting, drop pipe	97
velocities	96
Hydraulic ram	181
Hydrochloric acid	95
Hydropneumatic tanks	171
dimensions	174
Inches and fractions to centimeters	31
to decimal feet	30
Inches of mercury to feet of water	29
Incrustation, Ryznar stability index	287
Indicated vacuum	354
Infiltration galleries	304
Internal combustion engine operation	177

	Page No.
Iron and steel sheets, gauge	352
Irrigation, consumptive use	258
critical moisture	266
pipe (concrete)	121
soil intake rate	267
table	251
water quality	268
water requirement	258
Jet, hydraulic	154
Jetting, hydraulic, drop pipe	97
hydraulic, velocities	96
Kilowatt hours to lift water	176
Large diameter steel pipe	106
Latitude and longitude lengths	361
Length conversion table	3
Length of circular arc for radius one	59
Length of days	377
Linear drain, transient flow	302
Logarithms, base 10	53
natural (Napierian)	55
Longitude and latitude lengths	361
Lumber, conversions	349
dimensions	350
Manning formula	137
square root of decimal numbers	142
Map scales, English and metric	360
Materials, weight of	344
Maximum flow of water through pipes	150
Metric system multipliers	27
Miner's inch	129
Mineral hardness	342
Minor losses in pipe flow	144
Mohs scale	342
Moisture-holding capacity of soils	266
Monthly payments	38
Motors, efficiency	228
full-load current	227
Mud pumps, displacement of	67
Mud velocities, ascending	68
Mud weight, conversions	69
effect of solids content	70
Multipliers, metric system	27

	Page No.
Nails	346
Natural logarithms	55
trigonometric functions	48
Neat cement slurry	73
Net positive suction head	165
Nozzles, sprinkler	255
theoretical discharge	153
NPSH	165
Number of days of year	376
Numbering system, wells and springs	322
Numbers, five-halves power	58
three-halves power	57
two-thirds power	56
Numerical table	49
Ocean water, composition	356
Orifice bucket	196
flow table	212
Orifices, flow table	208
pipe	194
Oxygen, dissolved saturation	286
Pan evaporation, coefficients for consumptive use	264
consumptive use	260
Parallel pipes, flow in	144
Parshall flume table	204
Penetration test, standard	342
Permeability, estimating	307
Perpetual calendar	380
Pipe flow, friction loss	136
horizontal	215
minor losses in	144
vertical	217
Pipe orifices	194
flow table	208
Pipe sizes for reasonable friction loss	148
Pipe, A.P.I. drill	80
asbestos-cement	120
collapse pressure	118
concrete irrigation	121
ductile iron	119
freezing of water in	156
maximum flow of water through	150
PVC, dimensions	115
PVC, pressure ratings	117
PVC, thermal expansion	117

	Page No.
steel, dimensions	100
steel, large diameter	106
steel, thermal expansion	114
steel, thin-wall	107
Pits, suction, capacity of	66
Power conversion table	6
Power from watt-hour meters	226
Precipitation, effective	260
Present worth of annuity factor	42
Pressure tanks	171
Pressure, water, conversion	125
Properties of engineering materials	343
Pumping tests	311
bailer	317
slug	317
step-drawdown	317
Pumping vs. generation	155
Pumping, fuel requirement	177
kilowatt hours	176
theoretical horsepower	175
Pumps, air-lift	167
centrifugal	166
column friction loss	164
cylinder	183
efficiency	161
performance at different speeds	162
shaft and tube weights	163
shaft horsepower loss	163
suction lift	166
vertical turbine	159
PVC pipe, collapse pressure	118
dimensions	115
pressure ratings	117
thermal expansion	117
Ram, hydraulic	181
Reasonable friction loss, pipe sizes for	148
Rectangular (contracted) weir table	199
Rectangular (suppressed) weir table	200
Rectangular tanks, capacity of	66
Reinforcing bars, steel	353
Residual sodium carbonate	271
Right-angle gear drives	178
Rock types, weight of	340
Roller bits, rotary tricone	81

	Page No.
Rope drums, capacity	85
Rope, fiber, safe working loads	87
Rope, wire, breaking strength	88
sheave diameters	89
Rotary tricone roller bits	81
Ryznar stability index	287
Safe loads on chains	86
on steel hooks	86
Safe working loads on fiber ropes	87
Salt tolerance of crops	271
Salt-water drilling fluid	71
Sand settling rates	357
SAR	269
Screens	238
open area	240
slot widths	241
Screens development, acids or chemicals	91
compressed air	91
hydraulic jetting	92
surge-block	94
Sea water, composition	356
Seepage loss, canal	257
Settling rates, sand	357
Sheave diameters for wire rope	89
Sheet metal, corrugated	353
Sieve analyses	242
Sieves, U.S. standard	247
Sinking fund factor	43
Siphon tubes, discharge	256
Slot widths of screens	241
Slug tests	317
Slurry, bentonite cement	74
neat cement	73
Sodium adsorption ratio	269
Soil classification, particle size	342
graph	267
Soil classification, unified	341
Soils, moisture-holding capacity	266
Solid contents effect on mud weight	70
Specific capacity, theoretical	310
Specific conductance	279
temperature conversion	285
Specific gravity conversion	134
Speed of pumps, performance	162

	Page No.
Sprinkler irrigation, soil intake rate	267
Sprinkler laterals, friction loss	254
Sprinkler nozzles	255
Square root of decimal numbers	142
Standard penetration test	342
Standard time, world	378
Standards, drinking water	281
Steel and iron sheets, gauge	352
Steel hooks, safe loads on	86
Steel pipe, collapse pressure	118
dimensions	100
flanges	113
large diameter	106
thermal expansion	114
thin-wall	107
Steel plates, weight	351
Steel reinforcing bars	353
Step-drawdown tests	317
Stretch of tubing, drill pipe, or casing	111
Submersible pumps, cable selection	229
Suction lift of pumps	166
Suction pits, capacity of	66
Sulfamic acid	95
Surge-block screen development	94
Tanks, horizontal cylindrical	337
hydropneumatic	171
pressure	171
rectangular, capacity of	66
vertical cylindrical	338
Temperature conversion table	28
specific conductance	285
Temperature of ground water	321
Theis formula	294
Theoretical horsepower to lift water	175
Theoretical specific capacity	310
Thermal expansion, PVC pipe	117
steel pipe	114
Thiem formula	295
Thin-wall steel pipe	107
Three-halves power of numbers	57
Time scale, geologic	339
Time, decimal conversions	33
world standard	378
Total dissolved solids	279
Transient flow, linear drain	302

	Page No.
Transmissivity, estimating	307
Trapezoidal flume table	206
Triangles	46
Triangular weir table	202
Tricone roller bits, rotary	81
Trigonometric functions, natural	48
Tubing, A.P.I.	110
stretch	111
Turbine pumps	159
Two-thirds power of numbers	56
u vs W(u)	297
Unified soil classification	341
Uniformity coefficient	241
Unit conversions	7
USGS numbering system, wells and springs	322
V-belt drives	179
horsepower rating	180
Vacuum, indicated	354
Valves and fittings, friction losses	151
Vapor pressure of water	132
Velocity head	130
Vertical cylindrical tanks	338
Vertical pipe flow	217
Vertical turbine pumps	159
Viscosity and density of water	132
Volume between casing and hole	65
Volume conversion table	4
W(u) vs u	297
Water analysis, conversion factors	283
hardness conversions	285
Water hammer	154
Water intake rate, irrigation	267
Water level measurements	312
Water quality	277
alkalinity	280
electrical conductivity	279
hardness	278
irrigation	268
specific conductance	279
total dissolved solids	279
Water requirement, irrigation	258
Water use estimating	184

	Page No.
Water, freezing of in pipes	156
vapor pressure	132
viscosity and density	132
Watt-hour meters, measuring power	226
Weight adjustment of drilling fluid	72
Weight conversion table	5
Weight, drill collars	78
materials	344
pump shaft and tube	163
rock types	340
steel plates	351
Weir table, Cipolletti	201
rectangular contracted	199
rectangular suppressed	200
triangular	202
Weirs	191
Well development, acids and chemicals	91
compressed air	91
hydraulic fracturing	93
hydraulic jetting	92
surging	90
Well diameter, relation to yield	324
Well interference	296
Well screens (see Screens)	238
Well yield, relation to diameter	324
Wind chill factor	362
Wire gauges	231
Wire nails	346
Wire rope, breaking strength	88
sheave diameters	89
Wire, copper, properties	232
Wood screws	346
World climatic data	363
World standard time	378

NOTES:

NOTES:

INCHES

1 2 3 4 5 6 7

NOTES:

NOTES:

NOTES:

NOTES:

NOTES: